Power Electronics and Power Systems

Series editors

Joe H. Chow, Rensselaer Polytechnic Institute, Troy, New York, USA
Alex M. Stankovic, Tufts University, Medford, Massachusetts, USA
David Hill, The University of Hong Kong, Sydney, New South Wales, Australia

The Power Electronics and Power Systems Series encompasses power electronics, electric power restructuring, and holistic coverage of power systems. The Series comprises advanced textbooks, state-of-the-art titles, research monographs, professional books, and reference works related to the areas of electric power transmission and distribution, energy markets and regulation, electronic devices, electric machines and drives, computational techniques, and power converters and inverters. The Series features leading international scholars and researchers within authored books and edited compilations. All titles are peer reviewed prior to publication to ensure the highest quality content. To inquire about contributing to the Power Electronics and Power Systems Series, please contact Dr. Joe Chow, Administrative Dean of the College of Engineering and Professor of Electrical, Computer and Systems Engineering, Rensselaer Polytechnic Institute, Jonsson Engineering Center, Office 7012, 110 8th Street, Troy, NY USA, 518-276-6374, chowj@rpi.edu.

More information about this series at http://www.springer.com/series/6403

Pengwei Du · Ross Baldick
Aidan Tuohy
Editors

Integration of Large-Scale Renewable Energy into Bulk Power Systems

From Planning to Operation

Springer

Editors
Pengwei Du
Renewable Integration
Electric Reliability Council of Texas
Taylor, TX
USA

Aidan Tuohy
Electric Power Research Institute
Chicago, IL
USA

Ross Baldick
Department of Electrical and Computer Engineering
University of Texas at Austin
Austin, TX
USA

ISSN 2196-3185 ISSN 2196-3193 (electronic)
Power Electronics and Power Systems
ISBN 978-3-319-55579-9 ISBN 978-3-319-55581-2 (eBook)
DOI 10.1007/978-3-319-55581-2

Library of Congress Control Number: 2017934462

Printed on acid-free paper

This Springer imprint is published by Springer Nature
The registered company is Springer International Publishing AG
The registered company address is: Gewerbestrasse 11, 6330 Cham, Switzerland

Preface

Power systems are currently undergoing significant changes. Support for environmental protection and advocacy of sustainable economic development have led to a rapid growth of renewable energy-generating capacity in recent years, primarily in the form of wind and solar generation resources. Wind and solar power can provide huge benefits since they are plentiful, widely distributed, and clean. The massive deployment of these renewable energy resources also drives their per-unit cost down so that wind and solar power can now compete in the marketplace with conventional energy production. Power system operators will inevitably face challenges in adapting to this environment.

Wind and solar energy resources exhibit notably different characteristics from their fossil fuel-burning counterparts. Since the output of wind and solar generators is variable and uncertain across multiple timescales, they are often referred to as variable energy resources (VERs). Numerous integration studies and operational experiences have implied that as a large amount of VERs is integrated into an electric grid, it could fundamentally change how the grid is planned and operated. Thus, grid operators are concerned with the economics and reliability of a power grid where VERs are added. To overcome this concern, successful solutions have been developed to mitigate the adverse effects brought to the grid by the VERs. The operational experiences gained worldwide also provide useful guidance to renewable integration. However, most of the past efforts were focused on studies assuming a low or medium penetration level of VERs. Due to the assumptions made in these studies, some integration issues may not manifest themselves if the penetration level of VERs is not high enough. The issues associated with integrating VERs are also complicated and multifaceted, spanning from long-term planning to short-term operations. This requires comprehensive studies to be performed to ensure that the whole spectrum of issues is examined. Today, interconnection studies of VERs are evolving at a fast pace, and the focus is on scenarios with a high penetration level of VERs.

As we continue on the path of increasing the installed capacity of renewable electricity, many areas that have set goals to incorporate high levels of VERs in their power systems are quickly approaching those penetration levels. For example, electricity from renewable energy resources will account for 33% of total consumption in California by 2020. The Danish government aims to be fully independent of fossil fuels by 2050. When the penetration level of VERs is high, a large amount of conventional generators could be displaced from dispatch. As a result, if no significant changes are introduced to the grid, there may be a limit on the maximum amount of VERs that can be added to the grid while maintaining reliability. This limit depends on the size of the grid as well as the location and diversity of VERs. Beyond this limit, the installation of more VERs will not be effective or reliable unless those issues, which are an obstacle to the interconnection of VERs at a large scale, can be effectively mitigated. Some of these mitigations are physical in nature, while others are institutional.

As a first step to solve these issues, a thorough understanding of the technical challenges which could arise is a prerequisite to the successful integration of large-scale VERs into bulk power systems.

The first challenge in the integration of a large share of VERs into a grid is the need to upgrade the transmission network, i.e., to build long and often expensive transmission lines. These new transmission lines are necessary to transfer the power from wind/solar farms, usually in remote, low population areas, to high load locations. Otherwise, wind/solar power has to be constrained to below its full capacity, which leads to a waste of energy that could be produced. However, both the high cost and the long construction time could discourage such development of a new transmission network. Moreover, uncertainty of regulatory policies makes it hard to decide when to expand the transmission network's capacity and what the efficient capacity is. Since the installation of VERs can occur at unprecedented rates, building new transmission lines could significantly lag behind renewable resource development as has often been experienced in areas with high concentrations of VERs.

The second challenge associated with accommodating large-scale VERs is that it becomes necessary to increase system flexibility in order to maintain a generation–load balance in power system operations. Flexibility is the capability of a grid to respond to changes in load and variable generation. This flexibility difficulty arises because VERs possess two major attributes that notably impact the bulk power system's planning and operations. The first is variability—the output of variable generation changes according to the availability of the primary fuel (wind, sunlight, and moving water), resulting in fluctuations in plant output on all timescales. The second is uncertainty—the magnitude and timing of variable generation output is less predictable than for conventional generation.

Instantaneous electrical generation and consumption must remain in balance to maintain grid stability. Despite many years of experience that have been acquired by utilities to manage the variations from load, the variability, and uncertainty contributed from VERs will make the balancing task much harder. Numerous options have been proposed to increase the flexibility of a power grid, which fall into two categories: physical flexibility and institutional measures. For instance, more flexibility can be physically obtained by adding more responsive resources, using smart grid strategies, or reducing demand when wind production is high.

Institutional measures can include improving wind and solar forecasts, exporting and importing power to neighboring areas, and designing new market mechanisms to incentivize the provision of flexibility. Insufficient flexibility will result in an unsatisfactory frequency control performance or curtailment of power from VERs. Since a high penetration of VERs will eventually deplete the responsive capability of physical resources in a grid, adding more flexibility may come at a significant cost and it may take a long period of time to plan. A cost-effective portfolio for the provision of flexibility is a necessity in order to prepare for a future grid with a large amount of VERs.

The third challenge is the impact of VERs on the dynamics of a power grid where they are present. Different types of generators behave distinctly during grid disturbances. Today, most wind turbines use variable speed generators combined with partial- or full-scale power converters between the turbine generator and the collector system, which generally leads to more desirable properties for grid interconnection such as low-voltage ride through capability. Simulation results have shown that the transient stability for a power system with up to around 30% share of VERs is not a limiting factor. However, VERs have low short circuit ratio and do not contribute to the system's synchronous inertia since they are based on power electronics devices. As they displace a large share of synchronous machines in operations, it dramatically changes the dynamic characteristics of the grid, thus necessitating a need to study the dynamic impact of VERs on power systems. The low short circuit ratio could reduce the strength of the system where dynamic voltage support is crucial. Depending on the size of the system, frequency stability may be problematic if the frequency declines quickly enough to cause a cascading effect following the trip of a large generator unit. Therefore, all of these aspects need to be evaluated to ensure predictable, stable behavior during system faults when the penetration level of VERs is extremely high. The industry is also developing a new concept of essential reliability services as a necessary and critical part of the fundamental reliability functions to the grid reliability. Two examples of essential reliability services are voltage support and frequency support, which are both strongly affected by significant increases in the share of VERs.

The fourth challenge is managing large shares of VERs in a power market context. This requires coordination of scheduling processes across multiple time-scales and new market design to improve market efficiency and provide incentive to resources to provide reliability services. The VER forecast is one of the most cost-effective tools in system operators' control rooms to integrate renewable resources. The accuracy of variable generation forecasting has been steadily increased thanks to the improvement in both the numerical weather models and the statistical models used, as well as the geographic diversity of VERs which reduces the fluctuations in their power production. A power grid can see tremendous benefits in both economics and reliability from a well-functioning VERs forecast because it allows conventional resources to be committed and dispatched more efficiently.

In addition to the VER forecast, the reserves carried by the grid also play an important role in improving market efficiency. The actual power production from

VERs can sometimes deviate from their forecast due to their variations and uncertainties. When this happens, it can be managed through deployment of reserves allocated in advance. Reserves that are used in cases of large generating unit failures or load fluctuations may also be used to compensate for the variability and uncertainty of VERs. Determining required reserves highly depends on the operational policies and the response time of the reserve. It is generally agreed that more reserves are needed as the penetration level of VERs increases. As the economic implications of increasing reserves become more significant, some approaches have been proposed to maintain balance between reliability and efficiency. An excessive amount of reserves will reduce market efficiency while an insufficient amount of reserves is detrimental to reliability. Some approaches proposed include introducing a balancing market, sharing reserves among a larger footprint, and clearing the market through stochastic/robust optimization. The creation of a larger market can reduce the total requirements for reserve capacity due to geographic diversity of load and VERs, and thus saves operational cost. The exchanging and sharing of reserves between neighboring systems also introduces an increase in efficiency since less-expensive resources outside of a given balancing region is accessible by this region. The European Network of Transmission System Operators and the energy imbalance market (EIM) in the western US are such examples.

Recent work has shed some light on how to overcome these challenges in the context of a high penetration of VERs. From the encouraging results obtained from these studies, a high penetration of VERs is both technically feasible and economically viable for a future power grid.

This book will provide a thorough understanding of the basic aspects that need to be addressed for both system planning and operation at a high penetration of VERs as well as describes the most recent development of innovative technologies and cutting-edge research to address these challenges. Both system planning and operations are the key factors for a successful VER integration. Well-developed system planning can reveal the trend of the reliability issues that will grow over time, and it allows the most cost-effective long-term solutions to be implemented before these issues become prominent. Effective system operation can handle the challenges as they unfold and thus quickly improve the system's capability to integrate VERs. This book will focus on both and provides international experiences to demonstrate the advantages of the latest developments in system planning and operation in the areas of renewable integration.

This book covers a variety of subjects associated with the interconnection of VERs and presents a number of comprehensive and practical solutions which summarize the best practices and case studies for three power grids where a large amount of VERs are already present: Texas, Germany, and China.

This book also emphasizes the interrelation between the economic aspects, reliability, and policy development of renewable integration since any successful strategies that help improve the security of a future grid with VERs present have to be cost-effective, and the enhancement of these strategies need to be supported by the regulatory agencies in charge of the grid security and reliability.

This book could be useful for engineers and operators in power system planning and operation, as well as academic researchers. It can serve as an excellent introduction for university students in electrical engineering at both undergraduate and postgraduate levels. The dissemination of the knowledge contained in this book can stimulate more ideas and innovations to be developed and eventually help to facilitate the interconnection of more VERs in the future grid.

This book is divided into ten chapters.

Chapter 1: "Wind Integration in ERCOT" discusses the challenges and solutions associated with the integration of large-scale wind generation in the Electric Reliability Council of Texas (ERCOT) system. It details the factors that have contributed to the success of large-scale wind generation build-out in ERCOT, which include transmission access, an efficient energy market and ancillary services, and assessment of system frequency response performance.

Chapter 2: "Integration of Large-Scale Renewable Energy: Experience and Practice in China" focuses on the impact of integrating intermittent renewable energy on the security and stability of a power system. Through a recent event in which a large number of wind turbines were involved in serial trip-offs and went offline because of transient voltage problems; the lessons learned are summarized and three levels of defense in power system are described as countermeasures to enhance the stability of a power grid with a large amount of VERs.

Chapter 3: "The Role of Ensemble Forecasting in Integrating Renewables into Power Systems: From Theory to Real-Time Applications" presents a review of existing ensemble forecasting techniques to form a profound understanding of how ensemble forecasts is the key element to integrate VERs successfully into the power systems via increased reliability, early detection of risk, forecasting, and assistance in day-ahead and intraday balancing strategy.

Chapter 4: "Wind and Solar Forecasting" discusses other important aspects of wind and solar forecasting from the end user perspective. It introduces fundamental principles to guide how a forecasting solution should be optimized in order to provide maximum value to the end user. These principles consist of four components: sense, model, assess, and communicate. An in-depth description of the concept, significance, and connectivity of these four components in the context of wind/solar power forecast is provided.

Chapter 5: "Reserve Estimation in Renewable Integration Studies" analyzes the impacts of renewable energy on reserve requirements. It highlights and compares the methods for modeling required reserves, in addition to providing a description of the current policies for reserve provision.

Chapter 6: "Balancing Authority Cooperation Concepts to Reduce Variable Generation Integration Costs in the Western Interconnection: Consolidating Balancing Authorities and Sharing Balancing Reserves" explores the issues surrounding the consolidation of a larger operation area in order to maintain a better power balance. Through a case study of the Western Electricity Coordinating Council (WECC), the chapter demonstrates the benefits of a larger operation footprint through the savings in production cost and reduction in reserve requirements.

Chapter 7: "Robust Optimization in Electric Power Systems Operations" discusses one of the most important operations for independent system operators, unit commitment (UC), in the presence of VERs. Solving UC to optimality or near-optimality is crucial to reduce economic costs of operation, ensure a fair market outcome, and maintain a high security and reliability level in power systems. This chapter describes a robust optimization model which requires less accurate information on probability distributions of uncertain parameters, guarantees a higher level of feasibility (i.e., robustness) of the resulting solutions in the face of uncertainty, and leads to computationally more tractable and scalable models.

Chapter 8: "Planning of Large Scale Renewable Energy for Bulk Power Systems" introduces coordinated system planning to solve the reliability problem due to the high penetration of renewable power plants using asynchronous generators. This chapter specially discusses sub-synchronous resonance and voltage oscillations in addition to emerging techniques, such as parallel processing, that are used in these computation-intensive planning studies.

Chapter 9: "Voltage Control for Wind Power Integration Areas" describes the basic functions of a wind-automatic voltage control system (Wind-AVC) and its implementation to improve the grid's stability and solve the problem of voltage fluctuations caused by the intermittent output of the wind turbine generators and the relatively weak grid structure.

Chapter 10: "Risk Averse Security Constrained Stochastic Congestion Management" presents an innovative probabilistic security constrained congestion management (PSCCM), considering the probable outage of main elements of power systems as well as the uncertainty of wind power generation. The proposed approach is formulated as a two-stage stochastic programming problem, in which both of the base case (first stage) and all probable severe post-contingency states (second stage) are considered together. The control actions performed on the base case operation point are called preventive controls, whereas those activated following the occurrence of contingencies are corrective controls.

The future of VERs is promising as they offer many benefits to the grid and society as a whole. While this book covers the topics associated with the integration of VERs as comprehensively as possible, more work is still under development. The intent of this book is not only to help to adapt to a future grid with the benefit of installed renewable energy capacity fully explored but also to encourage more people to contribute to this dynamic field and enable further exploitation of new revolutions for renewable integration.

Finally, we acknowledge the innovative work contributed by all of the authors in this increasingly important area and appreciate the staff at Springer for their assistance and help in the preparation of this book.

Taylor, TX, USA Pengwei Du
Austin, TX, USA Ross Baldick
Chicago, IL, USA Aidan Tuohy

Contents

Chapter 1
Wind Integration in ERCOT

Julia Matevosyan and Pengwei Du

1.1 Overview

1.1.1 Texas Power System

The Electric Reliability Council of Texas is the electricity grid and market operator for the majority of the state of Texas. It manages the power flow to about 24 million of Texas customers, representing approximately 90% of the state electricity demand. As the independent system operator, ERCOT schedules power on an electric grid that connects more than 46,500 miles of transmission lines and over 550 generation units. The ERCOT's installed generation capacity is over 77 GW for summer peak demand [1]. This number accounts for summer ratings of thermal generation as well as capacity contribution from wind and solar generation, discussed later in this chapter. The most recent peak demand record of 69,621 MW was set on August 10, 2015.

The remaining part of the state of Texas is served from either from Southwest Power Pool (SPP) and Midcontinent Independent System Operator (MISO), which are parts of Eastern Interconnect (EI) or Western Electric Coordinating Council (WECC). ERCOT is connected to EI through two high voltage direct current (HVDC) ties rated at 220 and 600 MW. ERCOT is also connected to the Comision Federla de Electricidad (CFE), the electric grid in Mexico, through three HVDC ties rated at 36, 100 and 150 MW. As shown in Fig. 1.1, this arrangement makes ERCOT an electrical "island" with only limited asynchronous interconnections.

As a result of low natural gas prices and rapid development of wind generation, ERCOT's generation mix has dramatically changed in the past 20 years from being heavily dominated by coal and gas-steam generation in the late 1990s to now

J. Matevosyan (✉) · P. Du
ERCOT, Austin, USA
e-mail: Julia.Matevosjana@ercot.com

P. Du et al. (eds.), *Integration of Large-Scale Renewable Energy into Bulk Power Systems*, Power Electronics and Power Systems,
DOI 10.1007/978-3-319-55581-2_1

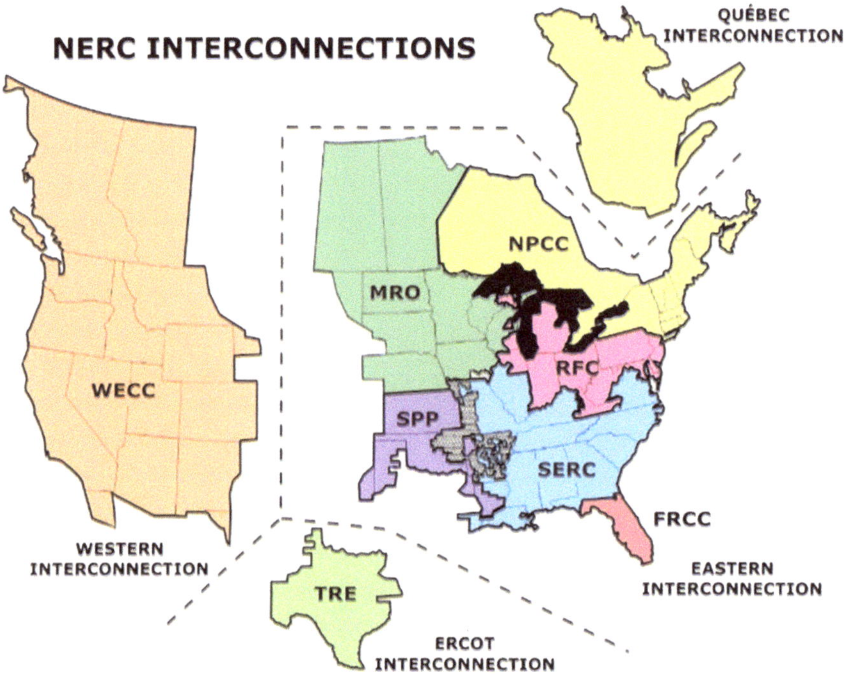

Fig. 1.1 NERC interconnections map [2]

having a substantial share of renewable generation and simple and combine cycle generation, as shown in Fig. 1.2.

The current generation mix consists of coal, nuclear, natural gas (steam generation, simple cycle and combined cycle combustion turbines), biomass, land fill gas, hydro, solar and wind generation. The ERCOT system is heavily reliant on natural gas generation both from an installed capacity and energy perspective, the latter of which is shown in Fig. 1.3.

1.1.2 Wind Generation Development in Texas

One of the reasons for rapid wind generation development is that Texas has rich wind resources—see Fig. 1.4. The best wind potential is in the Texas Panhandle[1] and West Texas as well as along the Gulf Coast. These areas have average wind generation capacity factor of 35–40%.

[1]The Texas Panhandle is a rectangular area of the state, consisting of the northernmost twenty-six counties in the state and bordered by New Mexico to the west and Oklahoma to the north and east.

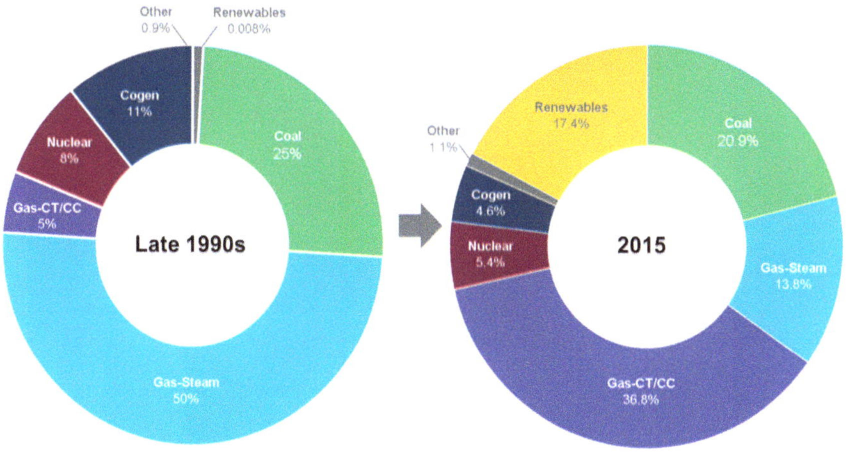

Fig. 1.2 Changes in ERCOT resource mix, in percent of installed capacity [3]

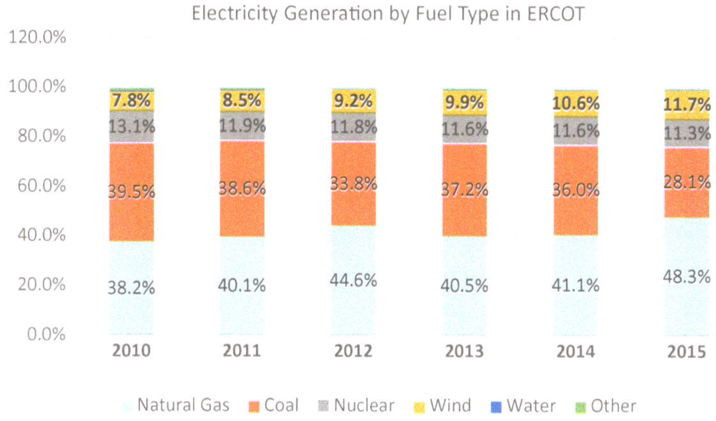

Fig. 1.3 Electricity generation by fuel type in ERCOT [4]

Apart from the availability of excellent wind resource, the Renewable Portfolio Standard (RPS) and Production Tax Credit (PTC) greatly influence the development of wind generation in Texas. An RPS is a mandate passed by the Texas Legislature to establish a minimum amount of renewable resources in the state's generation portfolio. Texas's RPS target was 10,000 MW of renewable generation by 2025. This target was exceeded by 2010. Additionally the federal Production Tax Credit (PTC) has been vital in maintaining economic viability of the wind generation projects. The PTC is a federal incentive that provides financial support for development of renewable energy facilities. It provides a 2.3 cent/kWh incentive for the first 10 years of a renewable facility operation. The PTC originally started in 1992 and expired at the end of 2013. Additionally, in December 2014 the PTC was

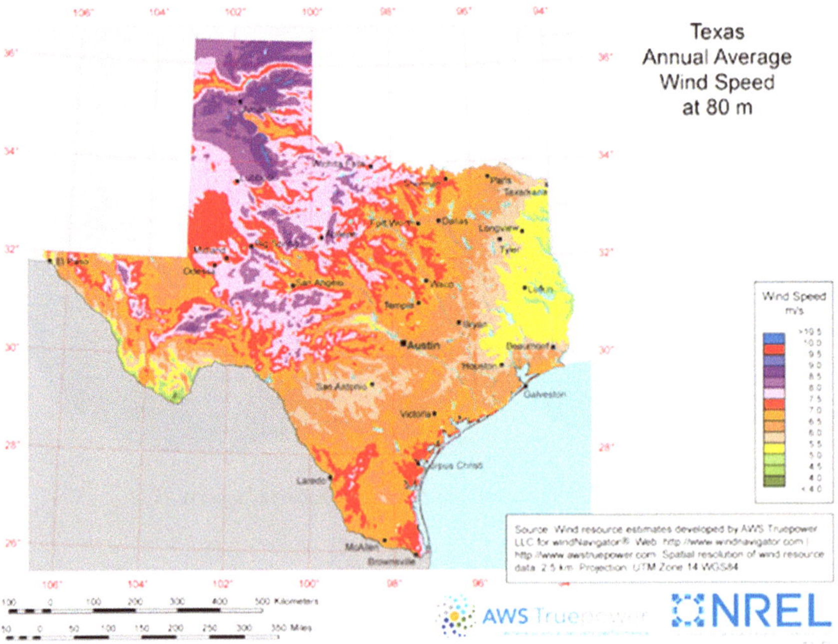

Fig. 1.4 Average annual wind speed in Texas at 80 m [5]

retroactively extended for the projects that were under construction by the end of 2014. In order to receive tax credits, these projects had to be operational by the January 1, 2017. These policy changes are clearly reflected in wind generation resources development in ERCOT, which has shown a steady increase of about 1,000 MW per year from 2008–2014 and a substantial leap in 2015 (during which over 3 GW was added) followed by an additional increase of more than 5 GW planned by the end of 2016, as seen in Fig. 1.5.

At the end of December 2015 another extension of PTC was signed. This 5-year extension will run through 2019. This new retroactive PTC extension allows developers to earn the full 2.3 cent/kWh tax credit for projects that meet "commence construction" criteria in 2015 and 2016 and are completed within two years. The credit drops by 20% in 2017, 20% further in 2018, and a final 20% more in 2019 before terminating on January 1, 2020 [6].

The installed wind capacity within the ERCOT footprint is 16,129 MW (as of June 1, 2016), making Texas the leading state for wind capacity in the USA. If Texas was a separate country it would be 6th in the world in terms of installed wind capacity as of the end of 2014 [8].

Most of the wind generation resources in ERCOT are connected at transmission voltage levels (69, 138 kV or 345 kV). Project sizes range from 1 to 250 MW (110 MW on average).

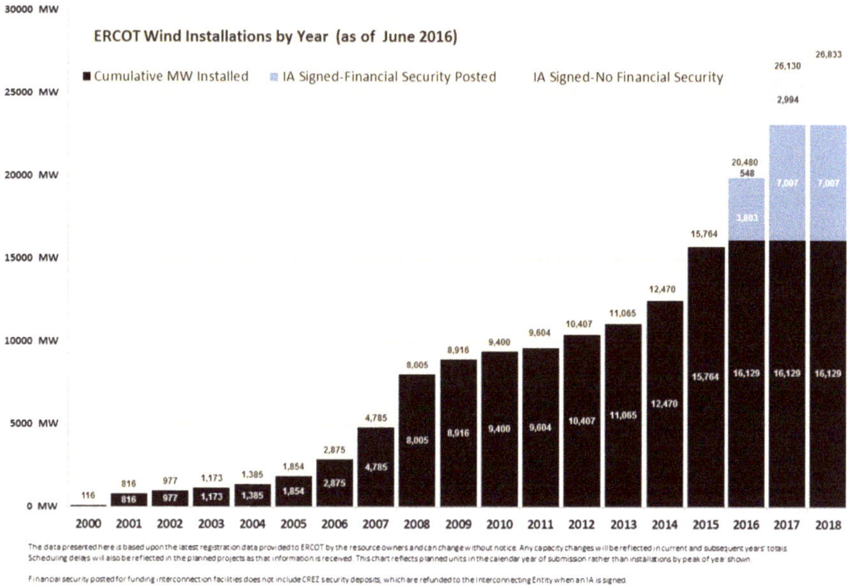

Fig. 1.5 ERCOT wind installations by year, [7]

The most recent instantaneous wind generation record of 14,023 MW was set on February 18th, 2016. The instantaneous penetration record was set on March 23, 2016 with wind serving 48.3% of electricity demand at the time (13,154 MW of wind generation and 27,244 MW load). A higher wind power penetration level could have been set at that time with wind generation potentially being sufficient to serve 51.3% of load; however, due to transmission constraints and thermal generators reaching their minimum sustainable generation levels, over 800 MW of wind generation was curtailed. With rapid developments of wind generation projects in 2015–2016 these records are likely to be exceeded in the near future.

About 89% of total installed wind generation capacity is located in the West Texas and Texas Panhandle. These areas are rich in wind resources (see Fig. 1.4), scarcely populated and thus are currently offering better economic opportunities for wind generation build out. About 11% of capacity is along the Gulf Coast near and south of Corpus Christi. This area has more favorable wind patterns well correlated with ERCOT load and, at times, high wholesale electricity market prices compared to West Texas and Panhandle. However, the coastal area has a lot of industrial and tourist activity which makes wind generation development there more challenging and expensive.

1.2 Transmission Development and Capacity Adequacy

1.2.1 Transmission Access

In ERCOT generation owners are not required to pay for transmission upgrades necessary to facilitate the interconnection. The generation owner only pays for the connection from their facility to the nearest point of interconnection on the transmission grid. Any other transmission upgrades necessary to accommodate new generating capacity are paid by the demand customers based on a flat rate approved by Public Utility Commission of Texas (PUCT). This reduces the cost of market entry for all new generation projects compared to some other systems where generators must pay for necessary transmission system upgrades associated with their project.

1.2.2 Transmission Reinforcement

As a result of the rapid development of wind generation in west Texas and very scarce transmission capacity in the region, there was initially insufficient transmission capacity to transfer wind energy to the large load centers in the north and central of the state (i.e. Dallas, Austin, San Antonio, and Houston). This situation could cause substantial wind energy curtailments when the penetration of wind generation was high. In 2005 the Texas legislature ordered the PUCT to determine areas where future renewables would likely to develop, so called "competitive renewable energy zones" (CREZ) as shown in Fig. 1.6. The commission designated ERCOT to develop a transmission plan to interconnect these areas to the existing transmission grid and elevate already-existing transmission congestion from the West to the North and Central parts of Texas. As a result about 3,600 right-of-way miles of 345 kV transmission lines were constructed with a capability of accommodating 18.5 GW of wind generation capacity. The project was completed in December 2013 with a cost of approximately $6.8 billion. The new lines are open-access and their use is not limited to wind generation. Some of these transmission lines also benefit fast growing demand from oil and gas industry in western Texas.

The CREZ transmission project also included new transmission facilities in the Texas Panhandle. Prior to the CREZ project, there were no ERCOT transmission lines extending into that area and therefore no load or generation in the area connected to ERCOT system. Furthermore in the beginning of the CREZ project there was no generation with signed interconnection agreements in the Panhandle area. The reactive equipment necessary to support the export of power from the Panhandle was implemented to accommodate 2,400 MW of wind generation capacity, even though the transmission lines were constructed for much larger generation capacity. This decision was made because the size and location of any

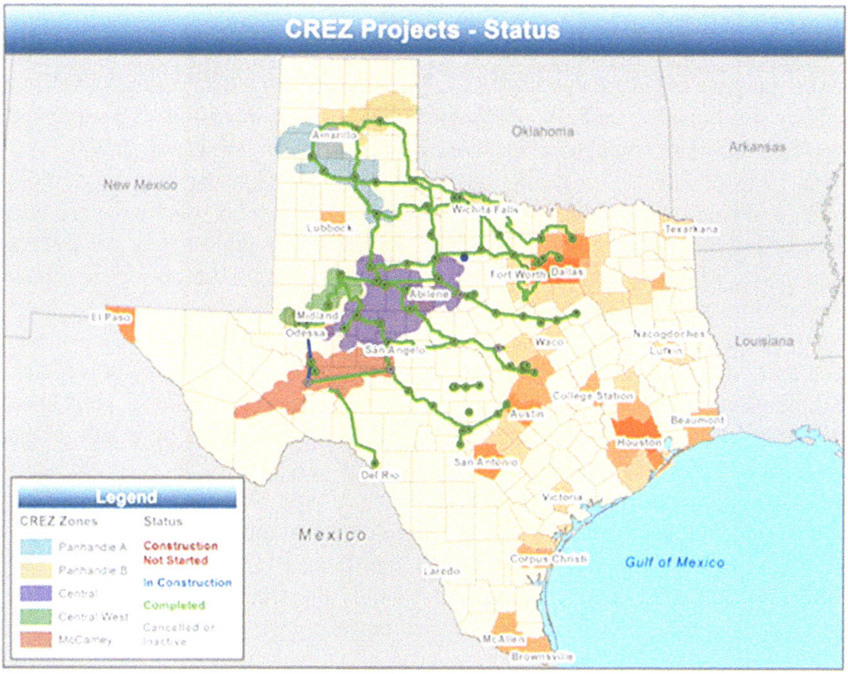

Fig. 1.6 Texas invested in transmission lines linking wind-rich "Competitive Renewable Energy Zones" to the state's largest cities [9]

additional equipment would be dependent upon the size and location of the wind generation that eventually would be developed in the area in the future. By the time CREZ transmission project was completed, there was only 200 MW of wind generation capacity installed in the Panhandle. However over the next two years (2014–2015) another 2.2 GW of wind generation capacity was built, reaching the limit that CREZ transmission facilities in Panhandle area were designed for. There is currently an additional 5.3 GW of wind, 0.2 GW of solar as well as 0.2 gas-fired generation capacity with signed interconnection agreements planned in Panhandle area in 2016–2017 timeframe [7].

The Panhandle part of the ERCOT grid is remote from synchronous generators and wind generation projects in the Panhandle are equipped with advanced power electronic devices that further weaken the system strength due to limited short circuit current contributions. As a consequence, voltage control becomes very difficult because a small reactive power change results in large voltage deviations. By the end of 2015 additional transmission reinforcement projects were proposed to provide reactive power support and allow further wind generation growth in the Panhandle area.

1.2.3 Capacity Adequacy and Wind Generation Resources

For the purpose of calculating the ERCOT generating reserve margin, ERCOT until recently was counting on 8.7% of installed wind capacity for all wind power plants regardless of their geographical location. This assumption was based on 2007 study using Loss of Load Probability methodology [10]. This capacity contribution number did not reflect subsequent increases in the number of wind generation projects and their geographic dispersion. Also, applying one capacity value for all wind power plants regardless of their location disregards the strong correlation between load patterns and wind generation in the coastal area.

In October 2014 a new methodology for calculating the capacity value of wind generation during peak load periods was approved. The new approach calculates average historical wind generation availability during the top 20 peak load hours, rather than the effective load carrying capability as determined by a loss of load probability studies. Wind generation availability is expressed as a percentage of installed wind capacity and evaluated over a multi-year period.[2] This methodology improves data and process transparency while also giving ERCOT the ability to update the wind capacity value on a more frequent basis [11].

This new approach also included determining the wind capacity values for all four seasons, as well as distinguishing between wind resources located in non-coastal and coastal wind regions, recognizing the differences in production patters between these two areas.

By the end of 2015 capacity contribution from wind generation, based on 6 years of historical data, was 12% for non-coastal wind and 55% for coastal wind across the summer peak and 18% and 37% respectively across the winter peak.

1.3 ERCOT Energy Market and Ancillary Services

1.3.1 Over View of Energy Market

The ERCOT electricity market is energy only, this means generators are getting their revenue from selling energy and ancillary services. There is no additional payment for long-term capacity availability as in some other regions, where capacity markets were introduced as additional tool to ensure long-term grid reliability by procuring the appropriate amount of generation capacity needed to meet predicted energy demand in a future year(s).

[2]The final value is the average of the previous 10 eligible years of Seasonal Peak Average values. Eligible years include 2009 through the most recent year for which data is available for the summer and winter Peak Load Seasons. If the number of eligible years is less than 10, the average shall be based on the number of eligible years available.

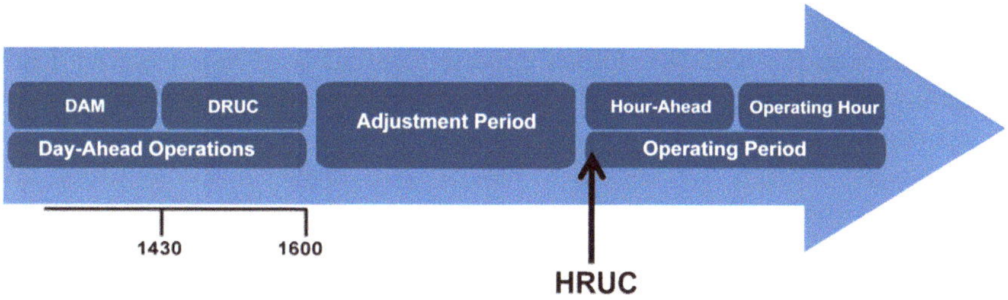

Fig. 1.7 Market operation in ERCOT

As detailed below, ERCOT market processes are designed to address uncertainty and variability of system load and wind generation at multiple timeframes—see Fig. 1.7.

Energy and Ancillary Services (Reserve) capacity are co-optimized in the Day Ahead Market (DAM). After DAM is cleared, day-ahead reliability unit commitment (DRUC) is carried out to ensure that sufficient generation is committed in the right location to serve next day's load forecast.

DRUC is re-evaluated every hour in hourly reliability unit commitment (HRUC), based on updated system conditions (e.g. generation outages, network topology changes etc.), as well as updated load and wind power production forecasts.

Security constrained economic dispatch (SCED) is executed every 5 min in real time and dispatch instructions are sent to generators to follow changing system conditions (load, renewable power production, unit outages, network topology changes). SCED uses real-time telemetry to receive up-to-date information from all generators.

Locational Marginal Prices (LMP) are calculated by SCED every 5 min based on submitted energy offers and system constraints. Generators are paid for energy based on the LMP at the generator's point of interconnection (weighted by time and volume over 15 min settlement interval). Loads pay for their consumed energy based on a 15 min average (weighted by time and volume) of 5-min LMPs within a predefined geographical area, called a load zone.

Within each 5 min interval, various types of reserves are deployed to keep system frequency close to the nominal value of 60 Hz.

1.3.2 Ancillary Services

Ancillary Services in ERCOT currently include Responsive Reserve Service, Regulation Service, Non-Spinning Reserve Service, Black Start and Emergency

Interruptible Loads. The latter two services are used in emergency and system recovery conditions, while the former three services are used to balance net load[3] variability and support frequency after generation outages. These services are described in more details below.

Responsive Reserve Service (RRS) bundles two distinct functions within one service. This reserve is used for frequency containment, i.e. to arrest frequency decline after generator trip, and as a replacement reserve to restore the depleted Responsive Reserves and bring the frequency back to 60 Hz. Until recently (June 2015) ERCOT procured 2800 MW of RRS for every hour in a year. Of this amount 50% could be provided by interruptible load resources with automatic under-frequency relays. The relays are activated within 0.5 s, if system frequency drops to 59.7 Hz or lower. These interruptible loads providing RRS are usually large industrial loads. The remainder of RRS is provided by generation resources and is deployed autonomously through governor response (as containment reserve) and/or through Security Constrained Economic Dispatch (as replacement reserve).

Recently, due to changing generation mix in ERCOT, the methodology for determining RRS has changed from procuring a constant amount of 2800 MW for all hours to determining necessary amounts of RRS dynamically based on expected system inertia conditions. The motivation behind this change is discussed in detail in Sect. 1.3.4.

Regulation Reserve service is a restoration service, used to restore frequency back to 60 Hz after a disturbance as well as to balance out intra 5-min variability in net load. Resources providing Regulation Service respond to an Automatic Generation Control (AGC) signal from ERCOT every 4 s. The Regulation reserve requirements are determined separately for Regulation Up and Regulation Down for each month, hours 1 through 24 (i.e. 24 values per month). The requirements are based on Regulation service deployments for the same month in the previous two years and a certain percentile for the 5-min net load variability for the same period. ERCOT also can increase Regulation requirements based on historic exhaustion of Regulation reserves as well as make adjustments to account for newly installed wind generation capacity that is not yet captured in the historic evaluation period [12].

Somewhat counterintuitively, Regulation requirements in ERCOT are trending down with increasing installed wind capacity, as shown in Figs. 1.8 and 1.9. This can be explained by the increase of geographical dispersion of wind generation resources as well as the continuous fine-tuning of AS methodology for determining the requirements.

Fast Responding Regulation Service (FRRS) was introduced in 2013 as a subset of Regulation service to allow resources to provide fast frequency support to the system. Resources providing FRRS are required to respond to a separate AGC signal as well as to a local frequency trigger (currently set at 59.91 Hz). Each resource with FRRS obligation should provide its full response within one second.

[3]Net load is defined as system load minus wind and solar power production.

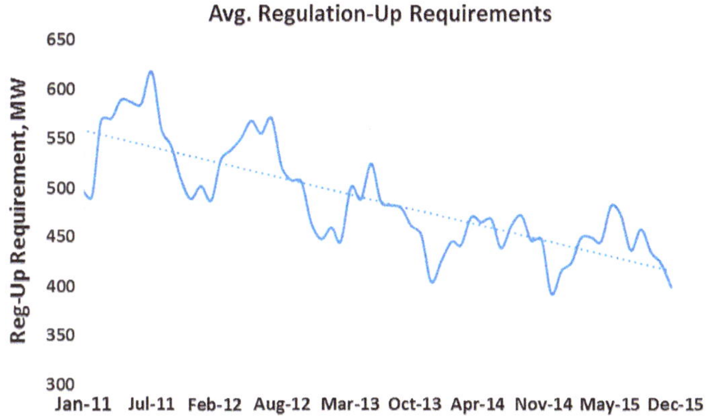

Fig. 1.8 Reduction in regulation up requirements between 2011 and 2015

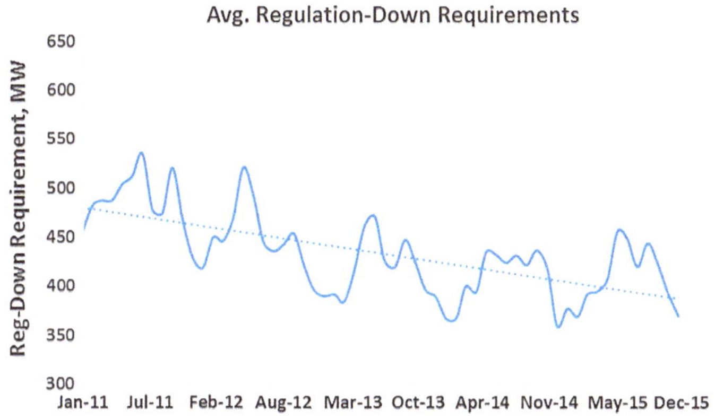

Fig. 1.9 Reduction in regulation up requirements between 2011 and 2015

Non-Spinning Reserve Service (NSRS) is a service to provide support within 30 min through online and/or offline resources. NSRS may be deployed to react to the loss of a generator, compensate for net load forecast error and to address the risk of large net load ramps or when low amount of generation capacity is available in SCED. Historically, the need for NSRS has occurred during hot or cold weather, during unexpected weather changes, or following large unit trips to replenish deployed reserves. The amount of NSRS is determined for every month in 4-h blocks (i.e. 6 values for each month). The NSRS requirement for each 4-h block is determined as a percentile of the 3-h ahead net load forecast errors in that block of hours from the same month of the previous three years. The percentile is chosen based on the risk of high net load ramps in the period under evaluation. If the risk of the net load ramp is high, the NSRS requirement is based on the 95th percentile of the hourly net load forecast error; if the risk is low, the 70th percentile applies. One example of the minimum requirement for NSRS in 2016 is provided in Fig. 1.10.

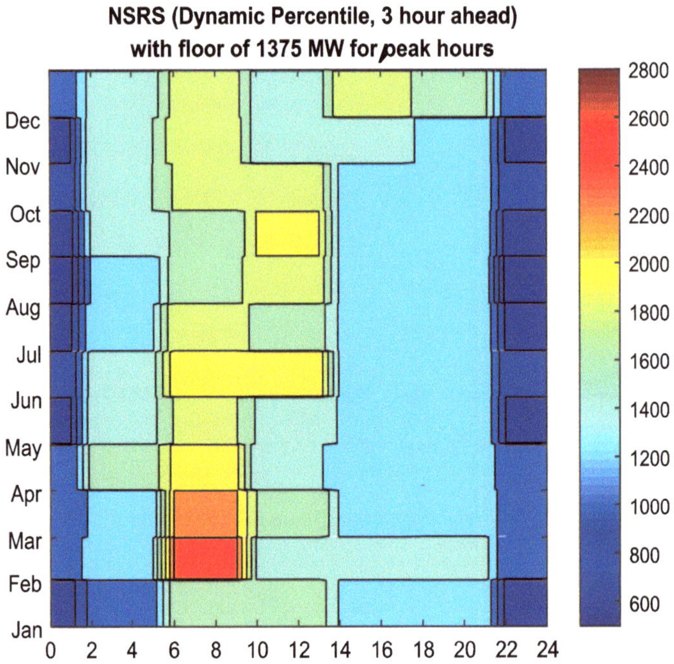

Fig. 1.10 Minimum NSRS requirement in 2016

1.3.3 Requirement for Primary Frequency Response

In ERCOT all online generation resources, including renewable generation, are required to provide primary frequency response through governor or governor-like action to the changes in the system frequency outside of the narrow deadband of ±0.017 Hz [13]. Primary frequency response shall have a droop characteristic of a maximum of 5%. However, only resources that are awarded Responsive Reserve in the DAM are required to reserve certain amount of capacity to meet their obligations. Other resources will provide primary frequency response only if they have available headroom. Figure 1.11 illustrates the response to an over-frequency event provided by wind generation resources through governor-like response.

1.3.4 ERCOT Frequency Performance

ERCOT's frequency performance is monitored by North American Reliability Corporation (NERC) through the Control Performance Standard (CPS1). CPS1 measures the quality of frequency control performance as follows [14]:

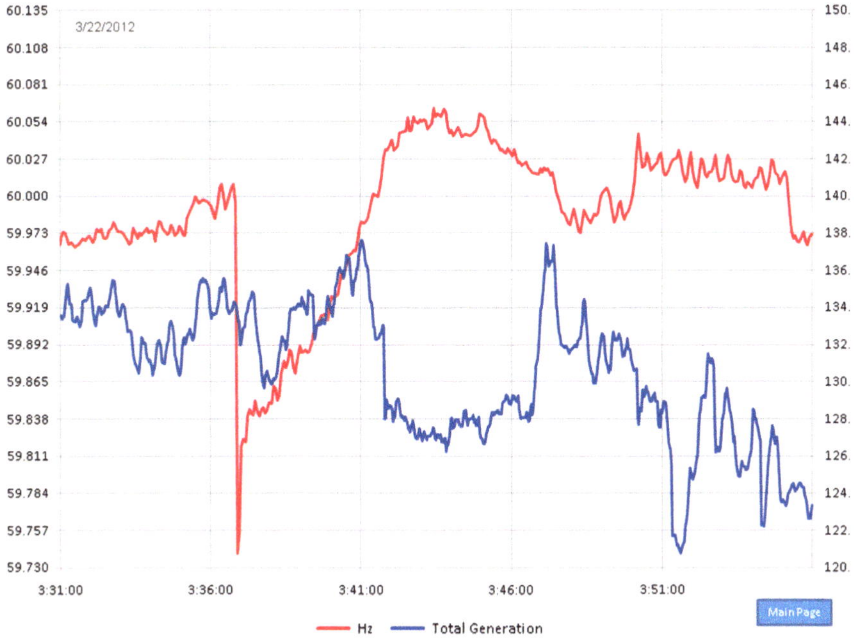

Fig. 1.11 Governor-like response from wind generation resources during an over-frequency event

$$CPS1 = 100\%(2 - \frac{\overline{\Delta f_{1m}}^2}{\varepsilon^2}) \quad (1)$$

where ε (Hz) is determined by NERC and represents the historical performance of frequency control (for ERCOT ε is 0.03 Hz), and $\overline{\Delta f_{1m}}$ is the 1-min average of system frequency deviations sampled every 4 s. The 1-year average of CPS1 must be larger than 100% for system frequency to comply with the standard.

ERCOT's frequency performance is continuously improving over the past years, Fig. 1.12, despite a growing share of variable generation resources and reduction in Regulation requirements, Figs. 1.8 and 1.9. This increase can be attributed to:

- Continuous fine-tuning of the procured AS amounts, based on historical data;
- Primary frequency response requirements for all generators, including renewable generation;
- Continuous performance monitoring by ERCOT after each significant frequency event;
- Generators preparing for implementation of BAL-TRE-001 standard that narrows governor deadband settings from 0.036 to 0.017 Hz.

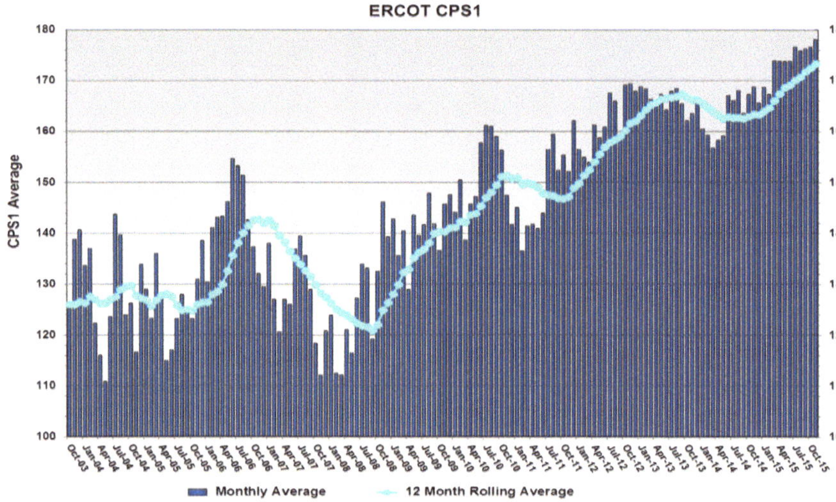

Fig. 1.12 Frequency performance in ERCOT from 2003–2015

1.3.5 Need for Modification of Ancillary Services

The increasing penetration of non-synchronous generation (wind and solar) in ERCOT leads to economic displacement of synchronous generators that would otherwise be committed to serve system load. This results in an overall reduction in the system's synchronous inertia (kinetic energy) within ERCOT Interconnection. Figure 1.13 shows the boxplots for the total system synchronous inertia for the past 3 years and the first quarter of 2016 (until 3/31/2016). The inertia is based on individual unit's power production. If a unit's power production is higher than a 5 MW threshold, the unit is considered online and its inertia contribution is counted toward the total system inertia. Individual unit inertia contribution is calculated as its inertia constant in seconds multiplied by its installed capacity in MVA). The corresponding lowest inertia in each year is provided in Table 1.1.

Table 1.2 provides additional information for wind generation penetration record in each year. These data points are marked with circles on the boxplots, Fig. 1.13. Notably, system inertia during the wind penetration record in each year is fairly close to the minimum inertia encountered in that year.

In the boxplots, Fig. 1.13, the median of the inertia in ERCOT is trending up even though the installed capacity of wind generation is increasing over the past few years. This can be explained by changes in unit commitment patterns. Lower gas prices in 2014–2016, Fig. 1.14, result in a change in the merit order between Combined Cycle units and Coal units, the latter of which have lower inertia contributions for the same unit size.

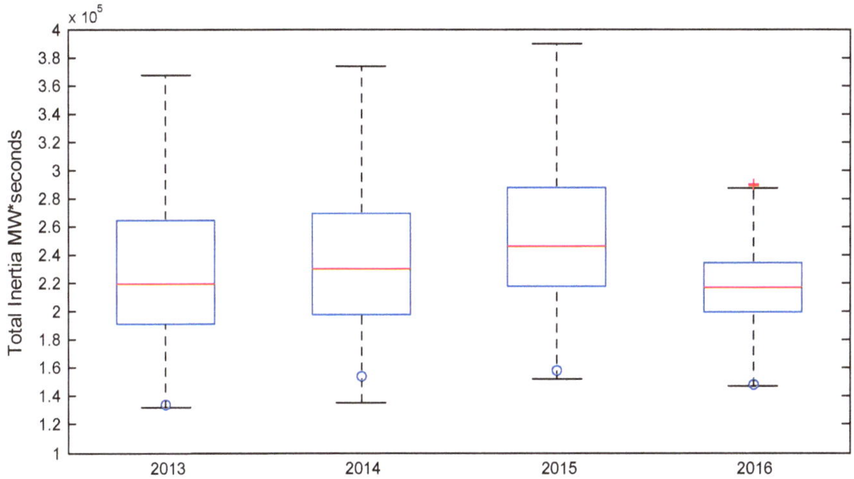

Fig. 1.13 Boxplot of the system inertia from 01/01/2013 to 3/31/2016 (On each box, the central mark (*red line*) is the median, the edges of the box (in *blue*) are the 25th and 75th percentiles, the whiskers correspond to +/−2.7 sigma (i.e., represent 99.3% coverage, assuming the data are normally distributed), and the outliers are plotted individually (*red crosses*). If necessary, the whiskers can be adjusted to show a different coverage.)

Table 1.1 Lowest inertia in different year (GW s)

2013	2014	2015	2016
132	135	152	147

Table 1.2 Wind generation power and load demand at the time of record of each year

	2013	2014	2015	2016
	3/9 3:15 am	11/3 2:28:56 am	12/20 3:05 am	3/23 1:10 am
Installed capacity (P_{wind_inst}) (MW)	10,570	12,527	16,170	16,547
P_{wind}/P_{load} (%)	35.8	39.93	44.71	48.28
P_{wind} (MW)	8,773	9,882	13,058	13,154
P_{wind}/P_{wind_inst} (%)	83	78	81	79
Net load ($P_{load} - P_{wind}$) (MW)	15,716	14,868	16,150	14,091
Inertia (MW * s)	134,196	154,599	158,970	148,798
Installed capacity (P_{wind_inst}) (MW)	10,570	12,527	16,170	16,547

1.3.6 Regulatory Requirement Changes

In January 2014 the US Federal Energy Regulatory Commission (FERC) approved the NERC BAL-003 standard which requires each interconnection to meet a

Fig. 1.14 Daily natural gas prices between 2010 and 2015

minimum Frequency Response Obligation (currently 413 MW/0.1 Hz for ERCOT), determined based on a simultaneous trip of the two largest generation units, i.e. 2750 MW. ERCOT also plans its operations to ensure that, for the instantaneous loss of two largest generators, the system frequency is arrested above the first stage of firm load shed, which in ERCOT is at 59.3 Hz.

In order to ensure compliance with the standard, the amount of frequency responsive reserve, i.e. RRS under the current set of Ancillary Services, has to be increased during low synchronous inertia hours. Since RRS currently also bundles several services together (fast frequency response from load resources with under-frequency relays, governor response from generation resources and frequency restoration provided by generation resources after a frequency event), it's becoming increasingly difficult to determine an adequate and efficient level of RRS to optimally serve system needs.

1.3.7 Changes to Responsive Reserve Requirement

A dynamic study was conducted by ERCOT staff in 2014 to examine the minimum primary frequency response requirement to prevent the frequency from dropping below 59.4 Hz (0.1 Hz above the prevailing first step of involuntary under-frequency load shedding) after the loss of two largest generation units (2750 MW). The study was conducted for thirteen scenarios with different system inertia levels. It was found that more Responsive Reserve Service (RRS) is needed for low-inertia situations to maintain the security and reliability of the grid.

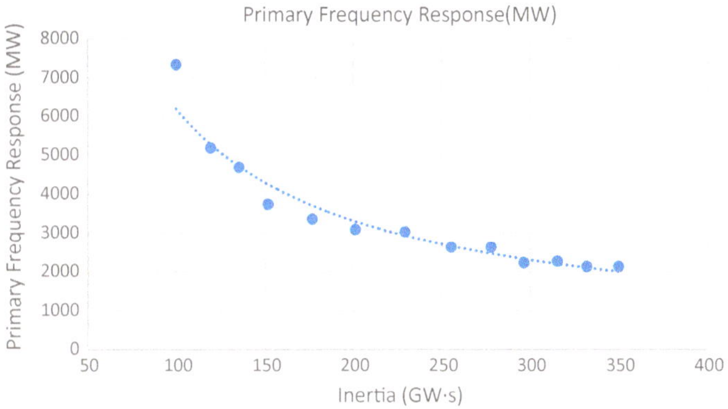

Fig. 1.15 Minimum primary frequency response requirement under different system inertia conditions

The RRS requirement, expressed in terms of primary frequency response (PFR)[4] is shown in Fig. 1.15. The study showed that when system inertia is extremely low, a large amount of incremental PFR will be needed, e.g., if the system inertia changes from 120 to 100 GW s, the minimum necessary quantity of PFR will increase from 5200 to 7350 MW. The study also showed that when the inertia of the ERCOT system is less than 100 GW s, the loss of the two largest units will cause voltage oscillations and voltage control issues. This issue cannot be mitigated solely by increasing frequency response reserves. Based on this study, in 2015, it was proposed to change Ancillary Service methodology and determine a minimum requirement of RRS based on expected system inertia conditions.

As described in Sect. 1.3.2, up to 50% of the total RRS requirement in ERCOT can be provided by load resources with under-frequency relays, delivering full response in 0.5 s after frequency falls at or below 59.7 Hz. Previous studies showed that during low inertia periods 1 MW of load resources is up to 2.35 times more effective than 1 MW of PFR from the generators, due to the fast speed of the response, as shown in Fig. 1.16. In the new AS methodology, the load resources are counted towards total RRS requirement with the equivalency ratios, based on inertia conditions that are seen in Fig. 1.16.

---

[4]PFR reserve is released from generation resources with RRS responsibility through governor action when system frequency falls outside of the pre-set dead band. While primary frequency response capability is required from all generators in ERCOT, as discussed in Sect. 1.3.3 above, only generators with RRS responsibility have to reserve capacity and be available to provide primary frequency response when needed.

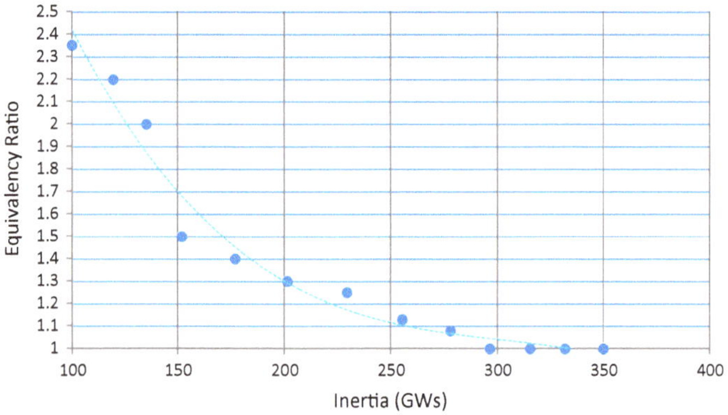

Fig. 1.16 Equivalency ratio between RRS provided by load resources and generation resources under different system inertia conditions

1.4 ERCOT Tools to Monitor and Predict System Inertia

1.4.1 Inertia Monitoring Tool and Dashboard

In order to facilitate the monitoring and analysis of total system inertia as well as contribution by individual generation types, a tool was developed in ERCOT for real-time synchronous inertia calculation performed once a minute[5] for each resource type as well as for the system total.

Additionally an experimental real-time inertia dashboard was set up to monitor the inertia in real-time as shown in Fig. 1.17. The dashboard also shows the inertia contributions by generator type for the previous 24 h. Monitoring by the generation type enables more granular analysis of inertia trends.

1.4.2 Inertia Prediction Tool

System inertia can be predicted using the Current Operating Plan (COP)[6] information submitted by generators to provide some foresight into what future operating conditions will be. The probability distribution function of a 3-h-ahead inertia estimation error using COP data in 2015 is shown in Fig. 1.18a. This prediction can give operators an opportunity to evaluate the sufficiency of procured Responsive Reserves as well as recognize the risk for extremely low inertia conditions ahead of time and prepare a mitigation plan if needed. The performance of 3-h-ahead RRS

[5]Currently being updated to calculate every 4 s.

[6]A resource at ERCOT should reflect in its COP the expected operating conditions.

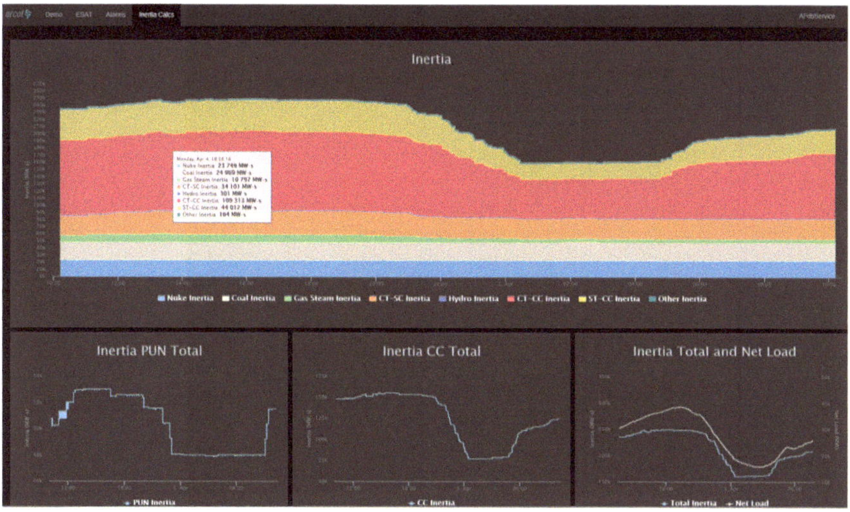

Fig. 1.17 A dashboard to monitor inertia in real time

estimation was evaluated based on 2015 data with the results shown below in Fig. 1.18b. In this evaluation, the 3-h-ahead RRS estimation produced a value below the actual RRS requirement (an under-estimation) 3.78% of the time and over-estimated the RRS requirement 10.1% of the time. One example of under-estimation of the RRS requirement is depicted in Fig. 1.19. On November 4, 2015, the system lambda (the system-wide price of energy without consideration of transmission constraints, shown by the dotted line in Fig. 1.19) dropped below \$10/MWh in the early morning. In response to this, some generation units which submitted "online" status in COP ahead of time were actually offline in real time, which resulted in an under-estimation of the system inertia. When the energy price has recovered, the generation units came back online, so the 3-h-ahead estimation of the system inertia matched well with the actual system inertia after 6 am.

1.5 Mitigation of Extremely Low System Inertia

Looking forward, the system inertia at ERCOT will further decrease as the wind generation continues to grow, and this poses a challenge to maintaining a satisfactory frequency control performance. Multiple technical solutions can be explored to mitigate the impact of low system inertia:

- Bring online additional synchronous inertia as needed by committing additional units, committing different units that have higher inertia, and/or using synchronous condensers.

Fig. 1.18 3-h-ahead estimation error of inertia and RRS requirement

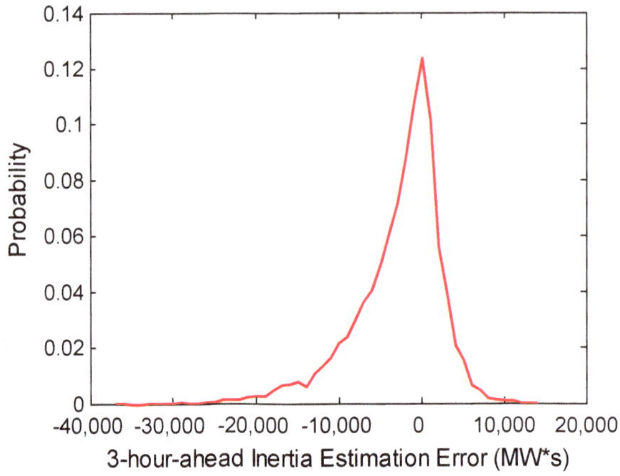

(a) errorin 3-hour-ahead inertia estimation

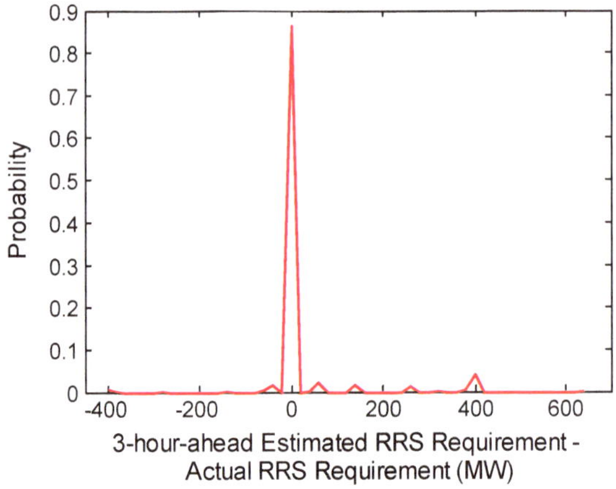

(b) error in 3-hour RRS requirement estimation

- Slow down the rate of change of frequency after an event (e.g. generator trip) by increasing rate of primary frequency response of the system in MW/Hz per second.

Slow down the rate of change of frequency after an event by increasing the speed of frequency response, i.e. add fast frequency response from load resources, storage, and synthetic inertia from wind generation.

These technical solutions can be implemented as described below and their effectiveness is summarized in Table 1.3.

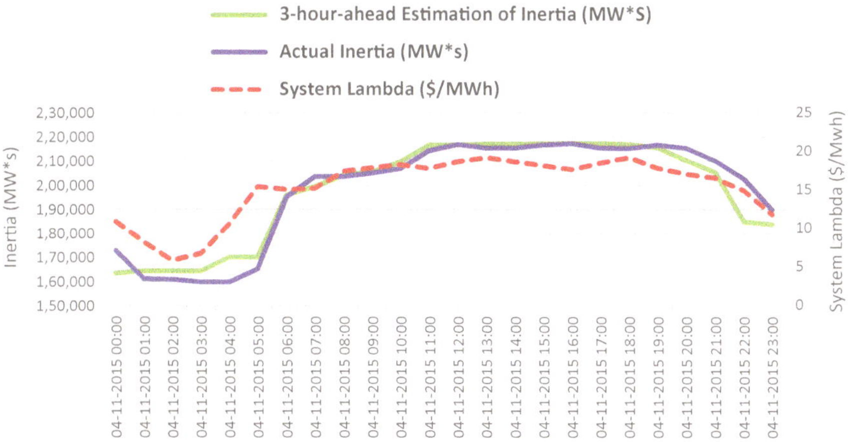

Fig. 1.19 System lambda and inertia on Nov. 4, 2015

Table 1.3 Options to mitigate impact of low system inertia

	Time horizon	Impact	Cost of implementation	Effectiveness
RUC more synchronous generators	Operation horizon	Increase inertia, wind curtailment, adverse effect on real-time market prices	Medium	Medium
Procure inertia as an ancillary service	Operation horizon	Introduce new AS, increase inertia, wind curtailment	High	Medium
Install synchronous condensers	Long-term planning horizon	Increase inertia, additional dynamic voltage support	High	High
Increase monthly RRS requirement	Operation planning horizon	Increase reserve and rate of response, wind curtailment	Medium	High
Fast frequency response (storage, load resources etc.)	Operation horizon	Introduce new AS, increase reserve	Low-medium	High
Synthetic inertia	Operation horizon	Introduce new AS or protocol requirement, requires implementation of centralized control	Low-medium	Medium

1.5.1 Bringing Additional Synchronous Inertia Online

- *Start up more generation units*

If system inertia is low, an operator may choose to bring more synchronous generation units online and thus increase system inertia. One benefit of this approach is that it would only be implemented during conditions when economically committed generation is not providing sufficient system inertia, and therefore it would not affect market solutions during other periods. The generating units, once started, would run at least at their minimum sustainable level, producing energy that is not needed otherwise. This may result in renewable curtailment and adverse effects on energy prices. The generators started up for inertia would need to be compensated (outside of normal market operations) for their startup costs and power production at minimum generation levels. It would be important to consider the relative inertia provided by the different units that could be brought online and their minimum generation levels, in order to get the maximum inertia at the least cost. This would only be recommended when the system inertia falls below the minimal inertia requirement, which causes a serious concern for reliability.

- *Procure inertia as a separate ancillary service product*

This mitigation measure would enable ERCOT to maintain necessary synchronous inertia levels through a market mechanism. Currently, inertia is provided by synchronous machines as the byproduct of energy or ancillary service and no direct compensation is offered for this service. Once historic analysis of system inertia shows that ERCOT more frequently experiences extremely low inertia conditions, it could be more efficient to incentivize synchronous inertia service as an Ancillary Service and to add a minimum inertia constraint into DAM clearing process. This may be the most efficient way to address low system inertia issues because it is implemented through a market mechanism. On the other hand, it would require significant changes to the current ERCOT market design and structure.

- *Include the need for inertia in the procurement of RRS*

The current procurement of RRS takes into account the expected level of inertia on the system to determine the quantity of RRS to be procured. However, the RRS procurement could be modified so that the different contributions to inertia from the resources offering to provide RRS could be considered in determining which offers are selected to provide RRS. In this construct, the RRS procurement would solve the need for RRS while simultaneously ensuring inertia sufficiency.

- *Install synchronous condensers*

A synchronous condenser is similar to synchronous generator but without a turbine; its main purpose is to provide dynamic reactive power support. Being a rotating machine, however, it can also contribute to total system synchronous inertia. This option may be expensive if used for inertial support alone, however recent studies for the Panhandle region also proposed the installation of synchronous condensers for voltage support. As one of the long-term solutions, use of synchronous condensers for multiple purposes may prove to be more cost efficient. Regardless, the contribution to inertia from synchronous condensers should be included in the assessment of total available system inertia.

1.5.2 Increase the Rate of Primary Frequency Response

- *Procure more responsive reserve service (RRS)*

As shown in Fig. 1.4, the need for RRS is increasing significantly during low inertia conditions. By procuring more RRS capacity spread over more units, a higher rate of response MW/Hz per second could be obtained during severe under-frequency events. This mitigation measure is easy to implement since it can be accommodated within the existing AS framework. However, this option would be inefficient to some extent, since a large amount of RRS would need to be reserved for the whole month but only used to hedge against a few hours when system inertia conditions are expected low. Additionally, if low inertia conditions are predicted ahead of time, additional RRS needs can be procured in real time through a Supplemental Ancillary Service Market (SASM), but the price of RRS may be extremely high as well. During low system inertia hours, high RRS requirements could also force more synchronous generation online to provide such a service, thus increasing system inertia and decreasing the actual need for RRS simultaneously.

1.5.3 Add Fast Frequency Response

- *Incentivize/use fast frequency response from other technologies*

What is critically needed at low system inertia conditions is a fast response to counteract high rate of change of frequency. Fast frequency-responsive load resources, e.g. large industrial loads, heat pumps, industrial refrigerator loads, and storage devices, can provide full response in a few hundred milliseconds to the under-frequency events. This type of response has been compensated equally as the generators within the current market framework and could be incentivized further in the future by introducing a fast frequency response Ancillary Service. This is one of the most viable solutions if it can attract a sufficient number of participants to best utilize their characteristics.

- *Synthetic inertial response from wind generation*

Another example of fast frequency response is synthetic inertia provision from wind generation resources (Type 3 and Type 4 wind turbines). When a wind turbine plant controller senses system frequency decline, it extracts kinetic energy from the rotating mass of a wind turbine, which is seen from the grid as an increase in active power injection. The effectiveness of the response and recovery of wind generation resource to its pre-disturbance state depends on operating conditions of a wind generation resource. Therefore this type of fast frequency response requires careful centralized coordination to enable reliable system operation. While synthetic inertial response capability is already included as a part of the interconnection requirements in Hydro Quebec [15], this technique has not been commercially utilized on a large scale.

1.6 Conclusions

As new wind generation was first beginning to come online in ERCOT, its impact on the daily system operation was insignificant. However as more wind generation was built, ERCOT and its market participants had to introduce a number of changes to the generation interconnection requirements, market rules and market design, ancillary services and operation practices to ensure continued system reliability with increasing amounts of variable non-synchronous generation. ERCOT is continuously working on further modification of ancillary services procurement methodology, tuning the AS amounts to satisfy ever-changing system needs. ERCOT is also researching possibilities to supplement diminishing inertial response from synchronous generation with the use of emerging technologies.

References

1. ERCOT, Capacity Demand Reserve Report, http://www.ercot.com/content/wcm/lists/96607/CapacityDemandandReserveReport_May2016.xlsx
2. NERC, Interconnection Map, http://www.nerc.com/AboutNERC/keyplayers/Pages/Regional-Entities.aspx
3. Bill Magness, ERCOT in 2016: Transition and Continuity, presentation to Golf Coast Power Association, Houston, Nov 2015, http://www.ercot.com/news/presentations/index.html
4. ERCOT News Release, Energy use in ERCOT region grows 2.2 percent in 2015, 15 Jan 2016, http://www.ercot.com/news/press_releases/show/86617
5. U.S. Department of Energy, Texas Wind Resource Map and Potential Wind Capacity, http://apps2.eere.energy.gov/wind/windexchange/wind_resource_maps.asp?stateab=tx
6. H.K. Trabish, U.S. Wind Industry Hits 70 GW Capacity Mark, Celebrates Tax Credit Extension, UtilityDive, 22 Dec 2015, http://www.utilitydive.com/news/us-wind-industry-hits-70-gw-capacity-mark-celebrates-tax-credit-extensio/411224/
7. ERCOT, Generation Interconnection Study Report, http://www.ercot.com/gridinfo/resource
8. Global Wind Energy Council, Global Statistics in 2014, http://www.gwec.net/global-figures/graphs/
9. W. Lasher, The Competitive Renewable Energy Zones Process, presentation at the Quadrennial Energy Review Pubic Meeting on Local and Tribal Issues, 11 Aug 2014, http://energy.gov/epsa/downloads/qer-public-meeting-santa-fe-nm-state-local-and-tribal-issues

10. Global Energy Decisions, ERCOT Target Reserve Margin Analysis, January 2007, http://www.ercot.com/content/meetings/gatf/keydocs/2007/20070112-GATF/ERCOT_Reserve_Margin_Analysis_Report.pdf
11. ERCOT, Nodal Protocol Revision Request 611: Modifications to CDR Wind Capacity Value
12. ERCOT, Methodologies for Determining Ancillary Service Requirements, http://www.ercot.com/mktrules/issues/NPRR611
13. NERC, BAL-001-TRE-1—Primary Frequency Response in the ERCOT Region, http://www.nerc.com/pa/Stand/Reliability%20Standards/BAL-001-TRE-1.pdf
14. NERC Resources Subcommittee, Balancing and frequency control, 26 Jan 2011, http://www.nerc.com/docs/oc/rs/NERC
15. Hydro-Québec TransÉnergie, Transmission Provider Technical Requirements for the Connection of Power Plants to the Hydro-Québec Transmission System, February 2009, http://www.hydroquebec.com/transenergie/fr/commerce/pdf/exigence_raccordement_fev_09_en.pdf

Chapter 2
Integration of Large-Scale Renewable Energy: Experience and Practice in China

Ming Ni, Feng Xue and Zhaojun Meng

2.1 Development Status of Renewable Energy in China

With the strong incentive and promotion of the utilization of renewable energy, the installed capacity of renewable energy generators in China is growing continuously, which results in the sustainable increase of electricity generated from renewable energy resources. Thus, the security of power supply of the whole society is ensured, and the structure of the power sources is further optimized.

By the end of 2014, the total installed capacity of the generators in China is 1360 GW while the installed capacity of renewable energy generators is up to 430 GW, accounting for 32% of the total installed capacity. Among the installed capacity of renewable energy generators, installed hydro, wind power and solar power capacity are 300 GW, 115 GW, and 28.05 GW, respectively. For solar power, a large portion of the installed solar power capacity are large-scale solar power plants which contribute 23.38 GW, and the rest of 4.67 GW is distributed generators. During 2014, electricity generated from renewable energy is 1200 TWh, taking up 22% of the total generated electricity, with 153.4 TWh of energy from wind generators and about 25 TWh from solar generators.

Figure 2.1 shows the installed wind capacity in China grown from 2004–2014.

Figure 2.2 shows the installed wind capacities in 30 of 31 provinces, municipalities, and autonomous regions that have wind farms (exclude Hong Kong, Macao, and Taiwan). The Inner Mongolia Autonomous Region leads the other provinces with 22.3 GW of installed wind capacity, followed by Gansu, Hebei, Xinjiang, and Shandong.

China is also the first country outside of Europe to have offshore wind farms connected to the grid. By the end of 2014, the installed capacity of offshore wind

M. Ni (✉) · F. Xue · Z. Meng
NARI Group Corporation, No. 19, Cheng Xin Da Dao,
Jiangning District, Nanjing 211000, China
e-mail: mingni2002@hotmail.com

P. Du et al. (eds.), *Integration of Large-Scale Renewable Energy into Bulk Power Systems*, Power Electronics and Power Systems,
DOI 10.1007/978-3-319-55581-2_2

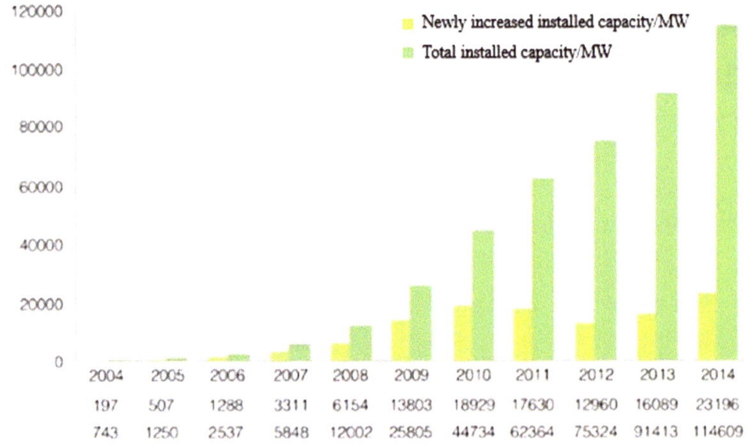

Fig. 2.1 Installed wind capacity in China

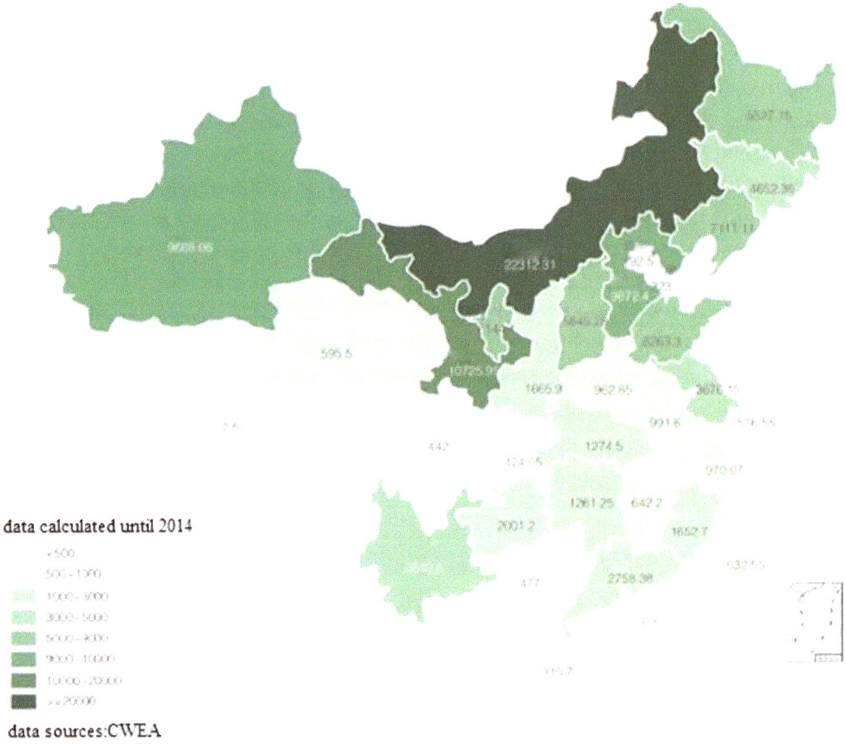

Fig. 2.2 Distribution of installed wind capacity in China

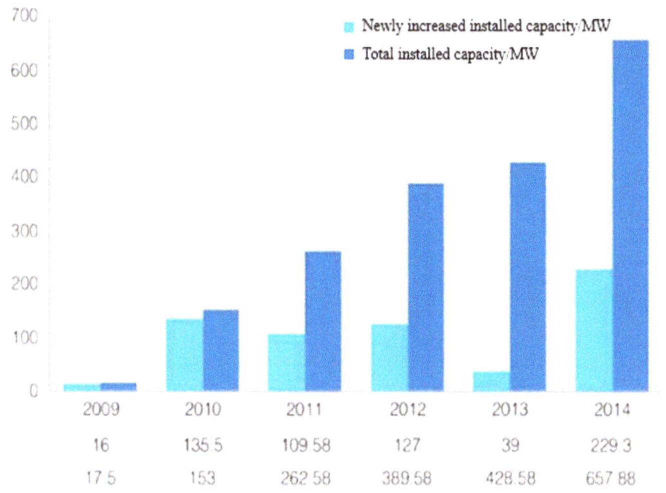

Fig. 2.3 Installed offshore wind capacity in China

power already built in China has summed to 657.88 MW. Among them, installed capacity of intertidal wind power is 434.48 MW, making up 65.6% of the total installed capacity of offshore wind power. Figure 2.3 shows the installed offshore wind capacity grown from 2009–2014 in China.

Figure 2.4 shows the distribution of effective wind power density in China. The north, northeast, and northwest areas of China, along with Tibet, are the wind-rich areas. However, they are far from the load centers (the North China Grid, Central China Grid, and East China Grid), which are also the most developed areas in China. Delivering wind power from wind-rich areas to the load centers is a big challenge to the integration of such a large amount of wind generation resources to the grid in China.

According to "The 12th five-year plan of renewable energy development" promulgated by the government of China in 2012, by the end of 2015, total installed wind power capacity in China will be approximately 100 GW, offshore wind power capacity will be 3–5 GW, and solar power capacity will be up to 21 GW. By the end of 2020, the total installed wind power capacity in China is estimated to be 200 GW, and solar power capacity can be 100 GW.

However, due to the rapid development of wind and solar power, and the commitment of the Climate Change Agreement, the government of china ought to adjust the development target of renewable energy in the 13th five-year plan to achieve the goal that non-fossil energy consumption accounts for 15% of the total energy consumption by 2020. At that time, installed wind power capacity is expected to reach 250–280 GW, and installed solar power capacity is expected to reach 150 GW or even 200 GW. By the end of 2020, there are projected to be eight wind plants with capacities larger than 10 GW in China, as shown in Fig. 2.5.

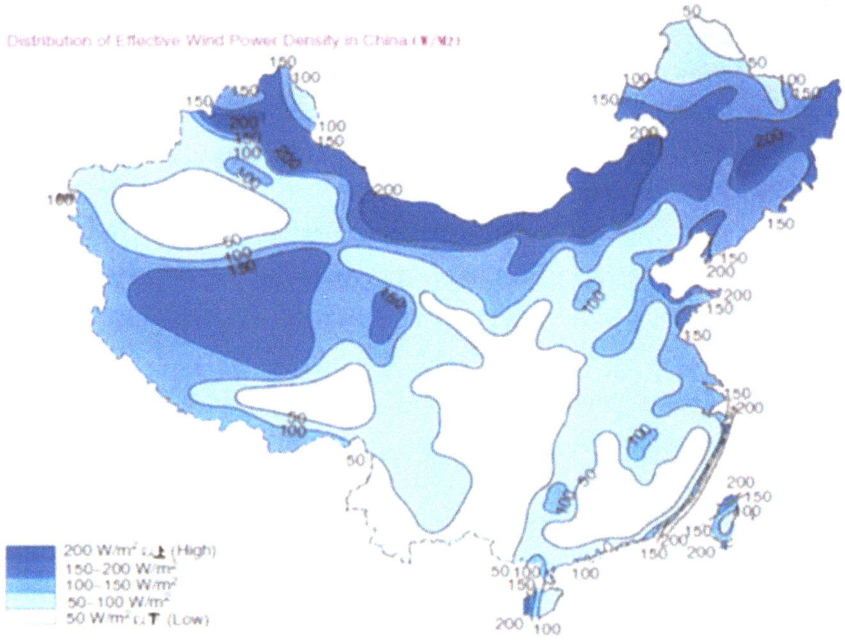

Fig. 2.4 Distribution of effective wind power density in China

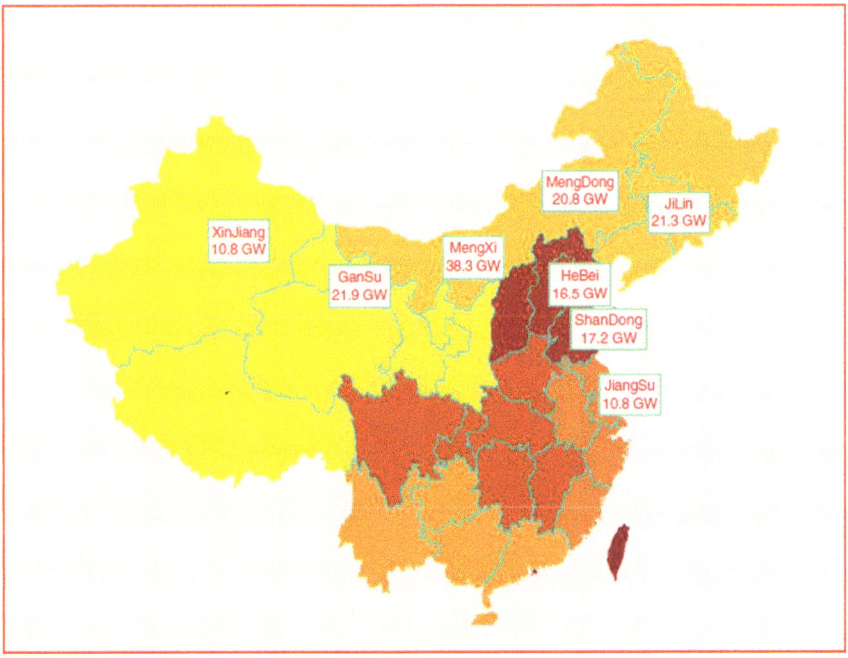

Fig. 2.5 Planned Chinese wind power plants with capacities of more than 10 GW in 2020

2.2 Problems in the Development of Renewable Energy

Renewable energy, especially wind and solar generators, has been experienced a dramatic growth. However, there exist some problems and factors that could restrain its further development in China.

A. ***Operation problem with renewable energy integration***

Wind and solar energy resources on land are mainly concentrated in three northern areas (northeast china, northwest china, and north china) and Qinghai-Tibet Plateau. Over 90% of national wind energy resource is located in three northern areas. The majority part of generators in three northern areas is thermal generators, with fast and flexible adjustable power source like hydroelectric power taking up merely 0.5–1.2%. Procurement of adequate peak regulation, which is pre-requisite to the accommodation of large-scale wind and solar generations, is especially difficult in heating period in winter.

On the other hand, despite the explosive increase of renewable energy, the renewable energy development is facing the following issues: (1) there are various types of wind turbines; (2) historical data of wind and solar resource is limited; (3) the development is mostly concentrated in three northern areas where the power system structure is weak; and (4) the majority power is thermoelectricity which lacks flexible adjustments, hence restriction of wind power output in some areas and large-scale trip-off accident of wind turbines happen from time to time. To maintain reliability and security of the power grid with increasing penetration of variable renewable resources has been one of the major obstacles to wind and solar power development in China.

B. ***Wind and solar energy curtailment***

In recent years, the curtailment of wind and photovoltaic generation has been the biggest problem in China's wind and solar farms. After 3 years of effort addressing this, the situation has been improved but still exists the problem. According to data from National Energy Bureau, the curtailed wind energy during the first half of year 2015 reaches 15.2%, increased by 6.7% compared to the same period in last year. Cumulative photovoltaic generation of the first half year nationwide adds up to 19 billion kWh, curtailed solar power is approximately 1.8 billion kWh, and curtailed solar power rate in Gansu and Xinjiang reach 28% and 19% respectively.

It's hard to predict whether the problem of curtailed wind and solar power can be solved soon, but the influence of wind and solar power curtailment on wind and photovoltaic generators industry is ineligible. The mostly affected one is investors of wind power, as many wind power enterprises in wind curtailment areas have undergone zero profit or even deficit. In addition, as the construction of power grid cannot catch up with the pace of wind power development, the risk of wind curtailment still exists.

Wind and solar energy curtailment has greatly affected the development of renewable energy industry in China. Therefore, the overall planning of power sources and power system must be strengthened, and smart grid technology must be developed. Peak regulation, frequency regulation and balancing energy storage should be arranged more efficiently.

C. ***Renewable energy pricing***

Currently, electricity price of renewable energy is determined based on market price plus government credit, ensuring the profit of wind and solar power industry. Not long ago, wind power price adjustment has caused an uproar in the industry. Many are worrying that the cutting down of electricity price may result in no profit in wind power industry, together with the influence of wind curtailment and national subsidy funds postponement, the enthusiasm of investors may be seriously set back. Therefore the problems should be taken into consideration in the thirteenth Five-Year Plan whether the goal of wind and solar power reaching 300 GW will be affected and how to organize future price mechanism without future price subside.

D. ***Renewable energy subsidy***

As for wind power, subsidy is an inevitable problem. With the future plan, if the present subsidy carries on, there will be huge deficiency in subsidy funds. Therefore there may be a limitation of total amount of renewable energy subsidy funds in the thirteenth Five-Year Plan, pressing a reduction of electricity price. Thus the trend in wind power industry is still under observation.

E. ***Development directions of renewable energy***

There are several disagreements in terms of development directions in renewable energy industry. Debates of wind power include the competition between large base in west, low-speed wind power development in the east and offshore wind power. Competition in photovoltaic power industry is between large-scale power station in west and distributed generation in east.

Although this year witnesses rapid development of low-speed wind generators in southeast area, some think the large base construction in west area remains the focal point according to wind resource conditions and technology mature level, as long as the problem of transmission pass way is solved. Low-speed wind power is the trend, requiring steady advance. Offshore wind power has great potential but should be developed in proper speed at current stage. According to the thirteenth Five-Year Energy Plan, the focus of wind power development is large base in west.

2.3 Typical Case of Large Scale Wind Generators Trip-off Accidents

Onshore wind energy resources in China are mainly located in three northern areas including Inner Mongolia, Gansu, Xinjiang and Heilongjiang. For example, six of the seven 10 GW wind power bases in national plan are located in three northern areas too. These areas are far from load centers, forming an opposing distribution of wind power resources and electric loads. Attributes of wind energy resources in China helped to adopt a main wind power development pattern, which combines large-scale centered exploitation and long-distant transmission. China's load centers are concentrated in populous southeast coastal areas, where the wind energy resources are less abundant and the development is also under restriction of land resources, environment and ecology as well, so that distributed generation cannot be deployed at large scale. Large-scale centered wind power development plan is more aligned with distribution of China's wind energy resources, and is preferred for speeding up wind power development, improving resource utilization, thus large-scale wind power base is the mainstream of China's wind power development manner. However, the wind power development pattern of large-scale centered exploitation and long-distant transmission are faced with various challenges in the frequency regulation and system stability.

From several wind turbine trip-off accidents in Jiuquan, Gansu in 2011, it has been learned that many wind turbines were involved in serial trip-offs and went offline because of transient voltage problems, which caused huge deficiency in active power and corresponding grid frequency decrease.

Jiuquan wind power base is located in the west end of Hosi Corridor in Gansu, and is rich in wind energy resources. By the end of 2011, installed capacity of wind turbines in Jiuquan is up to 5215.6 MW, which is collected by 750 kV Dunhuang Substation, 750 kV Guazhou Substation and 330 kV Yumen Substation into the main grid architecture consisting of 750 kV Dunhuang Substation, 750 kV Jiuquan Substation, 750 kV Hosi Substation, and 750 kV Wusheng Substation. Through this chained double circuit AC line transmission corridor, the power delivered by wind generation is integrated into main grid.

Four large-scale wind turbine trip-off accidents occurred continually on February 24, April 3, April 17 and April 25 2011 at Jiuquan wind power base. All of the accidents are developed from slight faults into serious faults, through the following stages:

(a) Initial fault stage

From the accident reports, the initial inducement is short circuits of electrical device near wind power plant. The first three accidents are induced by 35 kV cable head insulation breakdown causing 3-phase short-circuited, while the fourth one is induced by the short-circuited 330 kV bus in Jiayuguan Substation.

(b) Wind turbine generators low voltage tripping stage

Once there is short-circuit fault near wind power plant, wind turbine generator's terminal voltage drops drastically, which induces rotor current increase, and triggers Crowbar protection whose action time is shorter than double-fed induction generator stator clearing time. This results in temporary asynchronous operation which absorbs a lot of reactive power from grid and causes further voltage drop and more wind turbine trip-off. This is the main cause of a large number of wind turbine generators being tripped and going offline in the first stage. Secondly, some wind turbine generators do not have low voltage ride-through capability and the control strategy of wind farm's dynamic reactive power compensation device is not set reasonably, thus it doesn't respond in time and cannot compensate for reactive power deficiencies, which is also an important cause of instant trip-off accident of wind turbine generators.

(c) Wind turbine generators high voltage tripping stage

After a great number of wind generators are tripped off, transmitted active power in lines decreases, bus voltages in wind power plant recover. The nearby SVC and capacitors without automatic switching function continue running online and sustain reactive power output as is before accident. This results in large amount of reactive power flowing into grid and raising grid voltage. After fault clearance, the main grid voltage is over limit, which influences wind power plant close to initial fault point, and causes some wind power generators tripped by high voltage protection. This is the main cause of a large number of wind turbine generators going offline in the second stage. Also it exacerbates the excess of reactive power, which triggers a vicious circle of wind power generators going offline.

(d) Mutual interaction with main system stage

In "2–24" accident in Jiuquan, Gansu, a lot of wind power generators went offline and caused fluctuation in grid frequency, with the lowest frequency of 49.21 Hz and the highest frequency of 50.41 Hz. The wind generators that went through the low voltage and high voltage trip-off stages safely got tripped off because of high frequency protection. Therefore large-scale continuous wind generators trip-off phenomenon caused great active power deficiency, and resulted in disturbance in system dynamic frequency and more wind turbines being switched off. Thus the area of accident was further expanded and safe and stable operation of main grid was under threat. The typical trip-off accident described above is illustrated as the diagram in Fig. 2.6.

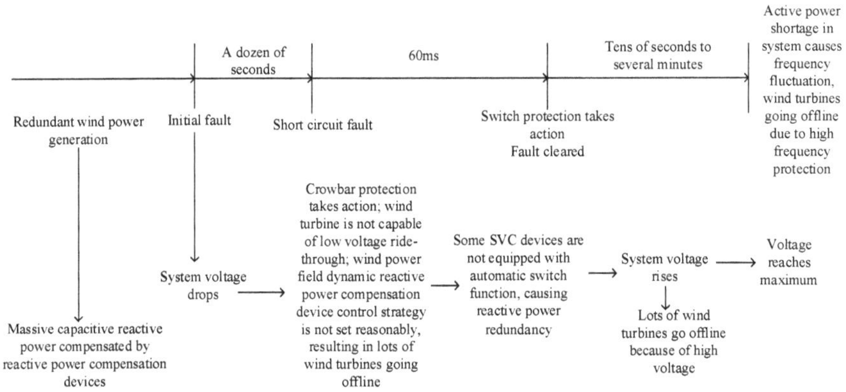

Fig. 2.6 A typical wind power generators trip-off accident process

Analysis of the four large-scale wind power generators trip-off accidents reveals the following common features:

① Jiuquan wind power base is located at the end of Gansu Hosi Corridor, and power is transmitted to main grid through double circuits 750 kV lines, which is a typical weak feeding end;
② Wind generators are with high output before accidents happen, and loads on transmission lines are too heavy;
③ Many wind turbine generators do not have low voltage ride through capability, and the control strategy of wind power plant's dynamic reactive power compensation device is not set reasonably, thus it doesn't respond in time, which is the main cause of instant trip-off accident of wind turbine generators;
④ All the wind power plants generate great power, and running SVC devices output great amount of reactive power while some SVC devices are not equipped with automatic switch function. This is the direct cause of high voltage tripping off;
⑤ Control manner of wind power generators is too simple, and reactive power regulation participation rate is low.

After the large-scale wind turbine trip-off accidents in Jiuquan, wind power companies, manufacturing companies and power grid enterprises respond quickly and pay great attention to this problem. Relevant parties have launched a comprehensive rectification work in Jiuquan wind power base with reference to State Electricity Regulatory Commission's accident investigation report and incident briefing. The first step is to launch rectification of wind power plant's reactive power compensation devices. Some wind power plants have launched rectification of 35 kV feed line small current fault line detection protection, changing the 35 kV feed line small current fault line detection protection from alarm to trip, which is helpful to avoid the consequence of feed line fault expansion. And all wind power plants have finished technically difficult but critical low voltage ride through rectification of wind turbine generators. Accident rectifications of Jiuquan wind power base are well implemented and main technical problems are solved effectively.

Analysis focused on such large scale wind power generators trip-off accidents indicated that great emphasis should be put on the impact of wind power centralized integration on power system security and stability, and the large-scale wind power base construction technique standards should be revised properly to ensure the wind power plants equipped with active power frequency control, reactive power voltage control and low voltage ride through capability. On that condition the pressure on power system security and stability brought by large-scale wind power centralized integration can be relieved, which will promote the continual stable and healthy development of large-scale wind power bases.

2.4 Impact of Integrating Intermittent Renewable Energy on Three Levels of Defense in Power System and Countermeasures

2.4.1 *Three Levels of Defense in Power System*

The ability for power system to stay secure and stable under large disturbance is classified into three-level standards in Chinese electric power industry standard DL 755-2001 (Guidance on Security and Stability for Power System):

(1) On the first-level standard, system should keep the stable operation and the normal power supply. When the disturbance caused by the failure of single component occurring in the power system under normal operating status, and protections, switches and re-closers act correctly, power system has to keep the stable operation and provide the normal power supply without taking any stability control measures; other electric components do not exceed their ratings; and cascading failures do not happen.
(2) On the second-level standards, system should keep the stable operation, but is allowed to lose partial load. When a serious disturbance happens to the power system under normal operating status, and protections, switches and re-closers act correctly, power system should keep stability stable. If necessary, stability control measures like generator tripping or load shedding are allowed.
(3) On the third-level standard, when the system cannot keep stable operation, it must prevent the system collapse and minimize the load loss. When the stability of power system is lost, actions must be taken to prevent system collapse, avoid long time blackouts in a large area, and minimize load loss as much as possible especially for the most important customers, including auxiliary power of the power plants. In a word, normal operation of power system should be resumed as soon as possible.

According to "Guidance on security and stability for power system", in order to ensure the security and stability of power system, primary system should establish the reasonable structure of power grid, be equipped with complete set of equipment, and arrange reasonable operation mode; secondary system should be equipped with highly developed relaying protection system and appropriate security and stability control measures. Thus, a complete defense system is formed, which is usually divided into three levels of defense.

A. ***The first level of defense: Security and stability control in normal operating condition***

In order to guarantee the normal operation of power system and satisfy the safety requirement under the first level of disturbance, the first level of defense to ensure the security and stability of power system should be composed of primary system, relay protection, and preventive control of security and stability which includes preventive control of generator output, additional control of generator excitation, compensation control of parallel and series capacitor, power modulation of HVDC (High-voltage direct current), and other flexible AC transmission controls, etc.

B. ***The second level of defense: Security and stability control in emergency condition***

In order to guarantee the normal operation of power system and satisfy the safety requirement under the second level of disturbance, the second level of defense to ensure the security and stability of power system should be composed of event-based control which can prevent stability disruption and components over-limit. Event-based control under that circumstance contains generator tripping, fast valving control, event-based control of generator excitation, braking control of dynamic resistance, forced compensation control of series or parallel capacitance, emergency power modulation of HVDC, and concentrated load shedding, etc.

C. ***The third level of defense: Security and stability control in extreme emergency condition***

In order to guarantee the normal operation of power system and satisfy the safety requirement under the third level of disturbance, the third level of defense should be composed of event-based control which can prevent expansion of accident and avoid system crash, such as system sectionalizing, resynchronization, event-based control of frequency and voltage, etc. At the same time, the third level defense should avoid mis-operation of lines' and units' protections when system oscillates, and prevent cascading trip of lines and units to ensure the security and stability of power system.

"Guidance on security and stability for power system" plays an important role in guiding the planning, constructing and operating of the power grids in China, ensuring security of large power grid, and avoiding the expansion of accident in power system.

2.4.2 *Impact of Integrating Intermittent Renewable Energy on Security and Stability of Power System*

Due to the characteristics of intermittent renewable energy, such as randomness, intermittency, and uncontrollability, it's hard for intermittent renewable energy to participate in the operation regulation in power system as conventional power plants, which will threaten the safe and stable operation of power system. For example, when a wind farm gets connected to a relatively weak node in power system, which means power grid's ability of controlling the region is weak, the operation department of power grid often hope that the connected wind farm can not only deliver power to the system, but also make a necessary contribution to the safe and stable operation of power system like conventional power plants, including participating in the system's active power and frequency regulation, and regulating the reactive power and voltage at points of interconnection. What's more, the operation department prefers necessary technical support provided by the wind farm during the system failure and recovery process. However, as mentioned above, it's hard for intermittent renewable energy to reach such operation goals, because of the randomness, intermittency, and uncontrollability of wind power. Therefore, when the total proportion of generating capacity of intermittent energy sources such as wind power in a power grid reaches up to a certain degree, the technical problems we face will be more serious, which will bring unfavorable influence to the security and stability of power system. Hence, it will endanger the power system and cannot guarantee the reliable power supply for system load.

Like wind generation, photovoltaic power stations and the photovoltaic power station groups start to operate, the scale of the installed capacity of intermittent energy can be compared to conventional energy units. Problems associated with integrating intermittent renewable energy, such as voltage problem, reactive power control, static stability, and dynamic stability, will become more and more serious when they are connected to transmission network directly.

Concretely speaking, the influences that integrated intermittent power generations brings to the safe and stable operation of power system are as follows.

A. ***Voltage fluctuation and flicker***

Most of the intermittent energy generators adopt soft grid-integration methods which will result in a large impulse current when starting to generate power. For instance, when wind speed is faster than the cut-out wind speed, wind power generators will automatically quit from the rated operation state. If the wind turbines in a wind farm quit simultaneously, it will result in the voltage fluctuation and flicker.

B. ***Harmonic pollution***

By utilizing power electronic convertors, intermittent energy generation will bring harmonics into the system which can cause distortion of voltage and current waveform. In sine circuits, harmonic voltage and current with the same frequency will produce active and reactive power with the same frequency which will reduce the voltage in power grid. Therefore, it's necessary to make constraints on the harmonic current injected into system by wind farms.

C. ***Frequency stability***

The influence of intermittent energy generation on the frequency of power system depends on the proportion that intermittent energy generation capacity accounts for relative to the total system capacity. When the capacity of intermittent energy generations has a large proportion, the stochastic volatility of the output power may have a remarkable impact on the frequency of power system, which will affect power quality and normal operation of some frequency sensitive load in the system. The scholars and engineers usually consider the effect of wind power on the system frequency through the index of limitation of wind power penetration. From the perspective of the whole power system, the limitation of wind power penetration represents the largest capacity of wind power that a size-given power grid can support. When the wind power capacity is within this limitation, the adjustable capacity in the power system should be able to ensure that frequency change of the power grid is within the permissible range. As the large power grid has sufficient reservation and adjustment ability, normally, the instability problems of system frequency don't need to be considered. However, as for the isolated small grid, the problems brought by intermittent energy generation such as frequency shift and stability problem cannot be ignored.

D. ***Voltage stability***

Challenges such as lower allowed short circuits current, fluctuation of system voltage and frequent drops of wind turbine generator are often encountered in wind power system, especially when more and more large-scale wind farms get connected to the gird. Most wind farms need reactive power support from system due to the utilization of induction generators, or it may cause voltage instability in small power grid. When using asynchronous generators, unless necessary preventive measures such as dynamic reactive power compensation are taken, otherwise the line loss will increase and voltage of end users in a long transmission line will decrease. When three-phase grounding fault happens, reduced stability of power grid will lead to the grid-wise voltage collapse.

Large-scale wind power and solar energy generation connected to grid will affect system stability, and its influencing degree is related to its proportion in the system. On one hand, when the proportion of wind power and solar energy generation capacity is small compared with the power grid capacity, its impact on power grid is very minor. So if grid control and distribution technology are taken good advantage of, the safe and stable operation can be ensured. On the other hand, when the proportion is more than a certain threshold, it will produce a significant impact on local power grid, which may lead to large-scale malignant accidents when serious.

Take an actual power grid as an example (as shown in Fig. 2.7). It can be seen from Fig. 2.7 that if the transmission lines between 330 kV Gan Jiayu 31 substation and 330 kV Gan Jiuquan 71 substation are disconnected, Jiayu region will become an isolated grid with a large number of centrally connected wind power.

Under that kind of circumstance, if no under-frequency low-voltage load-shedding measures are taken, to guarantee reasonable steady state voltage, and keep active power output and load level unchanged, different wind power penetration will relate to different frequency-response curves when accidents lead to a big power gap in isolated power grid (see Fig. 2.8). It shows frequency-response curves in an isolated grid only with grid inertia change, while other parameters of power system remain unchanged.

It can be seen from Fig. 2.8 that when Jiayu region becomes an isolated grid, if no under-frequency low-voltage load-shedding measures are taken, the system frequency will keep decreasing until frequency collapse, i.e., the system loses its stability. Generally speaking, when system frequency decreases to 49.0 Hz,

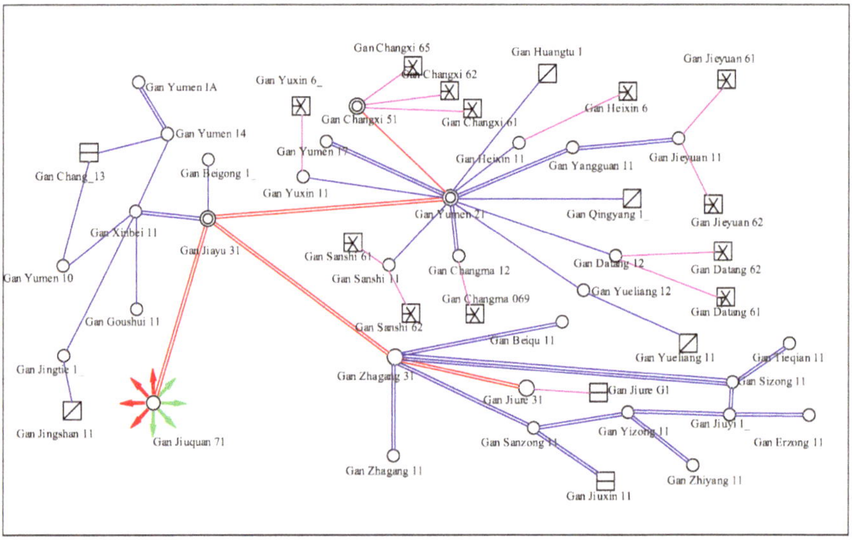

Fig. 2.7 Configuration of an actual power network

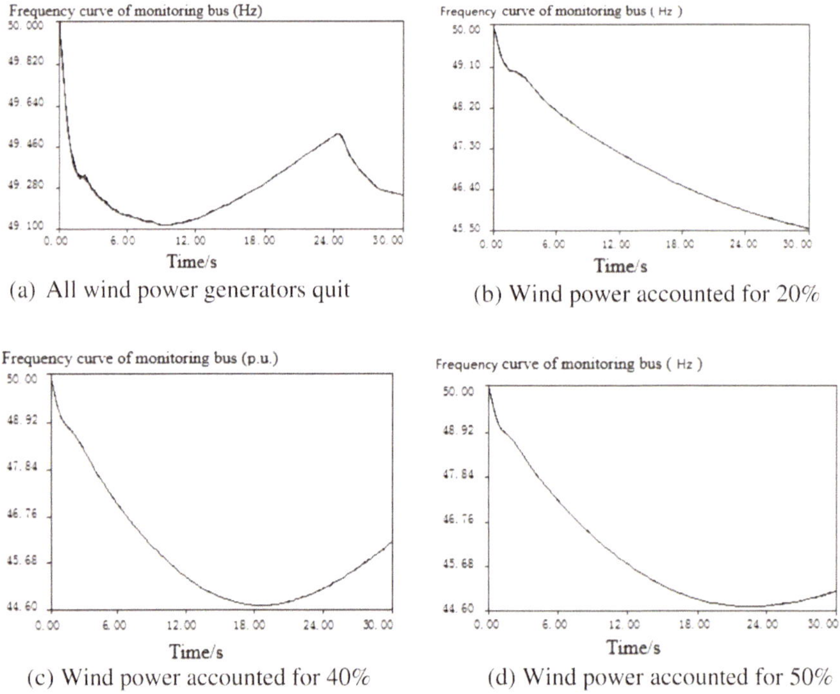

Fig. 2.8 Different frequency-response curves in isolated grid with different wind power penetration

Table 2.1 Frequency variation rate with different wind power penetration

Wind power penetration (%)	20	30	40	50
df/dt	−0.791	−1.044	−1.071	−1.026

security control equipment will start being activated. In this regard, in order to compare frequency decaying rate, we only need to pay attention to frequency change rate before it decreases to 49.0 Hz.

It is obvious that the frequency curve is nearly linear during its decrease from steady state to 49 Hz as observed from the simulation curves. Therefore, the slope of different curves which reflects various frequency drop process can be solved by adopting two-point method, namely df/dt. The results of df/dt are shown as Table 2.1.

The above results indicate that, on the fault condition of isolated grid, as the wind power output accounts for the increasing proportion of total energy output, frequency decaying rate increases within a certain proportion, and then slow down when exceeding that proportion.

2.4.3 *Impact of Integrated Intermittent Renewable Energy on Three Levels of Defense in Power Systems*

Three levels of defense in power system take some controllable and protective measures to prevent serious accidents according to the operating condition of system, and ensure security and stability of the system after fault happens. The traditional protection configuration and setting principle of three levels of defense, however, do not take the particularity of renewable energy integration into consideration. With the large-scale renewable energy integration, some corresponding changes must be made to the protection configuration and setting principle of three levels of defense. Therefore, the influence of renewable energy integration on three levels of defense in power system needs further discussion and studies, in order to make the defense system strong enough to keep the security and stability of the power grid.

A. ***Influence on relay protection***

If wind power or photovoltaic power account for a small proportion in power system, they will be simply regarded as a negative load, which has no effect on the configuration and setting calculation of power system relay protection. Nevertheless, with the rapid increase of the capacity of integrated intermittent energy, the simplified method mentioned above is no longer applicable.

In fault conditions, short-circuit current characteristics of wind power or photovoltaic generators has a direct impact on the setting and configuration of relay protection device. Configuration of relay protections, and their action coordination are strictly based on the correct and quantitative analysis of system failure. Only when voltage and current characteristics of power system with large amount of integrated intermittent energy are explicit in short circuit conditions, configuration of relay protection device and its function can be set correctly. Therefore, it is necessary to analyze its influencing factors through the characteristics of the short-circuit current provided by wind and photovoltaic generators, and determine the setting principle of relay protection accordingly. In addition, low voltage ride through control strategy of wind turbine may change the magnitude of the fault current and the phase relationship between voltage and current, which will have certain influence on protective elements that identify faults by the threshold value of current magnitude and directional elements that determine fault location by testing phase between voltage and current.

B. ***Influence on security and stability control***

It is common that wind power and photovoltaic generators are integrated in the regions where the power grid is weak. However, due to the weak grid structure of the end node, there might occur a series of problems after the intermittent power integration, such as voltage sag, transmission line overload, and change of system transient stability. As for that kind of situation, several security and stability control measures are needed to ensure the safe and stable operation of power system, and prevent consequences like voltage collapse, angle instability, and so on. Both trip-off accidents in wind power base and simulation analysis show that large-scale wind power and photovoltaic power integration has some special characteristics such as parameter settings of control system, crowbar line switching, low voltage ride through capability and uncertainty of resources. Therefore, fault characteristic and emergent switching feature of such integrated system are quite different from those of conventional power system. As for security and stability control device applied to renewable integration, the traditional criterion of stability control is no longer valid, a more appropriate stability control criterion must be put forward.

(a) Reactive power/voltage control

Wind power or photovoltaic power station cannot provide adequate reactive power support, that's why the reactive power required for its step-up transformer and sending lines needs to be transmitted over a long distance from the power grid, which increases voltage drop along the transmission line and makes the voltage stability of the region with wind and photovoltaic power stations worse. So the reactive power transmission is the main reason causing voltage problems in the system with large scale renewable energy integration.

Under the circumstance of weak regional grid structure, wind farms cannot re-establish terminal voltage after fault clearing, and active power cannot be transmitted out from the region with wind farms. Then, mechanical power is greater than the electromagnetic power, which will lead to instability of regional grid due to over-speed of wind turbine generators. At that time, wind turbine generators need to be shut down to ensure the security of regional power grid. Take the large-scale wind power trip-off accident happened somewhere in China for example, the system voltage dropped sharply during the fault, and a large amount of wind turbine generators without the low voltage ride through capability were taken off from the grid. In that accident condition, the mis-operation of reactive power compensation device in the wind farm, including not putting into service as requested, incorrect action of SVC/SVG, not meeting the requirement of control strategy, and response time problem, etc., brought out a large number of reactive power supply, leading to a voltage surge in power grid. The high voltage caused another set of wind turbine generators tripped off from the grid. Thus, the formulation and execution of wind power units shut-off and reactive power compensation strategy need further researches, in order to ensure the safety of the second level of defense in power system.

(b) Overload control

In normal condition, there is no overload problem when the large-scale intermittent generators and transmission system are in a stable operation state. However, in some other operation states where occur N-1 or N-2 failure in transformers or lines, it's hard to ensure that no overload appears in transformers or lines due to the uncertainty of their output. On that circumstance, actions such as limit of conventional units output or shut-off of part or all of the intermittent energy units in the grid need to be taken to avoid power flow switched in a wide range, and prevent the remaining lines from transmitting too much active and reactive power which can cause a more serious consequence.

(c) Transient stability control

Because the integrating power of large-scale intermittent energy changes the power flow direction of original system, intermittent energy generation replaces part of the conventional units which changes the inertia (idiomatic parameters) of the whole system and the transient stability characteristics of the system. The nature and degree of influence of intermittent energy integration on the transient stability have something to do with the integrated capacity, integrated location and the specific operation mode of the interconnected system.

C. ***Influence on under-frequency low-voltage load shedding***

When wind turbine generators get connected to the grid in a large scale, on the one hand, if they are lack of low voltage ride through capability, the impulse caused by short-circuit fault can easily result in generator tripping by low voltage protection of wind turbine generators, which leads to power shortage in the system, or even causes under-frequency load-shedding. On the other hand, when power system loses a large power supply and the frequency drops under 49.0 Hz, under-frequency load-shedding device will shed loads to recover system frequency, and then there is a possibility that wind turbine generators cause more serious power shortage due to its own under-frequency protection. If the wind power output takes a large proportion of the system total output, the sudden exit of wind turbine generators would exacerbate system frequency decline trend. Once the system frequency falls to the setting value of under-frequency protection for thermal power unit, power grid will face the danger of frequency collapse.

If a wind farm has the sufficient capability of low voltage ride through, that is to say, when voltage at the point of integration has dropped to 20% of rated voltage, wind turbine generators in the wind farm can continue to operate for 625 ms without taking off the grid. However, partial load at the point of integration will be cut off due to low voltage protection, resulting in power imbalance and over-frequency generator tripping. Source-grid interaction and the cascading reaction of different protection control devices will eventually have a great influence on the security and stability of power grid.

Take the actual system mentioned in Sect. 2.4.2 as example again. Wind power output accounts for 40% of total active power output, active power output of the isolated region is 1018 MW in all, active power back-up is 588.5 MW in total, and active load is 1251.3 MW. Power shortage caused by the disconnection of transmission lines after the accident is 255.6 MW + j44.8 MVar. In the simulation of the above accident situation, as the frequency fell to the setting value of frequency protection of wind turbine generators, both Gan Changma wind farm and Gan Changxi wind farm were taken off from the grid, losing 190 MW wind power totally. The simulation results are shown in Fig. 2.9, and action situation of under-frequency low-voltage load-shedding devices is shown in Table 2.2.

It can be seen from the simulation results that, when power shortage caused by wind farm trip-off accident in isolated regions increased up to 435.6 MW, the under-frequency load-shedding device in the original system would act 3–4 rounds to cut 435.2 MW of loads in total, which prevented the continuous decline of frequency in time, and regained the stability within a certain time.

When back-up active power is 488.5 MW in total, active load is 1251.3 MW, and power shortage caused by the disconnection of transmission lines after the accident is 261 MW + j44.8 MVar. In the simulation of the above accident situation, as the frequency fell to the setting value of frequency protection devices, both Gan Changma wind farm and Gan Changxi wind farm were taken off from the grid, losing 240 MW wind power in total. The simulation results are shown in Fig. 2.10, and action situation of under-frequency low-voltage load-shedding devices is shown in Table 2.3.

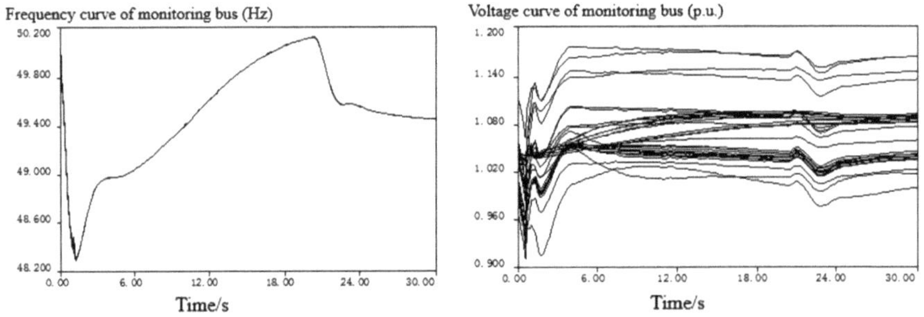

Fig. 2.9 Frequency-response curve (*left*) and voltage-response curve (*right*) of isolated grid with 190 MW wind loss

Table 2.2 Low-frequency low-voltage load shedding action

Bus name	Voltage classes (kV)	P (MW)	Q (MVar)	Action times	Type
Gan Jiayu 11	121.0	75.92	13.54	4	Low frequency
Gan Jiugang11	121.0	43.96	15.77	4	Low frequency
Gan Jiugang35	38.5	9.99	8.64	4	Low frequency
Gan JiuxinG2	13.8	2.78	0.08	4	Low frequency
Gan Jiuyi1_	121.0	9.99	3.08	4	Low frequency
Gan Sanzong11	121.0	25.97	7.64	4	Low frequency
Gan Sizong11	121.0	35.96	8.87	4	Low frequency
Gan Tieqian11	121.0	29.97	9.24	4	Low frequency
Gan Xinbei3_	38.5	14.40	3.64	3	Low frequency
Gan YangguanS1	38.5	4.80	1.02	3	Low frequency
Gan Yizong11	121.0	23.98	5.92	4	Low frequency
Gan Yumen11	121.0	0.64	0.04	3	Low frequency
Gan Yumen33	38.5	5.12	1.57	3	Low frequency
Gan Yumen32	38.5	29.76	10.24	3	Low frequency
Gan Zhagang11	121.0	35.96	8.87	4	Low frequency
Gan Zhiyang11	121.0	23.98	9.62	4	Low frequency
Gan Chang__12	10.5	3.03	0.71	3	Low frequency
Gan Chang__11	10.5	3.03	0.71	3	Low frequency
Gan Beiqu11	121.0	27.97	11.21	4	Low frequency
Gan Erzong11	121.0	27.97	14.18	4	Low frequency
Sum up		435.2	134.6		

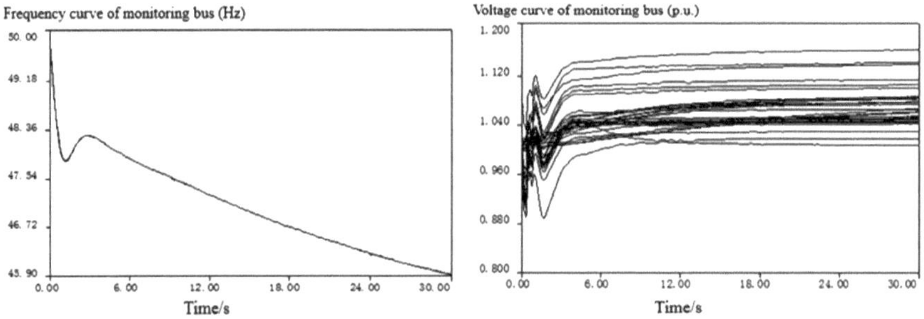

Fig. 2.10 Frequency-response curve (*left*) and voltage-response curve (*right*) of isolated grid with 240 MW wind loss

Table 2.3 Low-frequency low-voltage load shedding action

Bus name	Voltage classes (kV)	P (MW)	Q (MVar)	Action times	Type
Gan Jiayu 11	121.0	75.92	13.54	4	Low frequency
Gan Jiugang 11	121.0	43.96	15.77	4	Low frequency
Gan Jiugang 35	38.5	9.99	8.64	4	Low frequency
Gan Jiugang G5	10.5	1.11	0.03	4	Low frequency
Gan Jiugang G6	10.5	1.11	0.03	4	Low frequency
Gan JiuxinG2	13.8	2.78	0.08	4	Low frequency
Gan Jiuyi 1_	121.0	9.99	3.08	4	Low frequency
Gan Sanzong 11	121.0	25.97	7.64	4	Low frequency
Gan Sizong 11	121.0	35.96	8.87	4	Low frequency
Gan Tieqian11	121.0	29.97	9.24	4	Low frequency
Gan Xinbei 3_	38.5	16.65	4.24	4	Low frequency
Gan Yangguan S1	38.5	5.55	1.19	4	Low frequency
Gan Yizong11	121.0	23.98	5.92	4	Low frequency
Gan Yumen 11	121.0	0.74	0.05	4	Low frequency

(continued)

Table 2.3 (continued)

Bus name	Voltage classes (kV)	P (MW)	Q (MVar)	Action times	Type
Gan Yumen 33	38.5	5.92	1.86	4	Low frequency
Gan Yumen 32	38.5	34.41	11.84	4	Low frequency
Gan Zhagang 11	121.0	35.96	8.87	4	Low frequency
Gan Zhiyang 11	121.0	23.98	9.62	4	Low frequency
Gan Chang__12	10.5	3.50	0.82	4	Low frequency
Gan Chang__11	10.5	3.50	0.82	4	Low frequency
Gan Beiqu 11	121.0	27.97	11.21	4	Low frequency
Gan Erzong 11	121.0	27.97	14.18	4	Low frequency
Gan Jiugang 35	38.5	2.70	2.33	1	Low frequency
Sum up		449.6	140		

Figure 2.10 shows that when local region is operated in an isolated condition due to disconnection of inter-ties, wind farms in two locations would suddenly be taken off from the grid, resulting in the power shortage reaching up to 501.7 MW. Then, the ordinary round of existing under-frequency load-shedding devices were all activated, and so did the low-voltage devices in 35 kV Gan Jiugang bus. These two actions cut 449.6 MW of loads in all. However, it still failed to recover the system frequency and finally leaded to the instability of the system.

Comparing the above two simulation scenarios, it can be found that, on the one hand, the amount of load shed by under-frequency load-shedding devices in the first scenario can satisfy the power shortage requirement after the wind power trip-off accident, so system can be restored to the safe operation after a sharp frequency decline; on the other hand, in the second scenario, all actions of the ordinary round of under-frequency load-shedding devices cannot compensate the power shortage caused by wind power trip-off accident, so the system frequency kept falling till a collapse. Therefore, as for large-scale wind power integration, under-frequency low-voltage load shedding measures at the power grid side need to increase load-shedding amount based on the ordinary solutions. Meanwhile, only when conventional units add their output, can the increasing power shortage in local regions caused by a large amount of wind turbine generator systems trip-off accident be filled up.

Doubly-fed induction wind turbine generator can keep decoupled control of active and reactive power respectively, and realize the complete decoupled control of its speed and the grid frequency. However, they are unable to provide the frequency response of power grids when system frequency changes. Therefore, the inherent inertia of variable speed wind turbine generator based on doubly-fed motor becomes the implied inertia for the grid, which cannot help power grid to reduce the frequency change rate. When the inertia of wind turbine generator cannot perform in the transient process, that is to say, the inertia of whole system decreases, which has a negative impact on the frequency stability of the grid. In condition of large-scale integrated wind power replacing conventional power supply, not only does the system inertia reduce and the frequency variation rate increase, but also power system shortage increases caused by the tripping of wind power units due to the sensitivity of its protection action. Thus, the setting value of protection devices, as well as the amount of cutting load need further researches for the third level of defense in power system.

2.4.4 Countermeasures to Mitigate Effect of Integrated Intermittent Renewable Energy on Three Levels of Defense in Power Systems

Nowadays, large-scale intermittent energy is connected to power grid. In order to keep the security and stability of the system, reasonable grid structure must be set up first, which should be paid attention to in the system design and operation planning work. Second, the corresponding configuration, operation and control countermeasures should be put in place. It is likely that reasonable configuration and correct setting of system protection and automatic safety devices, the comprehensive utilization of on-line and real-time control, and practical application of energy storage in a large scale, etc. can mitigate the effect of the intermittent renewable energy on security and stability of power system, and the three levels of defense in power system as well, so as to achieve the goal of ensuring safe operation of power grid.

Main countermeasures to mitigate the effect of integrated intermittent renewable on three levels of defense in power systems are as follows.

A. ***Reasonable configuration of the power supply and power grid structure***

In order to satisfy the requirement of large-scale integration of intermittent energy, the existing structure of power supply and power grid should be optimized. On one hand, for the adjustment of power supply structure, first, controllable power source must be added, i.e. increasing the proportion of water, oil and gas units. Second, the scale of pumped storage power station should be expanded, as well as the new types of energy storage. Both of them will help to restrain the volatility and intermittency of intermittent energy. On the other hand, for the adjustment of grid structure to adapt to large-scale renewable energy integration, first, a variety of new power system pattern should be explored, including centralized and distributed integration, decoupling connection between regional power grids, and local consumption combining with long-distance transmission. Second, renewable power system structure with flexible control, high quality power supply, high reliability and scalability, should be set up, which will be based on the new types of transmission pattern, energy storage system, and new transmission technologies, such as flexible multi-terminal HVDC.

The new type of transmission and new structure of power supply and power system, which adapt to the renewable energy integration, transmission and energy distribution, will increase the system's capabilities in load shifting and complementing among conventional thermal power and other power sources in a big scope of space and time, grid resource allocation optimizing, grid controllability and renewable power integrating, which will enhance the security and stability of power system when large-scale renewable energy gets connected to the grid.

B. ***To strengthen the integrating test of intermittent energy units***

In the design stage of intermittent energy units integration, some technical requirements should be considered and checked, mainly including static/dynamic voltage and reactive power control requirements at the point of interconnection, high voltage and low voltage ride through requirements at the point of interconnection, frequency control requirements, the harmonic and power quality control requirements at the point of interconnection, and overvoltage and insulation coordination requirements, etc. If requirements for integration cannot be satisfied after intermittent energy units get connected to the grid, some necessary auxiliary control equipment should be installed to make sure that the integrated power system can still supply power to users safely and reliably.

Chinese national standard GB/T 19963–19963 (Technical Rule for Connecting Wind Farm to Power System), has stipulated technical requirements for new constructed or expanded wind farm interconnected through 110 (or 66) kV and above transmission lines, and also put forward some function requirements for wind farms, including active power/reactive power control, output forecasting, low voltage ride through, monitoring and communications.

The national standard GB/T 19963-19963 also stipulates the operation range of wind farm voltage, they are as follows.

(a) When voltage at the point of interconnection of wind farm is within 90–110% of the rated voltage, wind turbine generator system should operate in a normal state.
(b) When voltage flicker at the point of interconnection of wind farm satisfies the national standard GB/T 12326-2008 (Power Quality—Voltage Fluctuation and Flicker), the harmonics meet the national standard GB/T 14549-1993 (Power Quality—Harmonic in Public Power System), and the unbalance of three-phase voltages meet the national standard GB/T 15543—2008 (Power Quality—Three-phase Voltage Unbalance), wind turbine generator system should operate in a normal state.

Table 2.4 Rules for the operation range of system frequency in wind farms

System frequency range	Operation requirements
Under 48 Hz	According to the lowest frequency allowed in the operation of wind turbine generators in wind farms
48–49.5 Hz	Wind farms are capable of running at least 30 min whenever frequency is under 49.5 Hz
49.5–50.2 Hz	Continuous operation
Over 50.2 Hz	Whenever frequency is over 50.2 Hz, wind farms are capable of running at least 5 min, executing output down-regulation or high-circle cutter strategies assigned by power system dispatching center, and not allowing the interconnection of the offline wind units

The national standard GB/T 19963–19963 also stipulates the operation range of system frequency in wind farms, they are shown in Table 2.4.

In order to meet the demand on large-scale wind power integration, the national standard GB/T 19963–19963 must be polished up by including technology and management rules for operation control of integrated wind farms, and setting up standard system for wind farm integration, which can standardize the operation of wind power interconnection and guarantee its development in a healthy and ordered way.

C. ***To improve the intermittent energy generation technology and auxiliary equipment technology***

Dynamic reactive power compensation device can enhance reactive power compensation and voltage regulation capabilities of wind farms and photovoltaic power stations, and improve power quality. Series compensation/controllable series compensation technology is able to shorten the electric distance in transmission system and raise system's security and stability level. Controllable high-resistance technology is used to adjust the system voltage, so as to keep the voltage level during active power fluctuations in intermittent energy sending-out channel. In order to make some continuous fast adjustments to active power output of the whole wind farms, the existing control system, including background monitoring system of wind turbines, needs innovation, which means that except for wind speed and some other existing factors, wind turbine control system needs more constraints. Under that circumstance, the control system will become more complex, and it should ensure the reliable operation of wind turbines as usual.

(a) Variable-pitch control

It usually takes 2–3 s to see the effect of pitch control because of the existence of inertial element. Therefore, pitch control can only be regarded as the auxiliary measures of event-based control, or it can also be used as control measures for slow voltage fluctuation and voltage recovery.

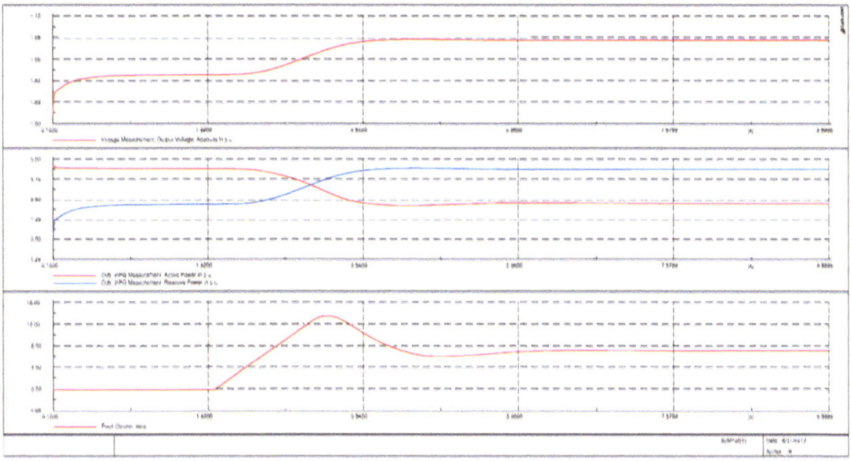

Fig. 2.11 Reactive power adjustment by variable propeller control

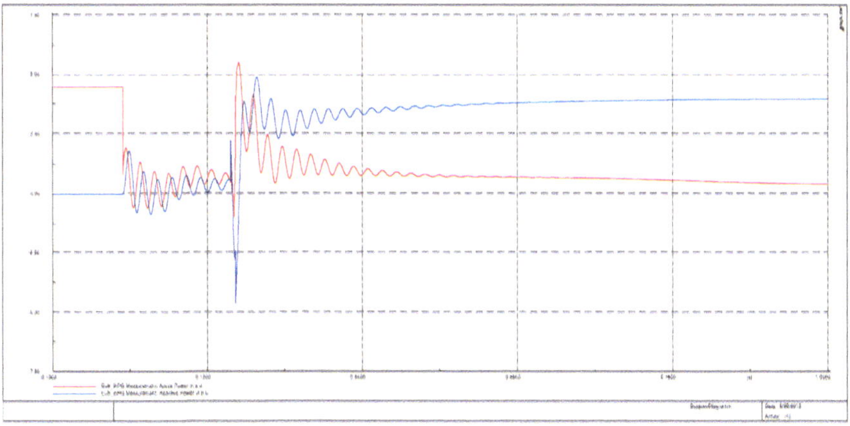

Fig. 2.12 Active and reactive power adjustment by frequency converter control

Figure 2.11 shows the variable-pitch control process of doubly-fed induction generator. It is obviously seen that it takes pitch control 2 s to reduce active power output, increase reactive power output and raise terminal voltage of wind turbines. As inertia element exists, it will take 2–3 s to see the effect of pitch control and 4–5 s for the whole control process. According to the related literature, the process of pitch control based on GE model takes about 5 s.

(b) Variable-speed control

Adjusting the control strategy of frequency convertor at the generator side can quickly modify the active and reactive power output of the units after the exit of Crowbar (as shown in Fig. 2.12), so as to achieve the purpose of the voltage control.

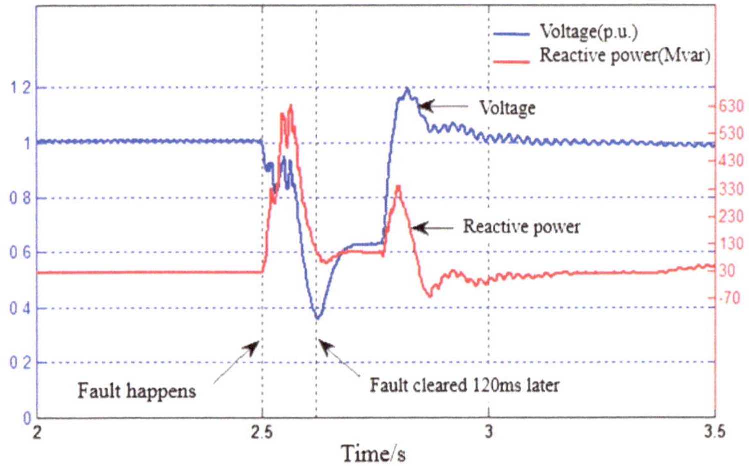

Fig. 2.13 Voltage and reactive power curves in SVC constant voltage control mode

Assume that wind turbine units are tripped and taken off from the grid when an accident happens at the high-voltage side of the transformer in the wind farm, then compare reactive-voltage characteristics of SVC working under three different control patterns, including constant voltage, constant reactive power, and constant power factor given. It is possible that all the working conditions under those three control patterns of SVC could lead to high voltage value (higher than 1.1 p.u. which is the threshold value of high voltage protection) at the point of interconnection after the fault in the wind farm is cleared. The highest voltage value can reach up to 1.2 p.u., which is likely to cause a secondary trip-off accident.

It is observed from the voltage curve and reactive power curve in Fig. 2.13, that in the early period of the fault, voltage dropped by a small magnitude and SVC injected a large amount of reactive power. However, with the falling of voltage at the point of interconnection of wind farm, reactive power output of SVC was reduced gradually. Once the fault cleared, voltage dropped to the lowest value, and so did the output of SVC. In the next 130 s of the low-voltage period, output of SVC remained unchanged until voltage rose back to over 0.6 p.u., and then its output increased at a rapid speed, resulting in voltage overshot with a maximum value of 1.2 p.u. at the point of interconnection. Hence, apart from the decline of reactive power absorbed by transmission lines and transformers, which is caused by the decline of active power output and line current due to trip-off accident in the wind farm, SVC is also one main reason for voltage overshot.

As for the low voltage problem which occurs in the fault period, SVC can operate in a constant-voltage pattern. As for the voltage overshot problem which occurs after the fault clearness, additional SVC locking control strategy should be adopted. The locking control strategy is to lock SVC in a delay time T1 once detecting the fault, and unlock SVC after a lasting time T2 to recover its voltage control ability. T1 and T2 can be obtained from the simulation of the actual power grid.

D. ***To strengthen researches on the control technology in integrating intermittent energy***

Take the event-based control of the second level defense in power system as example, one optimization rule of generator tripping sequence is put forward. As the execution time of event-based control is extremely short, considering the output change speed of wind turbines, in practical operation, overcut or undercut in event-based control will not happen in general. However, the fluctuations of wind farm output may result in complicate and changeable operation modes, which makes the formulation of control strategy more difficult. When it comes to the control strategy, researchers should make a concrete analysis of concrete conditions, take the influence of intermittent energy on the system stability into consideration, and give preference to intermittent energy power generators tripping or conventional power generators tripping.

Now, take conventional power generator and wind turbine generator as example. When the amounts of power to be cut from both of them are the same, compare the equivalent acceleration power of the system. Assume that the inertia of the conventional units and wind turbine generator system with the same capacity are M_1 and M_2. According to the characteristic of frequency-response curves of different generators, for the conventional unit and wind turbine generator with the same capacity, normally M_1 is larger than M_2, i.e. $M_1 > M_2$.

The equivalent acceleration power of the system can be calculated using the following formula.

$$P_{acc1} = \frac{M_a(P_{acc,s} - \Delta P) - (M_s - M_1)P_{acc,a}}{M_T - M_1} \tag{2.1}$$

$$P_{acc2} = \frac{M_a(P_{acc,s} - \Delta P) - (M_s - M_2)P_{acc,a}}{M_T - M_2} \tag{2.2}$$

Here, ΔP represents the cutting capacity of generators; P_{acc1} represents the equivalent acceleration power of the dominant reflection after the conventional generators being cut ΔP capacity; P_{acc2} represents equivalent acceleration power of the dominant reflection after wind turbine generator system being cut ΔP capacity; M_T is the equivalent inertia of all the generators; M_s is the equivalent inertia of critical generators; M_a is the equivalent inertia of generators excluding the critical generators; $P_{acc,\,s}$ is the equivalent acceleration power of critical generators; $P_{acc,\,a}$ is the equivalent acceleration power of generators excluding the critical generators.

$$P_{acc1} - P_{acc2} = \frac{M_a(P_{acc,s} - \Delta P + P_{acc,a}) - (M_1 - M_2)}{(M_T - M_1)(M_T - M_2)} \tag{2.3}$$

Because $M_1 > M_2$, $P_{acc1} - P_{acc2} > 0$, that is to say, the system acceleration power of conventional generators is larger, compared with the wind turbine generator system after they cut the same amount of power, which means the former system is more likely to be instable. In other words, shedding wind turbines in critical groups is more advantageous to the system stability.

E. ***To improve the performance of protection and automatic control device***

Due to the new electromagnetic transient characteristics of electric parameters in renewable energy unit during or after the faults, configuration principle and setting method of relay protection need to be further studied and reviewed. Phasor Measurement Unit (PMU) and Wide Area Measurement System (WAMS) can provide wide-area information for defense and event-based control of power system, and also be able to improve the performance of backup protection and automatic safety device that are less sensitive to time by utilizing the already built network. Therefore, through the analysis of relations between system fault characteristics and the system model and parameters, operation level and disturbance, and some other various factors, the new principle of wide-area protection in complex power grid is put forward, aiming at keeping system security and adapting to integrating renewable generators. This new principle can in some conditions detect system faults in time, and take measures to avoid serious accidents such as blackouts.

The national standard GB/T 19963-19963 set some requirements for relay protection and automatic control device, they are as below.

(a) Wind farm relay protection, automatic control device, design and arrangement of secondary circuit should satisfy the relevant provisions of the power system and the requirement of anti-accident measures;
(b) Under normal circumstances, for the grid-connected lines, only sectional phase-to-phase fault protection or ground fault protection are configured at the system side. When system meets the special requirements, the current differential protection can be configured;
(c) Substation in wind farm should be equipped with fault wave record equipment which have enough record channels and can record situations from 10 s before the fault to 60 s after the fault, and data transmission channels to power system dispatching department are needed as well.

For security and stability control system of large-scale intermittent renewable generators, such as wind farm and photovoltaic power station, in respect of control target, there is a trend that event-based control will consider the interaction between dynamic state of wind farms and power system. In respect of control measures, there is a trend that wind farm emergency power control will develop into fast control based on power electronics.

The formulation of under-frequency low-voltage load shedding scheme of system with large-scale intermittent energy generators should take intermittent energy particularity into consideration based on the traditional scheme. For instance, under-frequency protection of wind generators must coordinate with under-frequency load shedding scheme. When grid faults or abnormal wind speed result in system frequency anomalies, wind turbine generator should be able to restrict on the output and regulate the frequency, and setting value of wind turbine under-frequency tripping cannot be higher than setting value of the first round of under-frequency load shedding.

F. ***The development of large-scale energy storage technology***

Energy storage technology has become one of the core technologies of renewable exploitation. The current widely-used energy storage technologies are mainly pumped storage and compressed air energy storage. These two technologies are used in large centralized power stations with a capacity of more than 100 MW, and are widely applied to peak load shifting of large power grid. With the continuously increase of installed capacity and scale of intermittent renewable energy generators, adding energy storage device can provide fast active support and enhance the ability of power grid frequency regulation. As for renewable energy with volatility such as solar and wind generators, configuring large-scale energy storage devices with good dynamic response, long life and high reliability at the power supply side, can effectively solve the intermittent, uncertainty problems of large-scale renewable energy integration, and substantially improve the ability of power grid to accept renewable energy.

Energy storage contains heat storage and electricity storage. Wind/photovoltaic generators can be combined with energy storage battery, while solar thermal power technology can be combined with heat storage equipment. From the current situation, it seems that electricity storage technology fails to meet the ideal requirements on two issues, which are cost and capacity upgrade. Regardless of the above two issues, wind/photovoltaic generators combined with energy storage battery is still a promising research field, and it is possible that the combination can be widely applied if major breakthroughs are made in energy storage battery. Compared with electricity storage, the development of thermal storage technology is more mature, especially in the large-scale, high-capacity storage fields. Solar thermal power plants with heat storage tanks can fully satisfy technical property requirements for frequency regulation.

G. ***Implementation of online monitoring and control technology***

Large-scale integration of renewable energy generation such as wind or solar generators brings challenges to the safe and stable operation of power grid. One of the key problems is how to set up the renewable energy integration simulation model timely and accurately. The following two issues need to be addressed when building the simulation model.

(a) Model accuracy

Large intermittent renewable generation (such as wind farms, photovoltaic power stations) covers a large geographic area, and is affected obviously by external environmental factors such as geographic locations and weather change. How to effectively reflect these factors is the key problem which influences the calculation accuracy of the renewable energy integration simulation model.

(b) Model timeliness

The intermittent and volatility characteristics of renewable energy require that simulation model must track terminal power state fast to ensure timeliness of the calculation results. That is to say, on the premise of accurately reflecting terminal characteristics, the simulation model of intermittent renewable energy generation need to be simplified as much as possible. At the same time, its generator model should be forecasted in advance and calibrated on-line, through the weather forecast and online real-time state measurement.

To timely and accurately set up the renewable energy integration simulation model which reflects the external environment impact will be the basis of grid security and stability analysis for a power system with large amount of intermittent resources. Automatically adjusting model based on weather forecast information, and automatically correcting model parameters according to the online actual measurement of electrical values, will be the main process involved in on-line building intermittent renewable energy model adaptive to external environment, which can effectively reduce individual diversity and uncertainty of intermittent renewable energy generators.

Different from conventional generator which has relatively fixed steady and transient state model, and can be evaluated through offline model and online application, large-scale and high permeability of renewable energy generators with intermittent and volatility will make the traditional "offline calculation, online matching" power system security and stability control mode no longer valid in the future. Therefore, "online calculation, real-time matching" control mode should be put into effect. That is to say, in time period of 5–10 min, fast analysis and calculation of security and stability in power grid are made based on online system model which reflects running status of generators, power grid and load in computing time, so as to evaluate security and stability characteristics of power grid, and formulate the corresponding operation schedule and control strategy. Security defense system with the main characteristics of pre-arranged planning, static security analysis and passive defense, is now promoted to active security defensive system with the characteristics of dynamic security analysis, pre-warning, and online auxiliary decision making.

2.5 Application of Security and Stability Control Technique in Large Scale Renewable Energy Integration

2.5.1 Application Example 1—Wind Power Integration

A. ***Problems in wind power dispatching and operation in this district***

Wind farms in a certain province are concentrated in two districts. By June 2010, 10 wind farms were integrated into grid, with a total installed capacity of 1.05 GW. The power is transmitted to main grid through existing double circuit 330 kV line. The maximum transfer capability of the 330 kV grid is merely 700 MW. Even if the relevant power system's security and stability control devices are put into operation, the actual transfer capability is only raised up to 900 MW. Taken into consideration that local small hydropower takes up a part of transfer capability, especially in flood season and winter, when local hydropower, thermal units, and wind power will compete for the transmission corridor, wind power transmission capability could be limited. By the end of 2010, the first phase of this wind power development, which is also the first wind power plant of more than 10 GW in China, is integrated into the grid. The total installed capacity of wind power in this district reaches 5.16 GW, and as the power transmission mainly relies on a double circuit 750 kV line, wind generation is still under limitation in some periods of time. The situation will become worse as it's predicted that the installed capacity of the wind units in this district will reach 12.71 GW by the year 2015.

The traditional security and stability control system deployed in the district improves the power grid transmission limit, but it still cannot fulfill the need of transmitting all wind power produced. Grid operators should take preventive control of wind power in operation, in order to avoid the flow exceeding the stability limitation because of the stochastic and intermittent features of wind generators. When the grid operators performed preventive control, it initially took manual control. After a period of trial operation, the following problems of manual control are noticed:

(a) Dispatching center may not be able to send out adjustment orders in time, which may pose threat to grid security.
(b) Adjusting speed at field terminal is slow, which requires greater margin of power grid to ensure security.
(c) When maximum permitted wind output in the district is given, due to the stochastic and intermittent characteristics of wind output, real-time optimized control cannot be achieved by manual control, which could cause unfair allocation of wind output among farms, and result in not fully utilization of wind resources.
(d) The great number of wind power units put great stress on dispatchers.
(e) Each wind farm cannot learn the allocated output and actual output of other farms, which is hard for grid-source coordination.

To tackle the problems above, preventive control of wind power is required on the basis of the original security and stability control system, and is complement with event-based control function. Therefore, the security and stability control system, combining event-based control and preventive control for clustered wind power integration, must be realized. This control system can guarantee the stable and reliable operation of grid under all circumstances, improve and make the most usage of grid transmission capability, maximize the acceptance of wind power in the grid, determine the preventive control strategy automatically, and calculate the wind generators' output plan automatically. It can achieve the goal of minimizing the wind units' shut-off and full use of wind resource, even under fault condition.

B. ***Basic idea of preventive control***

The preventive control of wind power, unlike thermal power or hydropower, is hard to implement as the output of the wind unit cannot strictly follow the generation schedule set by grid dispatcher all the time. In order to take the full advantage of wind resources, wind generators should be allowed to produce as much power as possible. Thus there're two modes for wind power preventive control:

(a) Maximum output control mode. Under condition that grid is secure and stable, calculate the maximum generation output for each wind farms according to grid's wind power acceptance capability. When wind generator's output is below its upper limit, it is in the free generation state, while ramp-up rate must meet certain requirements. When the generator's output exceeds the upper limit, it can take up system resources of other wind farms according to other fields' availability, to achieve the goal of maximum generation and fully utilization of wind resources among wind farms.
(b) Output following mode. According to prediction of wind power of each wind farm, schedule generation plan for each wind farm after security evaluation, and each wind farm must follow the plan by making real-time active power output adjustment.

The two control modes are suitable for preventive control of several wind farms in a region, as well as coordinated preventive control of wind farm groups in different regions. As a special case, it can also be applied to the preventive control of one large-scale centralized wind farm.

To realize this preventive smart control, there are two main technique difficulties which need to be solved:

(a) System architecture design. That is to design the architecture of the whole system according to current communication channel condition, available device resources, and total investment, so as to guarantee the system reliability and feasibility, and expandability in future.
(b) System control strategy design. Control strategy design is the core of system design, which should incorporate the operators' control experience and methods in wind power farms' dispatching and operation. The control strategy of security and stability control system can be performed automatically in real-time instead of through manual actions by operators, which will release the operators from overwhelming work load. At the same time, a reasonable control strategy design can take full advantage of wind resources and transmission capabilities, accommodate more wind power, and strengthen management and control of wind farms.

C. ***System configuration and major function***

Clustered wind power security and stability control system in this district has a 4-layer architecture, respectively corresponding to control center station, control

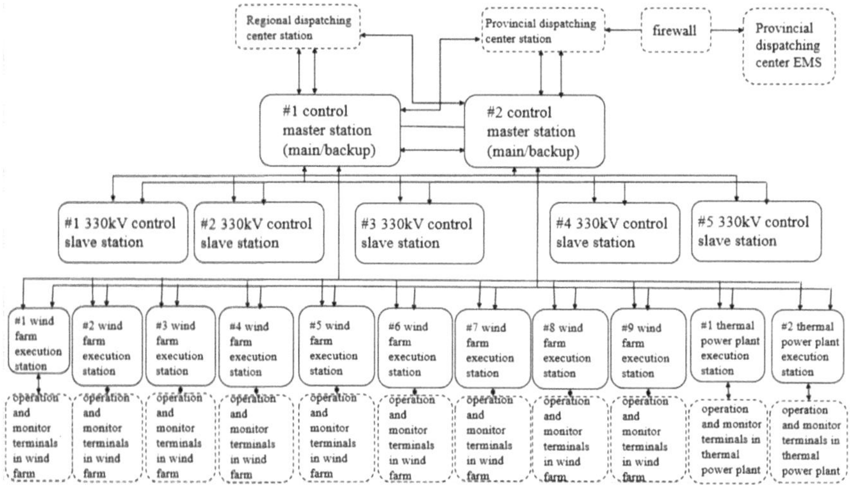

Fig. 2.14 Configuration of clustered wind power security and stability control system

master station, control slave station and control execution station. The details are demonstrated in Fig. 2.14.

(a) Control center station

There are two control center stations located in province dispatching center and regional dispatching center. One serves as main station, and another serves as backup. The major functions are (1) to perform real-time monitoring of the power grid, (2) to realize clustered wind power preventive strategy calculation, (3) to perform real-time generation plan calculation and publication, (4) to automatically reply to wind generator's request to increase its output, and (5) to switch between application algorithm and tracking algorithm, and between operation mode and control mode. Operators in province dispatching center and regional dispatching center can monitor the generation schedule of each wind farm, corresponding actual output, system reservations, 330 kV corridor's power flow, power flow of the step-up transformers in wind farms, and system margin data, etc., as well as the stability control devices' status, control modes, and action reports.

(b) Control master station

Two control master stations are located in two utilities in the area, responsible for information collection and exchange among 9 wind farms, 5 control slave stations and 2 control center stations. Their main functions are: (1) system status information exchanging between center stations and slave station, and (2) collecting real-time generation output plan of each wind farm from control center and sending them to slave stations. They also have the function of matching offline event-based control strategy table and sending the event-based control command to slave stations, according to system operation mode and fault information.

(c) Control slave station

Five control slave stations are composed with relevant slave stations in original grid security and stability control system, with the functions of monitoring power flow of every 330 kV interfaces in real-time, and uploading operation, fault situation and overload situations of monitored lines and critical interfaces to master stations and center stations, calculating the output plan for each units and sending the command to execution stations. They fulfill the coordination between clustered wind power preventive control and system event-based control.

(d) Control execution station

Nine wind farms in the area are equipped with stability control devices. As wind farm execution stations, they monitor and control real-time wind generators' output according to output schedule/plan set by control center stations under each operation mode. They can also produce over-capacity generation alarming and monitor units tripping under over-capacity and over-time generation.

Two traditional thermal power plants are equipped with stability control devices and added into the security and stability control system in order to improve wind power penetration capability as much as possible. With respect to the transfer capability margin in 330 kV interfaces, in a situation when wind farms' output need to increase and transmission interfaces' unused capability is limited, command will be sent to these two thermal power plants to reduce their generation and re-allocate some interface capacity to wind farms. In doing so, it will realize the coordinated adjustment of wind and thermal generation in this area.

Communication among control center stations, master stations, and slave stations is through 2 Mbit/s channel. To improve reliability, a redundant 2 Mbit/s channels are constructed. Control center station in the provincial dispatching center can also communicate with Energy Management System (EMS), and acquire data such as real time reservation, which will help peak regulation control. Furthermore, when communication between control center and slave stations is broken, control center stations of the provincial dispatching center can obtain the necessary data for control strategy calculation from EMS, to guarantee the correctness of control strategy and improve system reliability and availability.

E. ***Preventive control strategy***

(a) Overall Design

The fundamental purpose of clustered wind power preventive control is to take full advantage of event-based control and to improve the utilization of wind energy on the premise of secure and stable operation. General framework of control strategy is shown in Fig. 2.15. Control strategy is divided into two parts. The left part in Fig. 2.15 is inter-farm coordination control strategy, and the right part is wind and thermal power bundled export coordination control strategy.

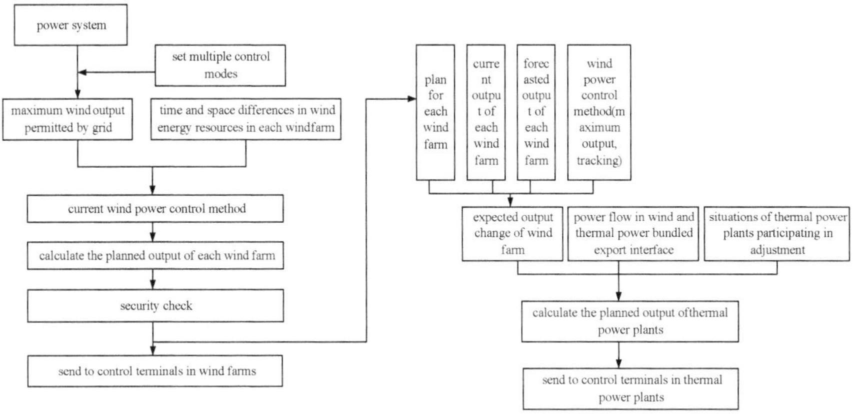

Fig. 2.15 Overall framework of control strategy

The general idea of inter-farm coordination control strategy is as follows. First, calculate in real time the maximum permitted wind power output in the system while respecting grid constraints; Second, coordinate and control output of each wind farm based on the maximum output change allowed by grid, and time and space difference of wind energy resources among farms. By using this control, it can be guaranteed that total wind output of the system is less than maximum permitted value, and the output of each wind farm is maximized. The keys to implement the method described above include control mode setting, maximum permitted generation calculation, and inter-farm coordination control according to time and space difference of wind energy resource among wind farms, which are detailed below.

(b) Control mode setting and maximum grid permitted wind power output calculation

In wind power dispatching operation, the major problem that the power grid is faced with is to handle wind power export and peak regulation. Control mode setting is mainly used in solving these two problems, and other control modes are set in order to adapt to grid emergency and system abnormality.

(1) Automatic adjustment control mode with reference to power flow margin of each interface.

- According to power flow margin changes in interfaces relevant to wind power export, increase or decrease wind output plan of each farm to ensure total power flow not to exceed interface permitted limits. This mode is set to tackle wind power export problem.

Limits for interfaces are different in different operation modes. To adapt to normal operation mode and various maintenance modes, various operation modes are set in the control mode. Set power flow limits on interfaces related to wind power export by offline analysis, and calculate transmission margin of each interface online. When grid operation mode changes, adjust generation plan of each wind power farm automatically according to operation mode changes. As wind farms cannot adjust in time, to prevent the situation that feed lines are cut off due to slow adjustment of wind output, cut off time of feed lines is increased to (planned change value/regulation rate provided by wind farms) + tolerated delay.

In this mode, maximum wind output permitted by grid is determined by current generation of each wind farm and transmission margin of each relevant export interface.

(2) Peak regulation mode.

- Increase or decrease current generation of each wind farm according to changes in peak modulation capacity of the grid, ensuring total wind outputs in the next time period not to exceed the sum of current wind outputs and peak modulation capacity that the grid can provide. This mode is mainly applied to the situation that wind power is needed to participate in system peak regulation in low load period.

In this mode, maximum wind outputs permitted by grid can be set by manual setting or automatic calculation. Automatic calculation is determined by grid peak regulation constraints.

(3) Other control modes

- Base point control mode. Generation output plan curves with a point every 15 min are set in the dispatching center, and distributed the plan to every farm. Once receiving the plan curve value, wind farms control output accordingly.
- Emergency derating mode. When the grid is in emergent condition and wind power output must be reduced to keep the grid secure and stable, the emergency derating mode can be selected. Dispatchers are only required to provide total amount of wind output that should be reduced, and the system can automatically calculate new generation plan of each farm and instruct each farm to carry out. This mode can immensely reduce the processing time when emergency happens. This function meets the following requirement in wind farm integration technique regulations set by State Grid Corporation of China: when power grid is in emergency, wind farms should control the active power output according to commands from grid dispatching department, and guarantee the rapidity and reliability of their active power control system.
- Dispatcher control mode. In operation, if relevant grid information cannot be acquired, dispatcher control mode will be activated. This mode is a manual mode.

Automatic adjustment of control mode with reference to power flow margin of each interface and peak regulation mode can be put into operation individually or simultaneously, while other mode can only be put into operation individually.

The control strategy has not taken into consideration of frequency regulation mode, but wind generation is not allowed to increase when system frequency is above 50.2 Hz.

(c) Inter-farm coordinated control

In this application case inter-farm coordinated control adopts two methods, with implementation details listed below:

(1) Maximum generation control

Procedure of maximum generation control is illustrated in Fig. 2.16.

Calculate total amount of wind generation plan that needs adjustment

$$P_{wPlanChange} = P_{\mathrm{wPlanMax}} - \sum P_{wPlan} \tag{2.4}$$

where $P_{wPlanChange}$ is total amount of wind generation plan that needs adjustment, $P_{wPlanMax}$ is maximum wind output permitted by grid, $\sum P_{wPlan}$ is the sum of expected outputs of all wind farms, while the expected output of farm with application is the sum of original plan plus the application value, and the expected output of farm without application is original planned value. Here, the maximum generation capacity provided by wind farm's centralized control system, and output increase application brought up by wind farm operators are processed uniformly. The difference between the maximum generation capacity and current generation is the reference of application value. If this value is negative, the value is set as 0, which means that wind farm will not submit output decrease application. Additionally, considering the ramp rate of wind unit, an upper limit of generation increase application is set. If upper limit is exceeded, then set the application value the same as upper limit.

According to the total amount of wind generators plan that needs adjustment $P_{wPlanChange}$, there're 3 conditions:

Condition 1: If $|P_{wPlanChange}| \leq P_{dz}$, there is no need to adjust plan for all wind farms.

Where P_{dz} is a constant set to avoid frequent adjustment of output plans when maximum output permitted by grid does not change drastically and no generation increase application is submitted by wind farms. When there are applications from wind farms, the value is set to 0, as the generation plan must be calculated again.

Condition 2: If $P_{wPlanChange} > P_{dz}$, it means output plan of the farms must be increased, and distribute $P_{wPlanChange}$ to each farm according to operation capacity ratio. If the plan for a farm exceeds its maximal capacity, the excess amount will be distributed to other farms.

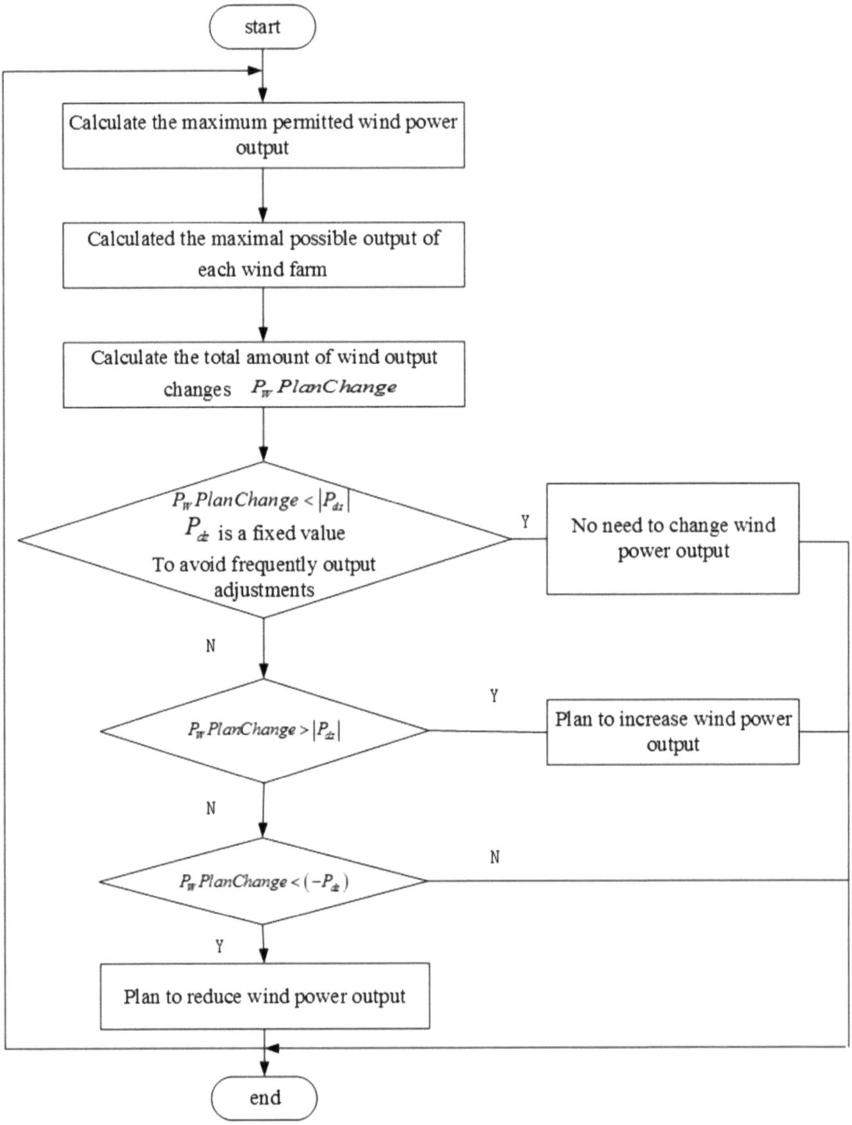

Fig. 2.16 Flow chart of maximum output control

Condition 3: If $P_{wPlanChange} < (-P_{dz})$, it means the plan for the farms must be reduced, and the total amount of wind output reduction should equal to $|P_{wPlanChange}|$. When determining the output reduction value for each wind farm, it should be fair and should consider the resource difference among wind farms. The implementation process is shown in Fig. 2.17.

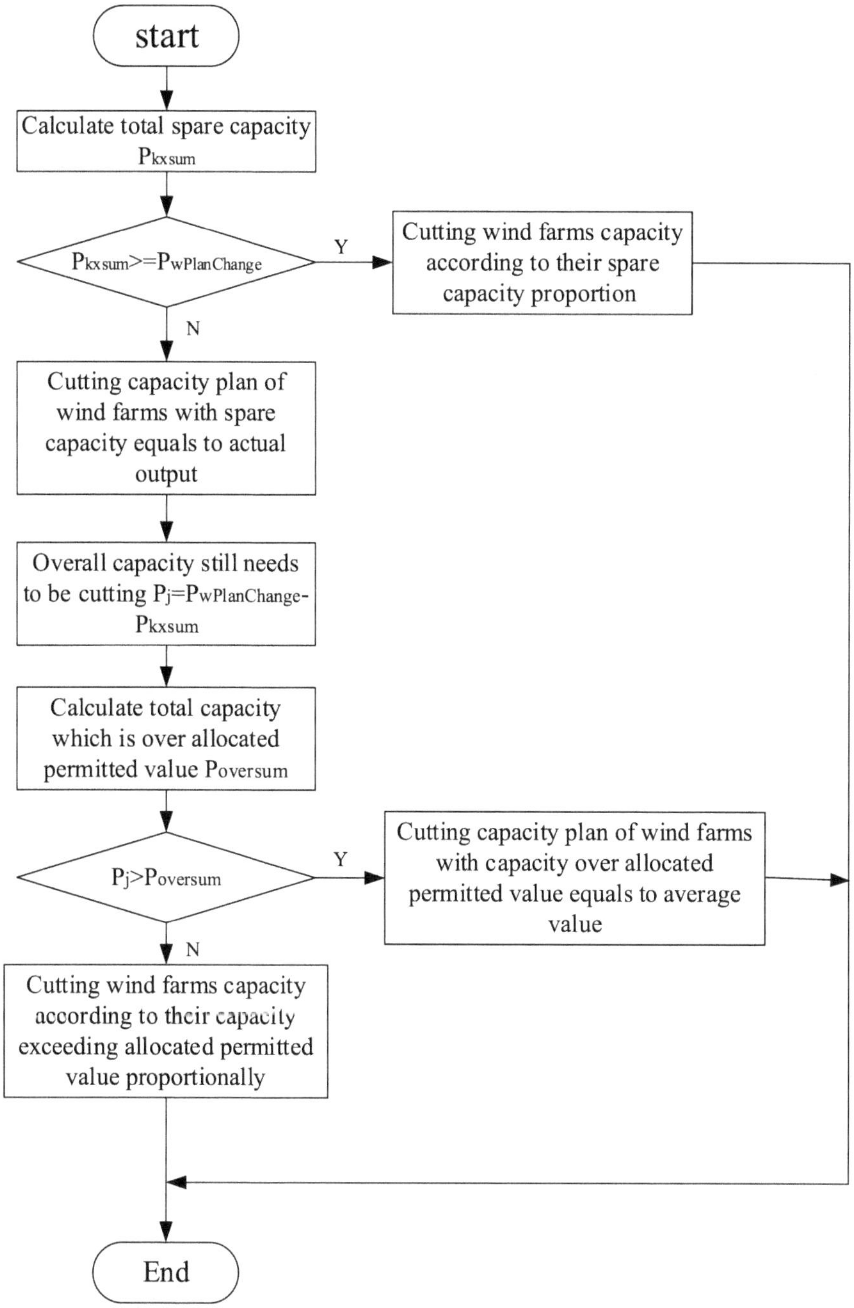

Fig. 2.17 Flow chart of wind output reducing plan determining process

In Fig. 2.17, spare capacity refers to the capacity which is the difference between current generation of the wind farm and its planned value. The fact that a wind farm has less spare capacity indicates that this farm takes up more shares of wind output permitted by grid. Thus the generation output of wind farms with small spare capacity should be reduced first, and the amount will be distributed to those wind farms with higher wind speed and smaller plan value. If a farm requests to increase its generation, the spare capacity of this field should not be reduced at this moment.

When spare capacities of all wind farms are reduced to 0, and if there's still need to reduce wind output plan, then reduction of the plan is done according to the ratio of the amount that planned value exceeds the allocated permitted value. Here the allocated permitted value refers to the amount of maximum wind output permitted by the grid that is distributed to each farm based on their operation capacity ratio. The control above considers the difference in wind resources in each farm, improves wind energy utilization, and preserves the principle of fairness of dispatching among wind farms.

(2) Generation tracking control

Under this mode, after wind output of each farm is predicted, and the security assessment is performed by the control center station, generation plan is sent out to each farm. All wind farms must track the plan and adjust their active power output in real-time.

Additionally, another objective is to avoid the situation in which the predicted wind generation is too high, taking up generation margin of other wind farms and making it difficult to fully utilize other wind energy resources. If the generation of a wind farm cannot reach the planned value (with tolerated mismatch) in the next 5 min, the output prediction of this wind farm will be set as its current generation for a given period.

F. ***Control interface with wind farm***

In this mode, every wind farm is equipped with stability control devices, which can collect real-time output of the farm and power flow on each 10 kV or 35 kV feed lines, send these information to control center station, and receive the real time generation plan data from center station. When heavier wind comes cross a farm and its output increases, in manual application mode, wind farm operators on duty can put forward generation increase request to control center station through control terminal provided by the devices, and control center station will automatically receive and process the application. If there's margin after calculation, then center station will reply automatically and send out new generation plan to the farm which it should follow. If the power generation of the farm is above the plan, stability control devices will alarm first. If the generation is not reduced below the plan within a certain time period, the devices will intelligently select some feed lines in the wind farm, and turn them off to cut off all wind power generators on these feed lines, as shown in Fig. 2.18a.

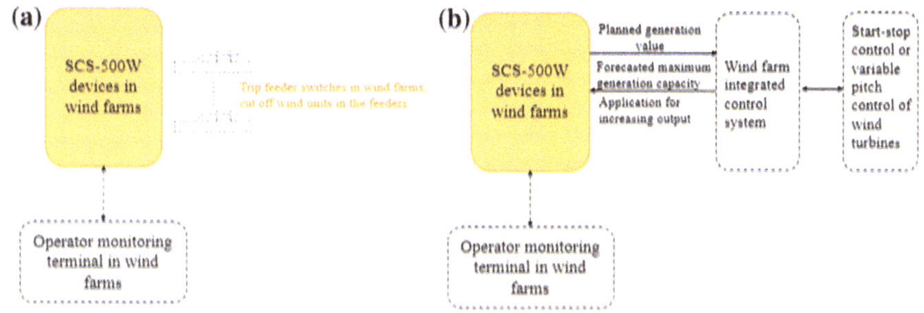

Fig. 2.18 Control interface with wind farms

To reduce the work load of wind farm operators, and to allow them focusing on monitoring wind turbines, the application for output adjustment is calculated and submitted automatically. That is implemented through the communication under ModBus/TCP protocol by using the Ethernet ports at wind farm's stability control devices and centralized control system. Wind farm centralized control system is further equipped with communication interface software with stability control devices, wind power super short-term forecasting and control software. Stability control devices receive the output plans from control center stations, and send them to wind farm centralized control system in real time. Then centralized control system will perform variable pitch control or starting/stopping control to adjust the output of the wind units according to plans. When heavier wind comes cross the farm, the centralized control system predicts the maximum power of the farm based on wind speed estimation, and sends the prediction to stability control devices through ModBus/TCP protocol. If it exceeds given plan, the centralized control system will send the generation increase application to stability control devices which will transfer the information to control center stations. After security check by dispatching center, control center station will automatically reply and send back new generation plan. Wind farm will follow the new plan and adjust active power output. This procedure realizes close loop control of active power output.

G. ***Field performance***

This system was put into operation in March 2010. Since then, the system has been running reliably, and performing very well. It improves wind energy utilization and ensures grid secure and stable operation.

The successful implementation of this system helps to maintain the secure operation of power grid in the early period of large-scale wind power development in the district, and provides solid base and stable operation and control platform for future wind farm construction in this district. It not only supports green energy such as wind power to be developed and utilized efficiently, but also guarantees the safe, stable and economic operation of power grid, which can result in great economic profits and significant social benefits.

2.5.2 Application Example 2—Wind Power Integration

A. ***Background***

This district is rich in wind energy resources. The wind power integrated into grid is over 1 GW by the end of the year of 2015. With the dramatic development of wind power, wind power in this district has turned from small-scale integration (low-voltage gird for local energy consumption) into large-scale integration (high-voltage-grid for long-distance transmission). A 220 kV substation in the district is a critical substation to support the enlarging construction scale of the wind power, to guarantee wind power export, to strengthen the connection of the northern part with main grid, and to keep safe and stable operation of main grid and wind farm.

Currently, the 220 kV substation is connected to the district main grid through 220 kV double-circuit lines. When wind output is high and one circuit line is tripped, another circuit line will be severely overloaded, and event-based control should be taken to cut off wind turbines quickly. If there is no event-based control, wind output should be limited. To maximize wind energy utilization, security and stability control is deployed to implement quick minimum wind turbine cut off in emergency state, which will help wind farms to produce at their full capacity.

B. ***System configuration***

The security and stability control system of this district with large wind power export has 2-layer structure, namely master station and slave station, as is shown in Fig. 2.19. Security and stability control system master station is installed in 220 kV substation, and slave stations are equipped in 8 wind farms.

General configuration principles of security and stability control system are:

(a) Devices in 220 kV substation have dual configuration. The devices determine the fault condition according to the collected local and remote information, and

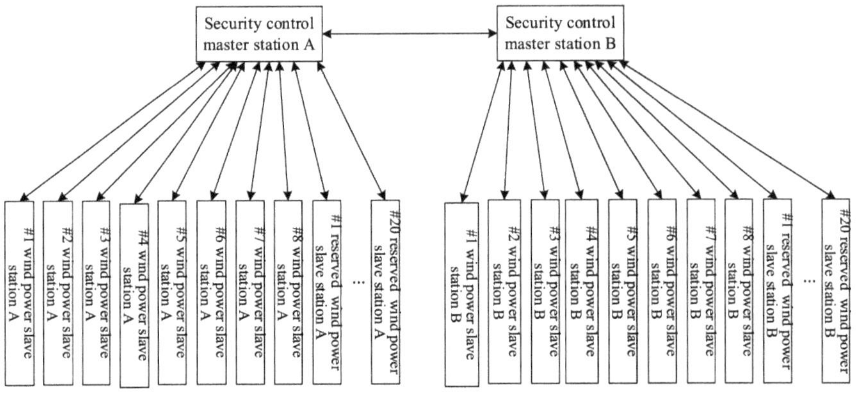

Fig. 2.19 Configuration of stability and security control system in a real district

send out wind turbine cutting-off command following pre-set strategy. Communication between master and slave stations are realized through reusing 2 M optical fiber.

(b) Devices in the aforementioned 8 wind farms have dual configuration. The devices collect wind generators information, send it to 220 kV substation master station, and receive the cutting-off command from the master station to cut off corresponding wind farm. All wind farms' slave stations communicate with the substation through reusing 2 M optical fiber, inter-station communication is through optical fiber double channel 2 M interfaces, where the two channels take different routes.

(c) Security and stability control system should have certain extensibility for the future systems. The security and stability control system may be commissioned before some wind farms being built. Thus the security and stability control system should be applicable as additional wind farms are put into operation, for example, 20 wind farms in the future. When new wind farms are added, there will be a minimum effort of upgrade, involving software update, which only needs adding more sites and communication plugins.

(d) The system can operate independently, or in a main-backup way, which can be switched manually.

C. ***Control strategy***

By now, there are three decision making methods for security and stability control system: offline strategy table or control logic determination; online quasi real-time control strategy determination; online real-time control strategy determination. This system in operation uses the first decision making method, i.e., which is to determine control strategy offline.

(a) Design principle

   (1) Strategy table in device should not be too complicated. Current consideration is to cut off wind turbines according to capacity, and select farm automatically. The major fault considered in the system is N-1 fault between any two districts.
   (2) Cutting off strategy for thermal stability problem should consider relatively fair shedding of units among wind farms. That means not to cut off a farm entirely as far as possible. Alternatively, it can be implemented by sorting the farms, and rotating the cutting off priority of each farm in a chronological order.
   (3) As the wind farms are coming into operation continually, with the commissioning of devices for new farms, guarantee that the commissioning debugging will not influence the operation of running devices through appropriate master station settings. Make sure that the commissioning debugging of new devices does not require the whole system to quit operation.
   (4) Open setting values should be as few as possible.

(5) If one round cutting off is not enough for thermal stability problem, a second round cutting off is allowed.

(b) Master station control strategy

Master stations in this system monitor the operation mode and power flow of two 220 kV substation double-circuit outgoing lines, receives power flow and operation data of transformers in step-up substations in wind farms. In the condition that one circuit line is tripped because of fault and the other is overloaded, master station's stability control device will instantly send out cutting off commands to relevant slave stations according to control strategy, and trip off transformers' low-voltage side switches or 35 kV collection line switches selectively, to ensure the electrical devices not to be damaged and guarantee secure and stable operation of power grid.

This system considers that in the case when any one of the substation's double-circuit lines is tripped off, the cutting off measure takes place only if another circuit is overloaded within certain time period Ttz. If the overloading of another circuit happens after certain time period Ttz, it indicates that the overloading is not caused by the tripping of the parallel line, in which case the dispatchers should take control. Additionally, if overloading happens without tripping, the dispatchers should take control as well.

After cutting off wind farms' transformers or 35 kV feed lines due to fault, if the overloading is not eliminated, the devices should take another round of cutting off after fixed time delay.

(c) Cutting off principle

Overloading cutting off should be based on the overloading amount and follow the over-cutting principle. The overloading amount equals active power P at the moment of action occurring minus the setting power Pdz triggering the overloading cutting-off action.

The process is detailed as follows. Set the cutting-off priority of each wind farm based on the setting value, meaning the farm with higher priority will be cut off earlier (the smaller the index, the higher the priority, and 0 means it cannot be cut off). In addition, set the cutting-off priority of each transformer in wind farm, and consider reserving one transformer with lowest priority in operation for each farm. Only if all transformers with high priority are cut off (except reserved transformer) and the farm still needs to cut off transformers, the transformer with next priority can be cut off, until the sum of active power of all the transformers cut off equals or exceeds overloading cutting-off amount. If the cutting-off requirements can be met by cutting off part of the transformers in a wind farm, the transformer cutting-off process should follow the order of from high priority to low priority.

If more transformers still need to be cut off when all transformers aside from reserved ones are already cut off, cut off reserved transformers in the priority order, until cutting-off transformers' capacity meets the requirements.

(d) Slave station control strategy

In normal operation, slave stations in wind farms measure power flow in transformers at step-up substations, and send these data to 220 kV master stations. On receiving cutting-off command from the master stations, slave station will trip off the switches on wind farm's transformer low-voltage side or switches of 35 kV collecting lines.

D. ***Reliability design***

In system operation, when a wind farm's slave station does not work due to a fault, but the master station still considers this farm during cutting-off calculation, it may result in under-cutting of wind turbines. To avoid this situation, wind farms do not send the power flow values of transformers under the following conditions: (1) general function pressure plate is not in service; (2) channel pressure plate is not in service; (3) devices are locked; or (4) master station receiving channel is abnormal. When master station cannot receive power flow information from corresponding transformer, this transformer is not included in the cutting-off candidate list.

In communication software designing, there are six levels of command code verification, e.g. CRC verification, address code verification, message code sum verification, positive and negative code of command verification, code value legal range verification, command code character word and information code character word switchover. A method of over three times of continuous confirmations of command is taken to ensure the devices not to take false action because of communication failures in channel.

E. ***Prospect***

The successful implementation of this system improves wind power export capability efficiently, ensures the grid safety, improves wind energy utilization, and has achieved significant economic and social benefits. Currently this system only solves the problem of overloading after one circuit is tripped off. Normally, this kind of control is done manually by dispatchers. Because of wind generator's inherent randomness and intermittency, transmission lines are easy to be overloaded. To avoid line overloading, in manual control mode, dispatchers tend to reserve relative large margins which is a waste of some transfer capabilities. To fully utilize grid's transmission capacity and improve wind output, it's required to establish a coordinated control system in normal and emergent situations. Instead of traditional manual control, control master station performs preventive control of each wind farm automatically based on the margin of export interfaces and the time and space difference of wind resources in each wind farm. Also, the control measure can be lowering wind output and cutting off wind units which is better than the manual control (only cutting off wind units). It will reduce the processing time in emergent state, reduce the reserved margin for export interfaces, and increase the wind outputs.

2.5.3 Application Example 3—Photovoltaic Power Integration

A. ***Background***

A province is rich in solar energy. By the end of 2011, the maximum photovoltaic power integrated in a district of this province reaches 900 MW. The power grid is at the end of Northwest power grid. Apart from a gas power plant, there is no large power plant to provide voltage support. This district is interconnected to 750 kV corridor through a single step-up transformer. Once the transformer is tripped, the district will lose the whole 750 kV corridor, which will have great impact on this district's grid. With large-scale integration of photovoltaic power station, when the photovoltaic station generates high power in the daytime, this district transfers from receiving end to sending end. The power in the export interface increases a lot which may result in transient stability problem.

The security and stability analysis shows that the grid in the district will be influenced by large-scale photovoltaic power integration. Thus it's necessary to add security and stability control devices in photovoltaic power stations and upgrade the original stability control system. Therefore, new stability control devices are installed in 4 collection stations and 30 photovoltaic power stations, and original stability control devices in two 750 kV substations and four 330 kV substations are upgraded.

B. ***System configuration and typical station function***

The configuration of security and stability control system with large-scale photovoltaic power integration is shown in Fig. 2.20. This system is installed in four kinds of plants or stations: ① 750 and 330 kV substations, shown with red and grey boxes; ② photovoltaic collection stations, shown with yellow boxes; ③ PV power stations, shown with green boxes; ④ traditional power stations, shown with white boxes. The following section will introduce the functions of the stability control devices in 750 and 330 kV substations, PV collection stations and PV power stations.

(a) Functions of stability control devices in 750 kV substation

Stability control devices in 750 kV substation monitor operation condition of four 750 kV lines, determine line faults, determine PV power station or loads cutting-off according to strategy tables pre-set in the devices. Meanwhile, the devices has the functions of determining bus overvoltage and sectionalizing 750 kV lines.

(1) Coordinated reactor cutting-off function

When the device identifies the tripping of the high voltage side of the step-up transformer, and the break of the 750 kV export interface, or receives the information of outage of other stations, the device will cut off the reactor in the 765 kV

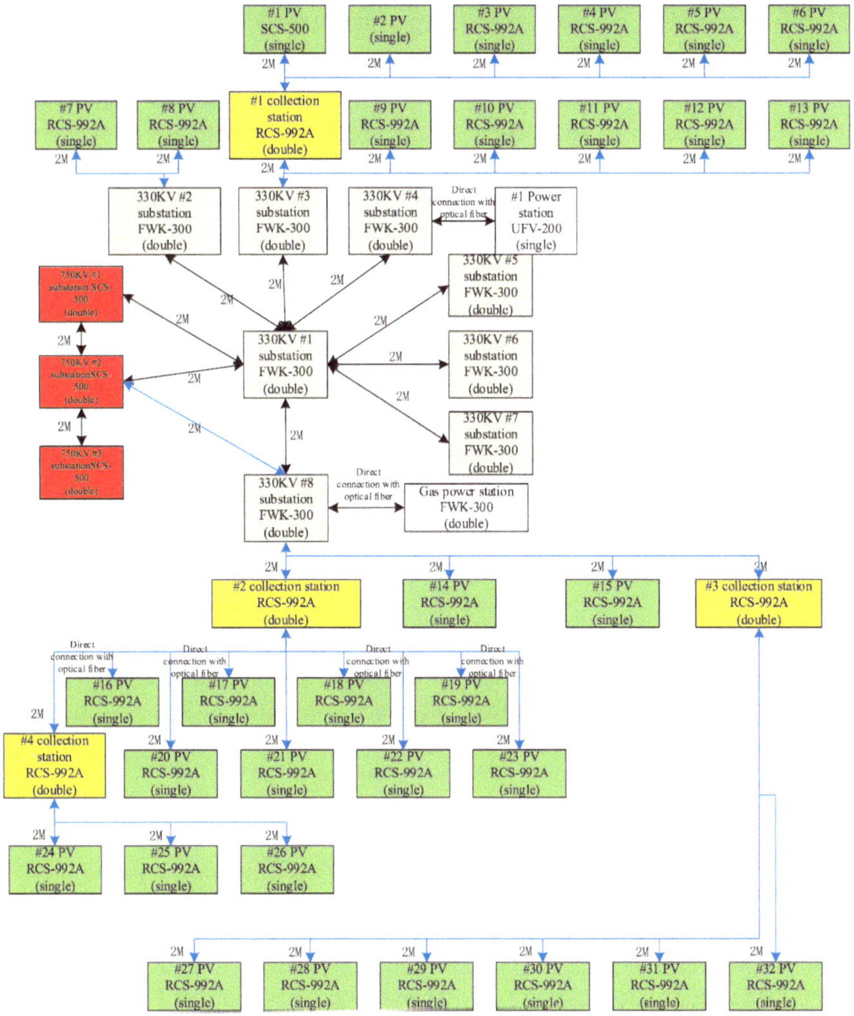

Fig. 2.20 Configuration diagram of security and stability control system with large-scale photovoltaic power integration

substation. On receiving the information of outage of other stations, device will wait Twait1 (delay) and then perform the coordinated cutting-off of the reactor in the station. On identifying the tripping of the high voltage side of the step-up transformer, device will wait Twait2 (delay) and then cut off reactor in the station. When device identifies the break of the export interface of this station, it will cut the reactor immediately.

(2) 750 kV interface overvoltage breaking control function

When the stability control device identifies the tripping of the high voltage or middle voltage sides of local transformers, or receives breaking information from other stations, and meanwhile detecting the action of protection for 750 kV line overvoltage, the device will trip the tie line.

(3) Function of receiving commands from other stations

When stability control device receives command from other stations to trip off tie line I(II), device will take action to trip off tie line I(II).

(b) Functions of stability control device in 330 kV #1 substation

Device's main cabinet collects operation condition of the 330 kV tie lines and signals of switches HWJ to identify failures of components. Device's auxiliary cabinet collects 110 kV load information. Based on the cut-off condition of 330 kV interface, the device determines the corresponding generator tripping or load shedding measures.

(c) Functions of stability control device in 330 kV #2 substation

(1) Information sharing
Collect information of 330 and 110 kV components in this substation, communicate with PV collection stations in the district, receive PV power station cut-off command from other districts and send it to PV collection stations.
(2) Control strategy when transformer interface is cut off
When the transformer interface is cut off, based on the power flow direction and magnitude in the interface before accident, if the interface is importing power, cut local 110 kV loads; if the interface is exporting power, cut off PV power (solar stations, collection stations), gas generators, etc.
(3) Control strategy when transformer is overloaded
When the transformer is importing and overloaded, cut local 110 kV loads. When the transformer is exporting and overloaded, trip off PV power (solar stations, collection stations), gas generators, etc.
(4) Control strategy for 110 kV bus low-frequency, low-voltage, over-frequency, over-voltage
Devices can identify low-frequency, low-voltage, over-frequency, over-voltage on 110 kV buses. When any of the aforementioned faults happens, devices take the measure of sectionalizing transformer's middle-voltage side. It should be noticed that after transformer sectionalizing, devices may sequentially identify the cut off of transformer interface, and take corresponding generator tripping or load shedding measures.

(d) Functions of stability control devices in photovoltaic collecting station

Devices in Photovoltaic collecting stations have several functions. Firstly, they monitor the operation state of double circuit out-going lines, two main transformers, and 24 photovoltaic feeders. Secondly, they send the information of total curtailment capacity of photovoltaic stations to stability control devices in substations of the region. Thirdly, they receive photovoltaic power station tripping command given by substation's stability control devices, and allocate the cutting capacity to photovoltaic feeders with photovoltaic power stations or collecting stations connected proportionally. Finally, they identify the overloading of local transformers, and allocate the cutting capacity to photovoltaic feeders with photovoltaic power stations or collecting stations connected proportionally according to the overloading capacity.

- Transformer overload action

Photovoltaic power control measures taken when transformer is overloaded should be processed by superposition rule. That is to say, total required cutting capacity = existing cutting capacity + overloading cutting capacity.

- Allocation measures of photovoltaic power cutting capacity

When total required cutting capacity is greater than or equal to total capacity that can be cut off in photovoltaic station, collecting stations trip all connected feeders; otherwise, stations cut photovoltaic power capacity in proportion of the cutting capacity that can be cut, which means $P_{\text{must cut in power station}} = P_{\text{must cut in collecting station}} * P_{\text{cuttable in power station}} / P_{\text{cuttable in collecting station}}$.

(e) Functions of stability control devices in photovoltaic power stations

The stability control devices in the photovoltaic power stations have the following functions. First, they monitor the operation state of 4 export transmission lines and 24 feeders. Second, they send the total capacity of photovoltaic power that can be cut to stability control devices in collecting station or master control station, and receive photovoltaic power cutting-off command. Third, choose to cut corresponding photovoltaic feeders or trip all export lines according to the priority level of photovoltaic feeders.

Photovoltaic power station will accumulate the curtailment capacity of each feeder, and send it to the collecting station or controlling master station.

If any of the following situation occurs, photovoltaic power station will send information invalid signal to stability control devices in collecting stations or controlling master stations, and then set the cuttable capacity to 0:

- Both Channel A, B for receiving information from collecting station (controlling master station) are abnormal.
- Pressure plates of both Channel A and B from photovoltaic power station to collecting station (controlling master station) are not in service.
- One of pressure plates of Channel A and B from photovoltaic power station to collecting station (controlling master station) works, another is not in service. And the channel in service is abnormal.
- Pressure plate of general function is not in service.
- Device is locked.

The principle for photovoltaic power cut-off is:

- If total capacity that can be cut off is less than or equal to total required cutting-off capacity, all export lines in power stations are tripped.
- If total capacity that can be cut off is larger than total required cutting-off capacity, then

  ① Check the photovoltaic capacity that can be cut off according to the priority order from round 1 to n, and accumulate the capacity;
  ② If accumulated capacity amount after n round is larger than or equal to the required cutting capacity, then feeders with these n round capacity will be labelled as waiting for cut feeders.
  ③ If all the cuttable feeders are in the list of waiting for cut feeders. Then trip all export lines; otherwise, only trip the feeders in the list.

C. ***Prospect***
Integrating large-scale photovoltaic power stations will have a great impact on the safety and security of regional power grid. Upgrades of current security and stability control system in regional power grid are necessary, which includes adding stability control devices in photovoltaic power station, monitoring active power output of photovoltaic power station, and sending or receiving photovoltaic power cutting-off command to/from master station. The application of this system in regional power grid provides technology and equipment support for planning, construction, and operation control of power grid after large-scale photovoltaic power integration. Also, this system plays a vital exemplary role, possesses wide popularization and application value, and has remarkable social and economic effect on clean energy exploitation and regional economy development in China.

2.6 Conclusion

By the end of 2012, China becomes the country in the world with the largest installed wind capacity and the largest annual electricity generated from wind power. The integration of large scale renewable energy brings tremendous challenges for the stability and security of the power system. From several wind turbine trip-off accidents in Jiuquan, Gansu in 2011, it has been learned that many wind turbines were involved in serial trip-offs and went offline because of transient voltage problems, which caused huge deficiency in active power and corresponding grid frequency decrease.

The capability for power system to stay secure and stable under large disturbance is classified into three-level standards in Chinese Electric Power Industry Standard, which is also called three-defense-line. Three levels of defense in power system take preventive control, event-based control and response-based control to prevent serious accidents according to the operating condition of system, and ensure security and stability of the system after fault happens. The traditional protection configuration and setting principle of three levels of defense, however, do not take the particularity of renewable energy integration into consideration. With the large-scale renewable energy integration, some corresponding changes must be made to the protection configuration and setting principle of the three levels of defense. In this chapter, the influence of renewable energy integration on three levels of defense in power system is studied, and the new designs of the security and stability control system for large scale renewable energy integration are introduced. These technologies help the power gird keep security and stability after integrating large amount of renewable energies.

Chapter 3
The Role of Ensemble Forecasting in Integrating Renewables into Power Systems: From Theory to Real-Time Applications

Corinna Möhrlen and Jess U. Jørgensen

3.1 Introduction

Nearly all decision making in business is associated with a cost or a loss. Many of these decisions are at the end of day weather related. If a decision is based on a deterministic forecast, then there is a risk of that the decision will be random, because the forecast used for the decision is likely to have a random error. The longer the forecast horizon, the more likely it is that a given decision turns out wrong, because the error grows with lead time and it is getting more unlikely that there is more than 50% chance of success. If a user of forecast detects that there is only 50% chance of success, then the user gives up and starts guessing instead, but guessing is not leading to any progress. Continued economic growth occurs, because humans act more clever by means of new and enhanced tools and because it is believed that "efficiency makes money grow". The alternative is recession in the economy and unemployment. It is therefore our obligation to assure that what we present as an improvement also can contribute to economic growth.

Today, it is no longer a sufficient condition to postulate that a given approach is better than another for a measure, which is not sufficiently hard linked to the economic value of the problem at hand. We need to conduct a proof in terms of resource usage and cost space. In fact, it is the latter that counts in a market economy. Legislation should then ensure that economic optimisation implicitly imply that resource usage is optimised as well, if necessary via fees and taxes.

Ensemble forecasts provide means to have an enhanced amount of information at hand in the decision process, but also enables evaluation of deterministic results.

C. Möhrlen (✉)
WEPROG GmbH, Böblingen, Germany
e-mail: com@weprog.com

J.U. Jørgensen
WEPROG ApS, Assens, Denmark
e-mail: juj@weprog.com

P. Du et al. (eds.), *Integration of Large-Scale Renewable Energy into Bulk Power Systems*, Power Electronics and Power Systems,
DOI 10.1007/978-3-319-55581-2_3

In this chapter, we will introduce the reader to ensemble forecasting in the context of the power industry. It is our objective to show that ensemble forecasting is today already a well proven methodology and will in the future be a pre-requisite to efficient decision making, risk analysis and a number of processes in the energy markets and grid management along with variable generation from wind and solar energy sources.

3.2 The Need for and Background of Ensemble Forecasting

Ensemble weather forecasts have become the established method to determine the uncertainty of the weather development since their development in the early 1990's (e.g. [17]). The main reasons for developing ensembles in meteorology has been to provide an objective value of the forecast uncertainty and also to be able to provide better warnings about extreme events. In fact, these two aspects are closely connected, because extreme weather traditionally has an implication on economic loss. The loss can be related to damages by not protecting something, which is too fragile to withstand extreme weather or by taking action that is inappropriate in extreme weather conditions. On the other hand, a loss can also be caused by invoking a protection mechanism based on a false alarm.

There are many decisions in which weather forecasts play a crucial role. The need for ensemble forecasts in that sense stems from the fact that the weather is a chaotic system, where small errors in the initial conditions can grow rapidly and affect predictability. Furthermore, predictability is limited by model errors linked to the approximate simulation of atmospheric processes of the state-of-the-art numerical models. These two sources of uncertainty limit the skill of single, deterministic forecasts in an unpredictable way, with days of high/poor quality forecasts followed by days of poor/high quality forecasts. Ensemble prediction is a feasible way to complement a single, deterministic forecast with an estimate of the probability density function of forecast states [20].

While a single deterministic forecast will be of sufficient quality most of the time, it is insufficient in extreme events. Since the power system needs to ensure stability and security also in extreme events by tradition, such conditions need to be covered in the forecasting process as soon as the penetration level increases over a threshold, where errors can exceed available reserve capacity. The trends to higher spatial resolution and deterministic numerical modelling are therefore two steps in the wrong direction in the context of forecasting of sudden extreme evolutions. In the future energy system, we can expect that a major fraction of the price volatility will be caused by special circumstances in the weather, where the demand is extreme. The total balancing costs will then be dominated by the peak prices for the largest renewable energy pools, because their forecast error is likely to be the root of the price volatility.

In the next section we will describe and examine different types of ensemble prediction systems (EPS) and their applicability in the context of using their output in the power industry.

3.2.1 Ensemble Prediction Methodologies

The pioneer work in ensemble forecasting stems from the 1990'ies (see e.g. [33, 46, 52, 60, 61]) and plays a central role in the society on risk management and in trading. The development of ensemble forecasting has been a funding resource ever since the development of the first operational ensemble forecasting methods and systems in 1993 and thereafter.

There is extensive literature about various ensemble developments and systems. For the reader new to ensemble forecasting, Cheung's review of ensemble forecasting techniques (2001) provides a comprehensive summary. The most important aspect to understand regarding ensemble forecasting is that an ensemble consists of a collection of forecasts that try to realise the possible uncertainties in a numerical forecast. A properly designed ensemble should be a finite approximation of the probability density function of the atmospheric state in phase space. In other words, each ensemble member is an equally probable state of the real atmosphere [62].

The method of producing perturbations depends on the particular system under consideration and its associated spatial scale. The simplest way is to add random noise to the original analysis (termed a Monte Carlo forecast, see next section), but this is not an optimal method, because the error characteristics in an analysis is often organised or correlated in some way. The ensemble approach requires the distribution of perturbations to be close to that of the initial state errors. This distribution in general is unknown and explains why the perturbation methodology is central to any discussion on ensemble forecasting [21].

We will therefore here focus on the aspects concerning the power industry and the challenges of applying various techniques for operational forecasting of wind power and PV at system operator level, at utility scale and for balance responsible parties.

3.2.1.1 Monte-Carlo Approach

The principle of Monte-Carlo forecasts is to produce random samples of an expected outcome, build a probability density function (PDF) and analyse the statistical behaviour of the outcome. Harrison [15] describes the overall intention of Monte-Carlo simulation as simulations that use random sampling and statistical modeling to estimate mathematical functions and mimic the operations of complex systems.

Monte-Carlo simulations have a long history in statistics and any type of description of uncertainty. In the context of weather forecasting they have also been the forerunner of today's ensemble forecasting systems. Murphy and Epstein [23, 48, 49] and Leith [42] were pioneers in meteorology to introduce Monte-Carlo simulations in weather predictions to generate probability density functions of the uncertainties. When developing stochastic dynamic methods in the late 1960'ies and 1970'ies, this was because computational effort for Monte-Carlo simulations were high.

However, at that time these techniques were considered encouraging and worthwhile following, because the lack of computational resources were expected to be the only real issue that would solve itself as new technology would become available. Nevertheless, as new technology became available, so did more sophisticated modelling techniques. Zhang and Pu [66] explain in their review on ensemble forecasting techniques the disadvantage with Monte-Carlo simulations being that the initial PDF needs to be known and that it is sampled randomly, which has its computational costs. In the 1970'ies where computer resources were very limited, this was an argument to develop other methods, where the sampling was less random [23]. But even 50 years later, the sampling of random uncertainties to form a PDF describing a realistic uncertainty of the atmosphere has still serious limitations, mainly because of the development of equations and their solvable methodologies that have increased with time and computer resources. Consequently, the difficulty comes with the size of the sample. Zhang and Pu [66] estimate for a common real-model with 10^7 degrees of freedom, a $10^7 \times 10^7$ dimension calculation for estimating the PDF will be involved. Parker [54] additionally mentions the uncertainty about which value should be assigned to each member. To explore this uncertainty requires a tremendous number of samples (and corresponding simulations) to generate a realistic PDF.

This deficiency of the Monte-Carlo simulations has caused that in the early 1990's other methodologies have been developed and brought into operation.

3.2.1.2 Initial Conditions Perturbation Approach

The ensemble approach following the Monte-Carlo simulations and other derived approaches thereof such as the "stochastic dynamic methods" with the longest history of operational use are those approaches, where the initial conditions are perturbed. While there exist a large number of ensemble prediction systems today (see e.g. the TIGGE project [14]), the EPS longest in operation for medium-range weather forecasting is the bred-vector perturbation approach at the National Center for Environmental Prediction (NCEP), introduced by Toth and Kalnay in 1993 [61]. This method is based on the argument that fast-growing perturbations develop naturally in a data assimilation cycle and will continue to grow as short- and medium-range forecast errors. A similar strategy has been developed and come into operation at the European Center for Medium-Range Weather Forecasts (ECMWF) in 1995. ECMWF uses a singular vector (SV) based method to identify the directions of fastest perturbation growth [18, 46]. Singular vectors maximize growth over a finite time interval and are consequently expected to dominate forecast errors at the end of that interval and possibly beyond. Instead of using a selective sampling procedure, the approach developed at the Meteorological Service of Canada (MSC) by Houtekamer et al. [30–32] generates initial conditions by assimilating randomly perturbed observations, using different model versions in a number of independent data assimilation cycles. This Monte-Carlolike procedure is referred to here as the perturbed observation (PO) approach [20].

Already in 1997, Palmer et al. [53] suggested that on the short-range (up to day 3), it may be possible to supplement the long-term singular vectors with vectors wherein the growth of initial perturbations are targeted for a specific area or lead time of interest such as maximised growth at day 3. This is worth noting, because for most applications in the power industry, especially wind energy and PV forecasting, there exist a discrepancy between such meteorological lead times of these ensemble techniques and the requirements of uncertainty in the power industry. While meteorologists focus on specific times such as 24 h ahead or 72 h ahead, the power industry has a need to know the uncertainty in a continuum, i.e. in every time step of the forecast. An introduction of calibration methods to circumvent such issues will be given in Sect. 3.2.4.

3.2.1.3 The Multi-model Approach

The only ensemble techniques that have inherent physical computations of uncertainty in every hour of the forecast lead time are the multi-model technique and the multi-scheme technique. The multi-model approach is a rather straight forward approach of using the output of many different deterministic forecast models to create an ensemble. The drawback of this approach is that a multi-model ensemble is likely to be rather under-dispersive, because deterministic models usually suppress extremes and thereby generate too little spread. There are other disadvantages with the multi-model approach such as the difficulty to maintain many different NWP modes or even collect output data from many different NWP models. However, the most critical issue with multi-model ensembles in operation is certainly the fact that deterministic models are tuned to provide best average forecasts, which often means that extremes are suppressed and therefore missed. Without extremes, the forecast spread however will not resemble a realistic uncertainty and the probability density function will be skew. To overcome such skewness, post-processing methodologies are necessary.

An example of an operational multi-model ensemble is the Poor-man's Ensemble System (PEPS), a project initiated by EUMETNET and operationally supported by the German weather service (DWD). EUMETNET started the project in 2006 by getting 20 European national meteorological services to participate, providing 23 forecast models. In the SRNWP-PEPS 40 deterministic and probabilistic forecast products are distributed to the contributing members on an operational basis. One of the main goals of the project has been the evaluation of PEPS to decide whether it provides a significant support and improvement of the warning process. In this system the single model forecasts are interpolated onto a reference grid, the PEPS grid. Exceedance probabilities are calculated at each PEPS grid point from the ensemble members using a nearest neighbour approach. Because the individual members have different resolutions and integration areas, the ensemble size depends on location [29].

3.2.1.4 The Multi-scheme Approach

The multi-scheme approach was developed in the early years of the new millennium, some years after the operational establishment of the global EPS', where the needs for limited areas and short-range probabilistic weather prediction were required for studies of precipitation uncertainty [25, 47, 59, 60]. The predictability horizon for precipitation, as well as for other surface or near-surface variables such as wind, radiation, temperature, is shorter than for more conservative parameters like mean sea level pressure, geopotential height or even upper level temperature. The reason for the start of utilizing LAM ensembles has been the interest in predicting and assessing the probability of occurrence of significant extreme events such as heavy rain fall, which can lead to severe damage (e.g. [12, 25]). This was the beginning of high-resolution LAM ensemble designs, which try to capture the comparatively rare events of surface variables.

At the same time, ensemble methodologies like the Ensemble Kalman-Filter (EnKF) for the assimilation were further developed and it was recognised that model errors can result in bias of the ensemble mean and insufficient ensemble spread due to its smaller projection onto the correct error growth direction. It is well known that those processes that cannot be explicitly resolved in the numerical models have to be approximated through different parameterization schemes. And it is exactly these schemes that are major sources of model error [43].

The success in form of ensemble spread, correctness and consistency of the probability distribution of the multi-scheme ensemble members is however closely dependent on the choice of processes that are computed differently within the numerical prediction model. If the chosen processes that have varied formulations are not directly connected to the variables that are under focus, the spread and hence uncertainty estimate and the skill of the ensemble to predict the uncertainty of specific variables looses its value.

The multi-scheme approach has later been demonstrated for other surface or near-surface variables to be the only approach that combines the advantages of the physical uncertainty computation with a reasonable computational effort. This is so, because the kernel of the NWP model is the same for all ensemble members. Specific physical and dynamical processes are then computed with different, but physically equivalent approaches [6–9, 11, 43]. Meng and Zhang [43] found in their experiments of a severe storm that a combination of different parameterisation schemes has the potential to provide better background error covariance estimation and smaller ensemble bias.

An example of an operational multi-scheme forecasting system is WEPROG's Multi-Scheme Ensemble Prediction System or short "MSEPS", which is a 75 member ensemble tuned with the focus on the fast surface processes [51]. The scientific background of the MSEPS system is described in [6]. WEPROG's MSEPS is a limited area ensemble prediction system using 75 different NWP formulations of various physical processes. These individual "schemes" each mainly differ in their formulation of the fast meteorological processes: dynamical advection, vertical mixing and condensation. The focus is on varying the formulations of those processes in the NWP model that are most relevant for the simulation of fronts and the friction between the atmosphere and the earth's surface, and hence critical to short-range meso-scale numerical weather prediction.

3.2.1.5 Ensemble Kalman Filter Approach

The ensemble Kalman filter (EnKF) first described by Evensen in 1994 [24] and later by Houtekamer and Mitchell in 1998 [33] is an approximation to the Kalman filter that has become feasible in the context of operational atmospheric data assimilation around 2005. The Ensemble Kalman Filter has its base in the computation of the background error covariance. The difficulty in the approach lies in that the true state of the atmosphere is unknown. This makes the estimation of the background error covariance difficult and expensive (e.g. [66]).

An example of an EnKF ensemble is the EnKF EPS at the Meteorological Service of Canada [34] that has been used operationally since 2005 as medium-range ensemble prediction system (EPS) and since 2007 as short-range EPS [35, 36].

In the Canadian EnKF algorithm, a strategy currently with four sub-ensembles [33, 45]) is used to preserve a representative ensemble. Consequently, in the absence of any differences between the model and the atmosphere or between the true and assumed observation- error statistics, the EnKF should maintain ensemble statistics that are representative of the actual error in the ensemble mean. It is thus possible to predict the analysis quality from the ensemble statistics for a hypothetical environment without model error. The negative impact of the different sources of model error on forecast quality can subsequently be quantified from the increase in ensemble spread as these components are added to the EnKF environment [36].

3.2.2 Evaluation of Quality and Value of Probabilistic Forecasts

Two decades after the introduction of the first operational ensemble forecasting systems from the European Center for Medium Range Forecasting (ECMWF) and the National Center for Environmental Prediction (NCEP) in 1993, it is still not common knowledge that there exist simple cost based algorithms on how to exploit probabilities, although research in defining quality and value with cost-loss methods of probabilistic forecasts has an even longer history (e.g. [28, 38, 49, 50, 65]).

The cost-loss method is a classical example of such an algorithm. Murphy and Ehrendorffer quote Winkler and Murphy [64] observing that the value of the imperfect forecast depends on the expected expense associates with the use of the information consulted by the decision maker in the absence of these forecasts. They concluded from this, that a realistic assumption then must be that since climatological forecasts would always be available to the decision makers, the value of imperfect forecasts can be defined as the difference in expected expenses between the situation involving climatological forecasts and the situation involving imperfect forecasts.

Apparently, this type of cost-loss evaluation is seldom applied in the power industry, most likely because the economic sensitivity related to marginal costs of certain actions is considered commercially sensitive. Another reason for the lack of decision making with the help of probabilistic methods has also been lack of short-range applicability of many of the ensembles generated world wide due to the approach used to generate them (see e.g. [20]).

Hagedorn and Smith [27] in fact showed with their *Weather Roulette* concept how we can evaluate whether investment in a probabilistic forecast can pay off. In their basic concept they show an application, where they find the true price for a climatology forecast (often also referred to as persistence) and a probabilistic forecast when measured over the success rate of the forecasts to predict the future.

3.2.3 Ensemble Predictions Versus Mixing of Multiple Deterministic Models

As discussed above in Sect. 3.2.1.3 there is an inherent risk in forming an ensemble with a number of deterministic models. With such an approach, the spread that determines the model error may not be consistent and rather small, because deterministic models often tend to suppress extremes, which are required to form a realistic description of model errors in the atmosphere.

Using an EPS formed with stochastic perturbations, a multi-scheme approach or an ensemble Kalman filter (EnKF) for wind power or solar power predictions is therefore fundamentally different from using one consisting of a few deterministic weather prediction models, because severe weather and critical wind power events are two different patterns [6, 12, 25]. The severity level increases with the wind speed in a weather context, while wind power has two different ranges of winds that cause strong ramping, one in the middle range and a narrow one just around the storm level (the cutoff level). Wind power forecasting models therefore have to be adopted to the use of the ensemble data.

In general, a wind power prediction model or module, that is directly implemented into an EPS is different from traditional power prediction tools, because the ensemble approach is designed to provide an objective uncertainty of the power forecasts due to the weather uncertainty and requires adaptation to make use of the additional information provided by an ensemble.

3.2.4 Ensemble Calibration Methodologies

It is only recently that the energy meteorology research community has developed ensemble calibration methodologies for the output from ensemble prediction systems using stochastic perturbation methods such as singular vectors (e.g. [52]) or breeding (e.g. [61]). The issue with these approaches is that they are designed for specific lead times, which are usually greater than 48 h. What that means in fact, is that the spread between the ensemble members is too small in the beginning of the forecast and often too large at the end of the forecast. In the scientific community these problems are referred to as under dispersiveness and over dispersiveness, respectively. The first showing so-called u-shaped histograms, while the latter shows bell-shaped histograms when the spread is evaluated objectively (see e.g. [28]).

To get around this issue, there have been developed calibration methods such as Ensemble Model Output Statistics (EMOS) methodologies (e.g. [63]) and derivatives thereof, such as

- BIAS corrections on each ensemble member
- Bivariate EMOS (e.g. [57])
- Variance deficit calibration (e.g. [16])
- Ensemble kernel dressing (e.g. [55])
- Bayesian Model Averaging (e.g. [56])
- Bivariate ensemble copula coupling (e.g. [37])
- Analogue ensemble calibration (e.g. [22])

Univariate or bivariate Bayesian model averaging techniques are theoretically also useful, but their high computational costs and similar performance compared to EMOS approaches means that they are not considered for practical applications at present [37].

3.2.5 The Importance of the Correct Choice of Ensemble Forecasts

It is well-known among meteorologists that forecast errors in real-world applications arise not only because of initial errors, but also because of the use of imperfect models. Representing forecast uncertainty related to the use of imperfect models is thought to be of an even greater challenge than simulating initial-value-related errors [20].

The deficiencies and benefits of the various approaches are hence an imperative aspect to consider when choosing probabilistic forecasts for real-world applications such as wind power or solar power forecasting. In fact, the success and the value of using probabilistic forecasts is dependent on the correct choice of methodology and requirements to fulfill a specific task.

Table 3.1 provides an overview over the various methods that have been described in the previous section and their specific characteristics in the context of applications for the power industry.

Table 3.1 Comparison of different ensemble forecasting technologies and their applicability in the power industry

Approach	Monte carlo simulations	Init. Cond. Perturb.	Ensemble kalman filter	Multi-model	Multi-scheme
Method	Statistical	Statistical	Statistical	Physical	Physical
Member differences	Random statistical perturbations of the initial conditions (Analysis)	Statistical perturbations from linearised equations singular vectors, non-linearised Lyapunov (Bred) vectors	Forecast error covariance	Every member is an individual NWP model	Computation of different processes inside one NWP model kernel
Application of the perturbations	Random simulations to form PDF	Differences are generated from perturbation of the initial conditions (Analyse)	Differences are a result of covariance error matrices of initial conditions	Models as well as initial conditions are different	Perturbations of the initial conditions and physical and dynamic processes inside the NWP model
No. of members	Limited	Limited	Computationally limited	Unlimited	Unlimited
Definitions and recognition of the differences	All differences are random and not unique	Can be defined statistically but are not unique	Can be defined statistically but are not unique	No differences can have a technical or physical reason	Yes well-defined
Expense (technically)	Large, because of random simulation of uncertainty	Reasonable because only 1NWP model is required	Low for small ensembles, large for larger ensembles	Huge maintenance of many models is required	Manageable 1 NWP model kernel with many schemes
Uncertainty validity	Dependent on uncertainty simulations	Predefined e.g. >3 days	Predefined in error matrix e.g. >3 days	Any forecast hour	Any forecast hour
Applicability in intra-day	Requires uncertainty computation to match time horizon	Requires calibration	Requires calibration	May miss extremes	Yes
Applicability in day-ahead	Requires uncertainty computation to match time horizon	Requires calibration	Requires calibration	May miss extremes	Yes
Applicability for futures (>2 days)	Yes	Yes	Yes	May miss extremes	Yes
Deficiencies	Large computational effort to create valid uncertainty	Requires calibration for many applications	Computationally expensive to avoid inbreeding	Members deterministic tuned, extremes suppressed	Computationally expensive

3.3 Best Practises on the Use of Ensemble Forecasts

It is soon 20 years ago, where it has been scientifically demonstrated that ensemble forecasting is the state-of-the-art method to find the instabilities in the atmosphere that causes most forecast uncertainty and error. It is therefore questionable, whether it is actually best practise to search for the best RMSE optimised forecast, because it has become apparent that optimisation on MAE/RMSE is often against the interest of the TSO and/or consumers.

Over time, it has also been shown that model changes that lead to better average deterministic forecasts lack internal perturbations that however seem to be required to forecast instabilities. Such instabilities are difficult to forecast and therefore it is often better to not predict them at all, if only a single value can be used and if low RMSE is the target. The likelihood of a forecast extreme coming out with an incorrect phase and thereby be hit by double punishment of errors is high. Thus, the stiff model system that may develop a small insignificant low may not be punished as much as the one that tries to develop a bigger and stronger low, which may be more realistic.

In a later section, it will be demonstrated with an example case how an instability did not take place in the RMSE optimised forecast, where the entire evolution of a low pressure system was in fact suppressed.

3.3.1 The Role of Stakeholders

It is a challenge to avoid price volatility, if dispersed wind power is handled in a liberalized market. The wind farm owners are not directly participants and would need to agree on a solidarity principle to protect themselves against other stakeholders. Nobody can expect that the balance responsible parties (BRP) have incentives to avoid high volatility events, unless they are penalised by the balancing cost or from the plant owners. The BRP may earn on one account and loose on another when prices change. Thus, the incentive on reduced volatility lies exclusively at the TSO among the direct participants in the market.

The TSO and consumers should essentially both prefer a stable energy price and a competitive market as this is the best basis to operate the grid in a stable manor. Price volatility is the result by not trying to tackle forecast uncertainty in time. Typical market stake- holders see volatility as an opportunity, thus it is only regulators and system responsible parties that have an interest to prevent that the price volatility grows to a level where the system becomes inefficient.

A direct parallel can be drawn to the economy leading to the finance crisis in 2008/2009. The crisis was triggered by volatility and resulted in years of artificial low interest, where the flow of money almost stopped through the banks. A similar situation could occur in the electricity market, which can for example be triggered by a physical handling problem or a market problem. Therefore, it is important to find methods to prevent volatility in a market compatible manor.

3.3.2 *How to Define a Forecast Optimization Criteria*

We therefore want to raise the question on the feasibility of MAE/RMSE optimisation on the basis of high forecast uncertainty. How should we take account for a minority of ensemble forecasts that lie at a very different place in the spread, far away from a majority of forecasts? Another question to ask in this respect is how to deal with strings of different outcomes?

It seems like the most appropriate step beyond the pure deterministic philosophy is to consider high and low uncertainty separate. If the uncertainty is low, then RMSE/MAE is a fair target. If the uncertainty is high, then the forecast should rather try to avoid the bigger mistakes and at the same time suggest an additional pre-allocation of reserve, which will prevent price volatility and increase system security.

Operational experience suggests that a good forecast lies approximately in the middle between the minimum and maximum of an ensemble configured to approximate a reasonable correct uncertainty of the weather situation. This type of *best guess* is also referred to as the mean or a weighted mean of the ensemble. A skewness adjustment may be considered, if the available reserve is asymmetric or the localisation of the data statistically suggests bias corrections. The forecast should follow a linear trend over several hours, unless it is very certain that the actual value will not follow this trend. In this way the choice of wind power forecast can contribute positively to increased efficiency of the energy system and because of the reduced volatility, there is reason to expect that the balancing costs will be lower, even if the MW deviation between forecast and actual generation will increase.

3.4 Forecasting Approaches for Large-Scale Wind and Solar Integration

The way forecasting for wind power and solar power has evolved in the various countries over the past 15 years can be categorised in 3 classes:

- Bottom-up forecasting approaches
- Top-down forecasting approaches
- Hybrid forecasting approaches

In the following 3 sections, the different approaches, their benefits and drawbacks in the integration of wind and solar power will be discussed.

3.4.1 *Bottom-Up Forecasting Approaches*

The *bottom-up forecasting approach* assumes that forecasts are generated for each unit and summed up over all units in order to generate area aggregated forecasts. In jurisdictions, where the development of wind and solar plants in general are of a size >10 MW, bottom-up approaches are common, also because larger plants always have individual SCADA systems and are of a size that even in fully liberalised markets, a detailed monitoring of the plant is carried out.

The advantage of the *bottom-up forecasting approach* is that detailed measuring information can be used to tune the forecast system and area aggregates are a direct sum of the forecasted generation. The disadvantage of the *bottom-up forecasting approach* can be that additional detail does not always seem to add to improved forecasts, but adds to the costs. Especially in large systems with many plants, this approach has been considered unnecessary and too costly. However, in recent years, where *BIG DATA* has become an abbreviation of the ability to analyse and make use of large amounts of data in order to increase understanding of specific processes, the *bottom-up forecasting approach* has become more interesting again, as its benefits of working with detailed data is obvious.

3.4.2 Top-Down Forecasting Approaches

The *top-down forecasting approach* is characterised by the simplification of a large problem size to a easy to handle process. That is, a *top-down forecasting approach* is using an up-scaling methodology to estimate an area aggregate from a number of representative units inside the area. This approach is useful in jurisdictions, where there are many smaller units that may or may not have SCADA or other monitoring possibilities for the system operator or unit managing party. In areas like Germany and Denmark, where wind energy has been a growing and developing industry since the 1980's, there are many small units that have no or non-standard monitoring system installed in order to be monitored and handled individually. Especially in the fast growing solar market in Germany, the regulators have for a long time considered small units as non-critical for system operation. With incentive schemes in place, the number of small units below 100 kWp however have exploded. At the end of 2014, over 1mio units were below this margin and are not monitored by the system operators. In such cases, the production can only be estimated with a top-down approach, where the generation is forecasted by up-scaling of generation units with a number of representative units. An example of such an approach is the inverse distance weighing, a two-dimensional interpolation function for irregularly-spaced data [58] that has been applied e.g. in Germany for forecasting of small roof-top solar plants by the system operators (e.g. [26]).

The advantage of the *top-down forecasting approach* is it's efficiency and simplicity. In normal weather conditions and over a longer period and large area, this approach delivers reasonably good results and is a useful tool. The disadvantage of the approach is that it is sensitive to the correctness, or better representativeness of the reference sites. As long as there are little changes, the algorithm, once trained, works quite well. However, in fast changing weather conditions or with continuously changing capacity that adds to the locations from where the forecasts are up-scaled, the algorithm can generate large and unwanted errors.

3.4.3 Hybrid Forecasting Approaches

The so-called *hybrid forecasting approaches* make use of both a bottom-up and top-down approach by for example summing up many small units in a regular grid that can then be summed up geographically correct. The way most up-scaling problems have been solved in the past is with a so-called distance-based up-scaling from reference sites as described in the previous Sect. 3.4.2. The disadvantage of such an approach can be mitigated by applying a regular grid for the so-called reference sites in an appropriate grid distance. By summing up the reference sites of such a regular grid, the power generated in each grid box is reflecting the weather related geographic differences much better. This type of up-scaling however requires a back-scaling of available measurements to the grid points in the regular grid.

Figure 3.1 shows how such a hybrid forecasting solution can be set up, where many small units ("sites") are associated via a distance methodology to the closest *grid point*. i.e. the respective grid point contains the installed capacity of all associated small units. Reference sites with measurements are marked black and are extrapolated to the grid points for example with an Ensemble Kalman Filter technique [9]. By using a correlation technique, the weather conditions can be taken into account spatially and temporarily and large errors at variable weather conditions and extreme events avoided.

3.4.4 Forecasting Approaches for Cross-Country Wind and Solar Integration

In order to fulfil Europe's targets of CO_2 reduction on energy production in the next 2 decades, the deployment of wind power throughout Europe has to increase significantly from today's installations. The targets set by most European countries can only be met with offshore wind power delivering the bulk of new production units. The coupling of electricity markets through-out Europe and the interconnection of offshore wind power in the North Sea, Baltic sea and parts of the Mediterranean sea will be a necessary step that eventually will lead to a centralized European Super-Grid connecting all European counties. This new grid structure will not only involve the enhancement of today's electricity network through more connection points, but will require focal points to collect, integrate and route the produced energy to the load centers and the markets with the highest demand. Such a European SuperGrid will also have to integrate all markets into one common market in order to enhance competition, the security of supply and reliability of the system for all countries in the EU.

An integrated European market requires large-scale forecasting of wind and solar power to ensure the efficient trading of electricity. By using ensemble forecasts of the large-scale wind power generation over entire Europe, the total reliability of the energy system can be maintained and further increased. The use of ensemble fore-

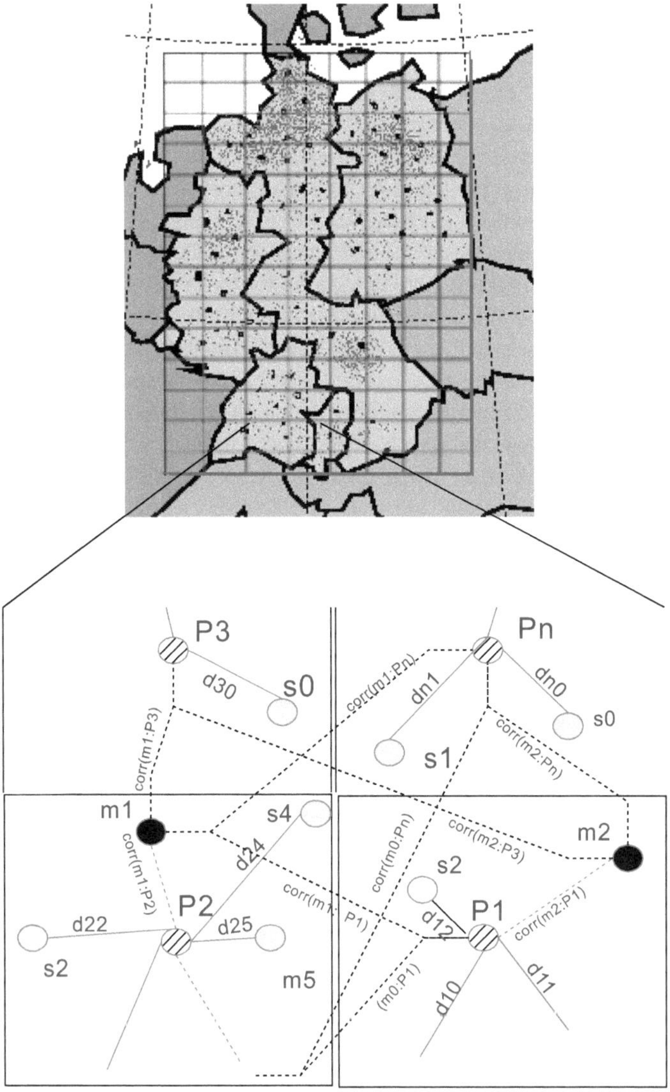

Fig. 3.1 Up-scaling principle for a hybrid forecasting approach of many small units summed up in a regular grid. The hatched points are the grid points in the regular grid, the *black* points are measurement points, correlated to all grid points for computation of the influence index, the *light gray* points are sites or units of wind power or solar power

casts in this context in fact will reduce price volatility, as excess wind generation from one region can balance missing wind generation in other regions. Additionally, the total wind generation can be estimated more accurately with the use of large forecast ensembles.

Möhrlen and Jørgensen [10] carried out a super grid study, where they found out that there will not be a likelihood for concurrent generation of more than 50% of the installed capacity and there is little to no risk for a complete lack of generation from RES. The statistical results for the day-ahead horizon showed that the SuperGrid produces nearly 30% less error than an average, which was computed as the weighted error for 8 countries assuming that no forecast errors are exchanged across inter-connectors between countries. In the "SuperGrid", errors were exchanged over the inter-connectors, whenever possible. The results from the study clearly showed that enlarging the grid intelligently, opens the possibility to save significant balancing capacity. Intelligently in this connection means that the grid is enlarged in such a way that it represents the possible physical flow of electricity instead of staying inside control areas. By considering physical bottlenecks rather than administrative areas, the handling of variable generation will be much more efficient.

Fortunately, the power society has plans to establish a fully integrated pan-European Electricity Market, which opens such possibilities. The cooperation of European transmission system operators (ENTSO-E) report about such plans in their 10-year Network Development plan, published on their web page (http://www.entso-e.eu). Similar to the findings of the "SuperGrid study" [10], there have been established different market coupling mechanisms for the north-western part of Europe and for the south-western part of Europe (see http://www.casc.eu). The North-Western Europe (NWE) Price Coupling, a project initiated by the Transmission System Operators and Power Exchanges of the countries in North-Western Europe including 17 partners from 9 counties (Finland, Estonia, Norway, Sweden, Denmark, Germany, France, The Netherlands, Belgium, UK). The South-Western Europe (SWE) Coupling Project is a joint project between the French, Spanish and Portuguese.

3.5 Challenges with Increasing Integration of RES Over Country Borders

The installed capacity in Europe and especially in Germany has been growing continuously and there is already over-capacity in some regions. Still, there will always be some congestion areas which will need special attention. However, the introduction of the so-called 10-year net development plans have been established in Europe by ENTSO-E, the cooperation of transmission system operators, to prevent major shortcomings in respect to congestion (see https://www.entsoe.eu/major-projects/ten-year-network-development-plan/).

For example in Germany, most of the energy from wind is generated in the north and north-east, while the load centers are in the south. This is causing more and more issues, as conventional and especially nuclear power plants in the south are being switched off. While the nuclear power plants provide cheap base load, many other power plants are no longer competitive and pushed out through the merit orders in the power markets. Nevertheless, the vast expansion of solar energy over the years from 2010–2015 to 37 GW of installed capacity, where 40% of the capacity is installed in the south, levels out a lot of the peak demand phases at midday.

3.5.1 *Lack of Extreme Weather Events Reduces Awareness of Risks*

As a consequence of the different generation pattern from wind power and PV and the fact that PV's generation curve follows naturally the demand on the morning and midday peaks, the amount of large generation units are in the process of reduction. This in return reduces the capability to balance severe forecast errors from wind power and PV generation. Grid security may therefore need to get more attention again over the next years, also because there have not been severe weather events that brought attention to extreme weather forecasting for over 5 years.

There has been made progress with a number of grid initiatives to couple control areas and to couple markets. In Germany, 2009 marked the year where the balancing area in Germany was coupled among the 4 control areas in the so-called "grid control cooperation" [5]. This grid control cooperation (GCC) is a network control concept, where the four German TSOs optimise their control energy use and the control reserve provision, technically and economically through communication between the load-frequency controllers of the TSOs. The horizontal structure of the control areas in the European interconnected system, offers the GCC the possibility to exploit synergies in terms of network control like in a single fictitious control area, without giving up the proven structure of control areas. It also enables a flexible response in case of network bottlenecks. (see https://www.regelleistung.net/).

This concept was extended further into a "international grid control cooperation" (iGCC), where the coupling of Germany to the northern counties took place at the end of 2009 with the establishment of a first-inter-regional market coupling office, the so-called European Market Coupling Company GmbH (EMCC). The EMCC launched its volume coupling system to connect the Nordic and German electricity markets. In November 2010, EMCC and Central-Western European (CWE) Transmission System Operators and Power Exchanges established Interim Tight Volume Coupling (ITVC). The 16 ITVC stakeholders utilised EMCCs coupling as a regional link between CWE price coupling and Nordic-Baltic market splitting [3].

EMCC's market coupling system used its own hourly calculated price information from 23 bidding areas and hourly flow calculation from 32 inter-connectors, managing an electricity volume representing about 55% of European consumption. The algorithm represented the different rules and products of the coupled regional day-ahead power markets, even with regards to special regulations, e.g. curtailment rules, grid loss factors and clock change procedures.

The next natural extension to the EMCC was then to the west and led to the establishment of a "North-Western European market coupling" (NWE) in the beginning of 2014, now including four Power Exchanges and 13 TSOs in the North-Western Europe (NWE) day-ahead price coupling. The NWE region is stretching from France to Finland and operates under a common day-ahead power price calculation using the Price Coupling of Regions (PCR) solution. The same solution is also used at the same time in the South-West Europe (SWE) region in a common synchronised mode, with France being part of both the NWE and the SWE region.

The development of market coupling aimed to reduce congestion and extreme prices by preventing that reserve requirements are treated separately in neighboring control areas, which often leads to a lack of competition. Market coupling has in fact started the evolution of new techniques to forecast reserve requirements and thereby enhanced competition on the reserve market.

Seen in that light, it can be concluded that there exists a kind of predecessor of such an European SuperGrid environment in central Europe as described in the previous section (Sect. 3.4.4). One of the ideas behind the SuperGrid is that the predictability of wind power and PV increases by increasing the region, i.e. aggregating over more and better dispersed units, where the forecast levels out errors automatically. This argument is valid to a large extend with such market couplings that reduce congestion and price volatility. However, besides the technical and physical flow constraints in the grid, the scale of weather systems that may span over the entire area should also not be neglected.

3.5.1.1 Errors in Large Systems from Weather and Climate

On the weather and climate side, the development of enlarged areas look slightly different, because of the generally unknown long-term and short-term development of weather patterns and climate change.

We will therefore present a case below, where area aggregation even over the very large area of 7 countries is insufficient. We will also demonstrate that even in very large areas there can establish errors in the day-ahead forecasts of significant size. In order to study this event in detail we have setup 10 different forecasting system configurations focussing on different aspects of the modelling chain leading to uncertainty of the end result. Table 3.2 shows the model configurations and their differences.

Figures 3.2, 3.3 and 3.4 show the outcome of the different forecast system runs for our extreme event case on June, 11, 2010.

The first figure shows the day ahead forecast valid at 5pm over Denmark. It shows that a large group of ensemble members have almost no generation from wind symbolized by the dark blue colour everywhere. A few forecasts show almost full generation in a large region of Denmark. From this forecast it is evident that the uncertainty at this time is extremely high. The second figure confirms that the uncertainty is rather high even on the 9 h forecast horizon. In fact, this would be the last weather forecast before the event in a 6 h schedule. The figures therefore illustrate that there is extreme uncertainty in the weather, but high likelihood of no generation in Denmark. The likelihood is higher for high generation in Germany than in Denmark.

Table 3.2 Description of boundary model configurations for the test case on the 10th June 2010. Abbrev. Meaning: OPR = operational model setup, T = test setup, NM = nested inner Model, BM = boundary generating model, ANM = nested inner assimilation model, ABM = boundary assimilation model, Std = standard, XL = eXtended large, Glb = global, VL = very large, XS = small, SL = Semi-Lagrangian, ES = use of either Euler, Euler Upstream, Semi Lagrangian advection, Ana-SST = SST analysis scheme, Pcr = reference power curve scaling used to increase the responsiveness of power curves

Description	OPR	EU	T01	T02	T03	T04	T05	T06	T07	T08	T09	T10
No. of members in NM	75	75	75	75	75	75	75	75	75	75	75	8
No. of members in BM	8	10	8	8	11	1	8	8	11	8	8	1
No. of members in ABM	8	10	8	8	11	n/a	8	25	11	8	8	8
No. of members in ANM	75	75	75	25	75	75	75	0	75	75	25	8
No. of nesting levels	2	2	2	2	2	1	2	2	2	2	2	2
Model Area of NM	EU	EU	EU	EU	EU	VL	EU	EU	XS	XS	XS	Ext
Model Area of BM	Std	Ext	VL	Ext	Ext	Glb	Ext	Ext	Ext	Ext	Ext	Hem
Spatial resol. of NM	0.45	0.45	0.45	0.45	0.45	0.45	0.45	0.45	0.45	0.22	0.22	0.22
Spatial resol. of BM	0.45	0.45	0.45	0.45	0.45	0.4	0.45	0.45	0.45	0.45	0.45	0.6
BM Boundary update freq.	6	6	6	6	6	n/a	6	6	6	6	6	6
NM Boundary update freq.	1	1	1	1	1	n/a	1	1	1	1	1	1
BM Advection scheme	SL	E	SL	SL	ES	SL	SL	SL	ES	SL	SL	SL
Iterations of Ana-SST	1	1	1	3	3	3	3	3	3	3	3	3
Power curve method	Std	Std	Std	Pcr	Std	Pcr	Pcr	Pcr	Std	Pcr	Pcr	Pcr

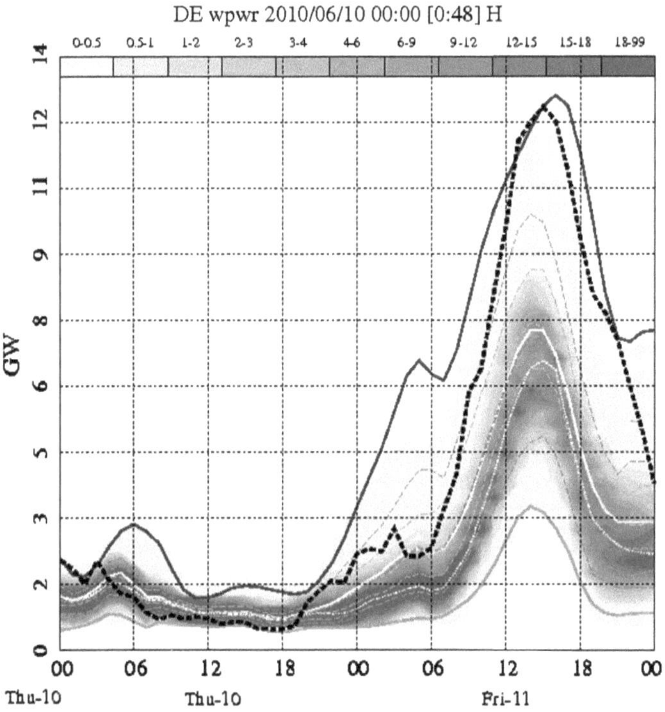

Fig. 3.2 Probabilistic 48 h wind power forecast for Germany on the 10th June 2010. The *dark gray fat line* and the *light gray fat line* are the maximum and the minimum, respectively, the *white fat line* is the best guess, the *white thin line* the highest probability, the *dashed dark gray line* close to the *white fat line* is the mean, the 60, and 80% percentiles are the *gray dashed lines* above the mean and the *gray dashed lines* below the mean are the 40 and 30% percentiles

The third figure shows the probability distribution of German wind power generation. All figures are taken from the forecast setup named T04 from Table 3.2, which performed considerably better than the T10, but also considerably worse than the operational/backup setups. Table 3.2 shows how T04 differed from the other systems.

The 11th of June 2010 was an unusual day seen from a wind power perspective, because there was no sign of a significant low pressure system at the beginning of the forecast on the 10th of June. The low pressure system developed almost instantly during the forecast as a result of an instability in the large scale flow. The large scale weather pattern was dominated by two stronger low pressure systems in the Atlantic, respectively north of Scotland and west of France. There were several other very weak low pressure systems, some of these were located around Denmark in a northerly flow of colder air. None of these small low pressure systems were located over Denmark and none seemed to cause any wind in Denmark, because they were small and insignificant. The weather had been warm up to that day in both Denmark and Germany. This meant that the northerly flow from Norway in the middle tro-

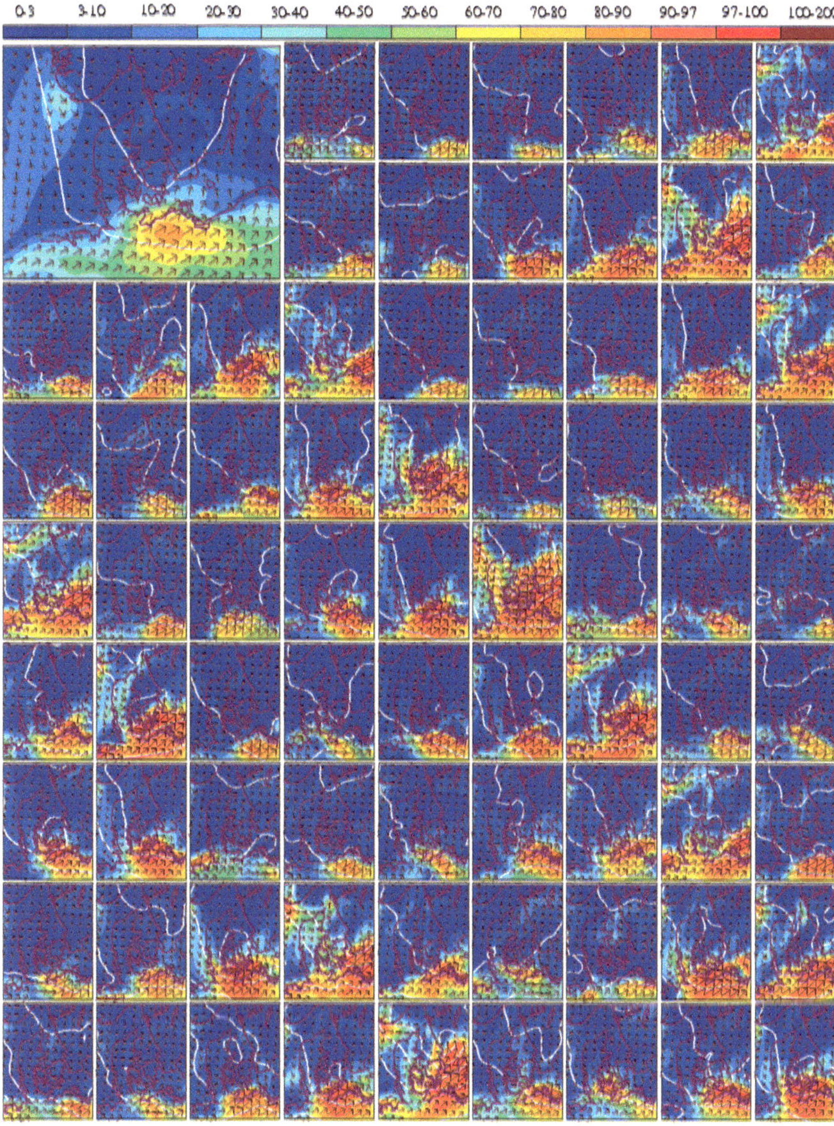

Fig. 3.3 Short-term Ensemble Forecast (9 h) from the 11th June 2010 displayed in horizontal plots of wind power load factor, inclusive wind speed arrows and isobars. The large figure is the mean of the 75 forecasts

posphere would lead to a situation, where cold air would lie on top of warm air with continued heating from below in the afternoon. One could therefore expect strong rain in the late afternoon, but the formation of a low pressure system with strong wind would still be an unlikely evolution. However, if the afternoon convection would be simultaneous over a large region, then a low could develop. Without a large ensem-

Fig. 3.4 Ensemble Forecast (39 h) from the 10th June 2010 displayed in horizontal plots of wind power load factor, inclusive wind speed arrows and isobars. The large figure is the mean of the 75 forecasts

ble of forecasts one could not estimate the probability of the low pressure systems development.

Figure 3.5 shows the typical output of combined forecasts (also referred to as meta forecasts). They are all dampened to take account for the uncertainty in the

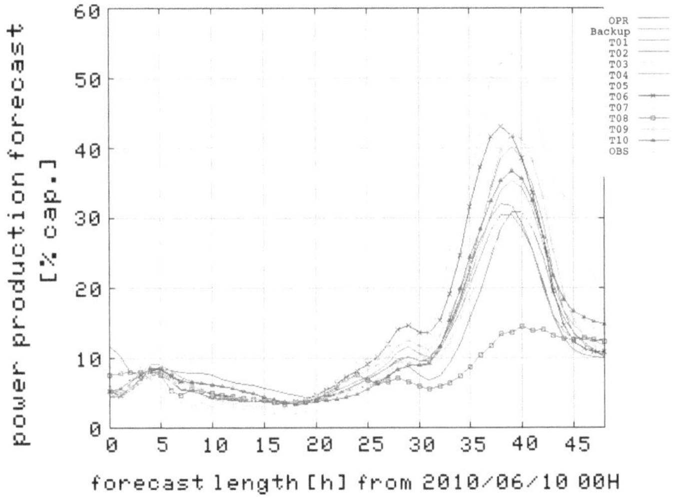

Fig. 3.5 Wind power forecast at the 10th June 2010 and 48 h ahead from different forecast systems. *Operational* stands for operational setup, *Backup* for backup setup, *T* for test setup. See Table 3.2 for details regarding the differences of the various systems

evolution and suggest therefore values in the middle of the possible range of values, only partially influenced by the probability. Figure 3.2 on the other hand shows the probability of the generation, which is not represented by the different meta forecasts on Fig. 3.5.

The typical pattern is that the meta forecasts are MAE or RMSE optimized, because those measures are by tradition used to evaluate, if forecasts are good or not. A collection of meta forecast therefore does not represent a likely physical outcome, but a defensive guess on what is expected to give the least error. Their spread or rather difference to each other is therefore suppressed, because of RMSE optimisation. One could now argue that the forecast with the best MAE/RMSE measured over 18 months (T10) turned out to be a good minimum forecast even though it is a combined forecast. However, there is no combined forecast, which represents the actual generation. Only the percentiles of the ensemble and the individual members above P80 provide a good estimate for the maximum.

A study of the operational day-ahead forecasts showed that they had the center of the power generation 100–200 km too much to the East. Because Germany is regarded as a SuperGrid, such an error is not counted even though it is an error. If we would verify all forecasts in smaller regions on distribution system operation (DSO) level, then T10 on Fig. 3.2 could probably be defined to be the best forecast, because no meta forecast would have the power generation at the right location. That is because T10 is not double punished anywhere for having generation at the wrong time.

In this way we can argue for that the absolutely worst SuperGrid forecast may be the best DSO forecast, if RMSE is used as an error measure on both. The fact that

Percentile 85 was a good forecast shows that the event was difficult, because the further out in the percentile range we find the best forecast, the more difficult the event, if the distance to P50 in MW is simultaneously large. It is not unusual that the P85 forecast is the most accurate. Thus, such an event could easily be overlooked as being extreme, because there have been events with more imbalance. What is interesting in this event is that the bulk of the ensemble forecasts were extremely different and the ensemble spread was large. This indicates, that if the final forecast coming out would have been wrong, whether as a meta forecast of different deterministic forecasts or the best guess of an ensemble, the event could have been causing a major issue on the grid. Because of the potential security issue on the grid, we decided to carry out forecasts in high spatial resolution, as this is the expected trend for forecasting in the future. From the results of the experiments, we conclude that the forecasts could have resulted in an imbalance of 12 GW of wind power, if the Netherlands, Denmark and Germany would already have been coupled together in the NWE market coupling.

An interesting aspect was that the more we increased the spacial resolution of the model system, the worse the result became regarding the likelihood of high power generation. The fact that the power system did not experience an imbalance of 12 GW could be regarded as luck. It is difficult to assess what the total imbalance was in this case. However, the answers of four of the balance responsible parties in Denmark and Germany indicated that all forecasters had forecasted well under the actual produced power. It was only WEPROG's P90 and maximum forecast that over-predicted the production. One of the test setups (model T10 in Fig. 3.5) had been showing very promising results over entire 2009 and the first 5 months of 2010 with improvements of 25% of the total error in Germany and 15% in Denmark. This is an exceptionally strong improvement for this type of work even though the model configuration is fundamentally different than the other setups in several stages of the model chain. However, in this particular case, we noticed a complete failure to produce the correct power production. The average load factor was 10% in the forecast, while it was estimated to reach 55% of the installed capacity at the peak. The operational systems forecasted 42–44% generation at the peak correctly timed but spatially shifted (see Fig. 3.5).

In T10 only one ensemble member produced a forecast, which was close to the actual. However, the other forecasts had so little wind power generation that the final forecast was seriously under-predicting the event. It has been found that the forecast, which performed best, was one of those with the poorest long term statistics. The influence on the combination forecast was then minor. The event emphasizes that extreme care is required when evaluating forecasts, because T10 could match the other systems measured over the month on MAE/RMSE, but made one extreme error during the month, which is much more serious than any of the other system's smaller errors during the month. A similar event without impact on Denmark took place only 30 days later, this time on the border between Germany and Poland. The relative performance of all systems were similar to the 11th of June case. In this time, a low pressure system developed over Poland in southerly flow, where again only 1 forecast caught the event in the T10 setup. The two events were similar in one way, but different in many other ways. This time the critical spot was in Poland, just outside the SuperGrid and the high density of the wind generation, and the T10 had plenty of wind. The root of the problem is however the same, i.e. the low came from a mountainous region. Therefore, the two events complement each other in the study of the cause of the error. Given the success of T10 it would be extremely convenient for the future development, if those two events could be handled with just the same quality as the average of all the other systems. Therefore, extremely many model formulation changes have been applied to T10 to search for the crucial parameters. So far, only softer Norwegian mountains and coarser model resolution seem to have positive impact, but such changes make MAE/RMSE scores worse measured over long time. From the study of more than 1000 test forecasts for the two events (see Table 3.2), it has in fact become apparent that typical future forecasting techniques with very high spatial resolution seem to have a potential risk to fail to forecast such events on the day-ahead horizon. What happens is that several small low pressure systems develop independently in a region of 400×400 km. For dispersed wind power this would mean almost no generation, but in fact one large low pressure system developed with significant power generation as the result.

Even if we consider the operational forecasts in the two events, it was found that they under-predicted strongly, while the percentile P85 was in both cases nearly perfect. The fact that the operational systems did warn about the event was positive, but it is not positive that increased spatial model resolution reduces the forecast skill in such events. Knowing that the distance between P10 and P90 was extremely high, it would be desirable to already on the day-ahead horizon trade this uncertainty into the market in competition. The benefit of such a possibility becomes evident, when we note that the low pressure system seemed to give concurrent uncertainty corresponding to 40% of full capacity in Denmark, Germany and the Netherlands, which was equivalent to 12 GW. This amount of uncertain generation could otherwise give rise to concerns and price volatility considering the relative small volume in the intraday market.

From this experience we can therefore summarise our learnings as:

- A continued focus and evaluation criteria being to find the best deterministic forecast, where "best" is evaluated by a single error measure such as MAE or RMSE, may lead to severe problems on the grid now, if a forecast fails during a low demand period and even more so in the future
- The warnings provided by ensemble forecasts are becoming more important, because the risk of a major error is increasing, especially with increasing capacity in combination with changing predictability of the weather
- It is not trivial to state what a good forecast actually is. Is this the one that rarely or never makes serious mistakes or is this the one that makes the lowest average error? The results suggest that this is not the same model formulation

What we want to emphasise with this analysis, is that extremely much care has to be taken, when evaluating what is a good and a bad forecast. Over several decades, a core experience that developed in meteorology was that all information is good information, also if it is uncertain or wrong information.

This is state-of-the-art practice particularly in data assimilation and will also need to be adopted in other applications using weather forecasts, if the weather forecast is the key input to the application. Users of weather forecasts automatically experience that finding sooner or later, if they analyse the performance of their forecasting system regularly.

The wind power forecast is more or less a direct output of weather forecasts and there is no other possibility to forecast wind power than by using weather forecasts. Demand forecasting and hydro energy are not in the same degree driven by weather forecasts. In wind and solar power context it is therefore required to adopt the experience from meteorology, which is the ability to use information that is only maybe correct. The 11th of June 2010 is a proof that there are events that can only be predicted by using uncertainty forecasts. By not doing so, there is a risk of extreme price volatility and possibly also grid security issues. We have demonstrated with future compatible forecasting techniques that this risk is enhanced, and with increasing installed capacity of wind power and PV this risk will further increase over time. Especially, when the penetration of RES increases over 20–30% of energy consumption, such errors have significant impact in terms of MW on the grid.

The solar eclipse on March, 20, 2015 has nicely demonstrated the impact of uncertainty that was suddenly at up to 17 GW of renewable energy feeding into the power grid. Such events are rare and are known very long in advance. For this reason they are not dangerous, as everybody can plan well in advance and it is a pre-defined time span of not more than 1–3 hours. The TSOs and DSOs can plan reserve capacity and curtail strategies to ensure grid stability and the market participants and balance responsible parties can schedule and bid into the market with their units accordingly.

It is therefore important to note that the extreme event example from June 10, 2010 has been chosen to demonstrate that the evolution of the weather depends on factors that cannot always be predicted explicitly. Small disturbances in a complicated system can lead to instabilities that may have major impact on human life and in less dramatic ways on the energy system, both direct and indirect. If we accept that wind power and PV should be major contributors to the energy generation in the future, then it will also be imperative to adopt the electricity markets in a way that makes handling of Renewables not only efficient, but also possible at all times. The feasibility of wind power and PV will level out, if they are always operating in competition, while many other generators operate at very little competition at times, where the wind is not blowing, the sun not shining and the demand is high.

3.6 A High Level Picture of the German Solar Power Integration Experience

The German solar energy experience has shown that it is possible to control the evolution of the energy system with simple financial instruments. About 830.000 independent solar systems with varying sizes from typical roof top 6–8 kwp to 80 MWp were built within 10 years from 2000–2010. In the 5 years following 2010, 630.000 units have been built, summing up to 1.47 mio at the beginning of 2015. This means that in average more than 300 systems have been connected each day of the year during 15 years. With the target of 56 GW solar power in Germany by 2020, new installations are connected with unchanged pace.

Table 3.3 shows the distribution of PV installations at the beginning of 2015 and Fig. 3.6 shows the growth rate over the past 15 years. The evolution can be compared to how Denmark reached a 20% penetration from wind already before the 2 MW turbines became commonly available. The Danish and German capacity has been built on the basis of fixed price tariffs for a long enough period to guarantee the return of the investment. Both countries have been socializing the additional cost of the renewable energy in the form of kWh based tariffs. That meant that all consumers contribute to the renewable energy proportional to their consumption. In Germany, the renewable energy law, abbreviated "EEG" (from German "Erneuerbare Energien Gesetz", 2012) allows for exceptions for the heavy industry in order to not loose competitiveness. The exceptions are regularity evaluated to ensure fairness.

It is difficult to summarize in a fair manor the status of the German solar energy system at a level of almost 40 GW, because there are different opinions to the subject. On the one hand the 40 GW would not have been built without socializing the costs and providing the investors security of investment. Building 40 GW requires an industry and creates a lot of employment on all levels in the society. Also a long term export capability of the technology, project planning tools and installation experience has been developed. Meanwhile the existing power generators have felt competition and difficulty in staying competitive against the socialized solar energy. A long term insurance against high energy prices has been established and this will ease strategic planning, increased economic stability and generally increase the long-term

Table 3.3 Distribution of German PV installations at the end of 2014

Time valid units	Plant size (kWp)	Inst. Cap. (MW)	Inst. Cap. (%)	No. of plants (count)	No. of plants (%)
2014–12	<30	13,775	36	1,308,157	89
2014–12	<100	6,573	17	125,575	9
2014–12	<1000	7,626	20	31,043	2
2014–12	>10000	8,218	22	3,073	<1
2014–12	>10000	1,635	4	98	<1
SUM		37,827		1,467,946	

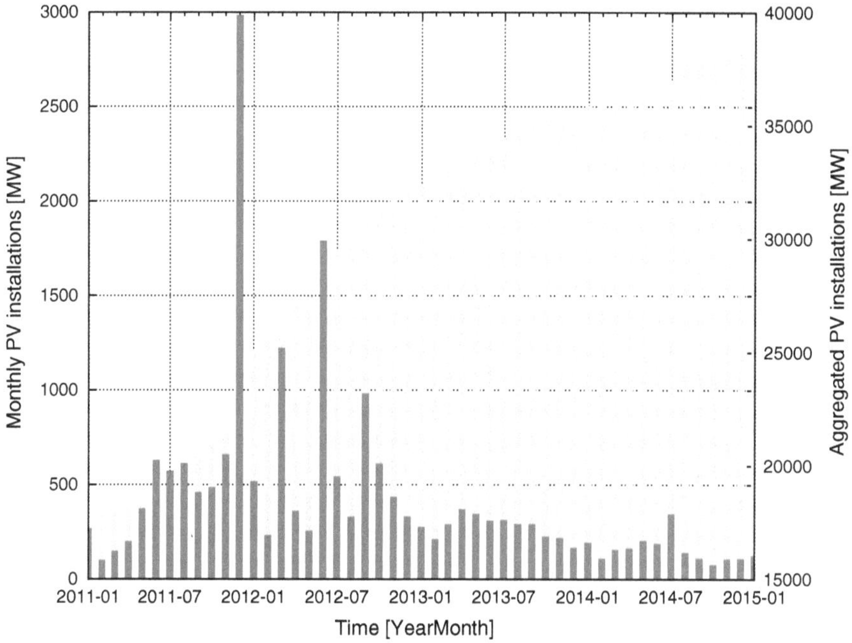

Fig. 3.6 Development of the German PV installations since 2011. The *black boxes* show the monthly installations in [MW] and the *gray boxes* the aggregated installation in [MW]

investment level for renewable energy, making energy from fossil fuels less and less feasible.

In the short-term there is however an energy market close to despair, because existing fossil fuel capacity is under pressure and the various socializing support mechanisms are gradually reduced. As a result, high efficiency is required to survive on market terms. This is a change, because a major fraction of the energy has by tradition been sold over the counter at prices above the current market prices. With increasing renewable capacity and lower market prices, it is becoming increasingly profitable to buy energy at market prices instead.

The bottom line is that although the German solar system is dispersed and on many hands today, it has a huge impact on the economy and society. In fact, it has impact world wide, because the experience gained in Germany has—despite socializing costs—contributed to encourage the development of PV on a global scale. Without the German experience, it would certainly have taken longer to reach a competitive cost of solar energy at other places in the world.

It is also certain that the most liberal economists will be able to argue against the German "solar project". They would argue that too strong socializing lead to inefficiency. We will describe examples of such inefficiency later. In systems, where costs have been socialized, this is what can be expected. Nevertheless, efficiency can in such cases be improved with political initiatives. This is also what is happening in Germany to address this issue of reduced efficiency due to socialized costs by generating increased competition.

3.6.1 *What Has Triggered the Solar Boom?*

The relative growth of solar capacity in Europe has been higher than in Asia and Northern America. This pattern is driven by a mixture of political decisions, modest space for wind turbines, visual impact concern, a strong political support to build wind power offshore and a shrinking onshore wind resource potential for hub heights under 100 m. The future prospect of being energy dependent on what is today considered unstable political systems also encouraged initiatives to increase self-supply and self-sufficiency of energy. On large-scale industrial as well as consumer level.

From a consumer perspective, the additional solar capacity has generated an unexpected price drop of energy. Around year 2010, it was predicted that the energy prices would continue to increase for a decade. Just 2 years later, the price forecast trend had changed sign and has decreased continuously for several years now.

The long term price forecast was a catalyst for increased small-scale solar energy, because consumers had an incentive to combine heat pumps and solar system to achieve some independence on electricity prices. This was most profitable, when heating systems should be exchanged. For new buildings, the German government disallowed installations of oil heating in the renewable energy act. Approximately 5% of the German households have a solar system on the roof suitable to supply enough energy for average self-supply.

Consumers export and import energy more or less permanently, but it is worth while to note that the export of energy does not utilize transmission capacity. Given that still a small percentage of the households produce energy, a high fraction of the energy is effectively utilized locally in the distribution net. Because of the fairly good correlation between peak demand and solar generation, it has so far been decided to keep the production incentive for small-scale solar on a high level, while large scale solar has been reduced. This reflects that larger solar systems may need grid reinforcement, while the smaller systems do not.

3.6.2 *What Happened to the Energy Price Forecast?*

The unexpected drop in the energy price was not only driven by growing renewable energy capacity. *Fracking*, the technology to exploit large gas resources increased the amount and the expectations of available gas on the market and initiated decreasing oil prices at the end of 2014.

The change of the price trend was accelerated, because of a now critical mass of solar energy penetration that produces energy during the hours where peak prices were previously highest, especially caused by high day-time demand during periods of low wind penetration so far.

Yet another initiative contributed to increased competition and lower market prices. The market coupling as described in Sect. 3.5.1 was introduced in 2008 to level out prices and optimize inter-connector usage between the Nordic counties and central Europe, mainly Germany, Austria and France. It was expected to also prevent that excess renewable energy caused negative prices.

So far the impact of market coupling has been that price takers produce more energy and price makers produce less energy, because the complexity of the energy system grows and it is getting increasingly difficult to predict the price. The interconnection capacity is mostly high enough to avoid congestion. A price forecasting system therefore needs extremely much information about the market and the weather. Consequently, the average monthly prices level off over time with increasing amounts of Renewables. Figure 3.7 shows the price development at the European Power exchange (EPEX) between 2010 and 2014.

It can be seen that generation capacity with high marginal costs is pushed out of the market for primary power. The result is a competitive market, where consumers benefit from low prices even though the fraction of renewable energy increases and continues to receive production incentives. This means that a part of the benefit is reinvested into the development of Renewables and the prices do not drop as much as

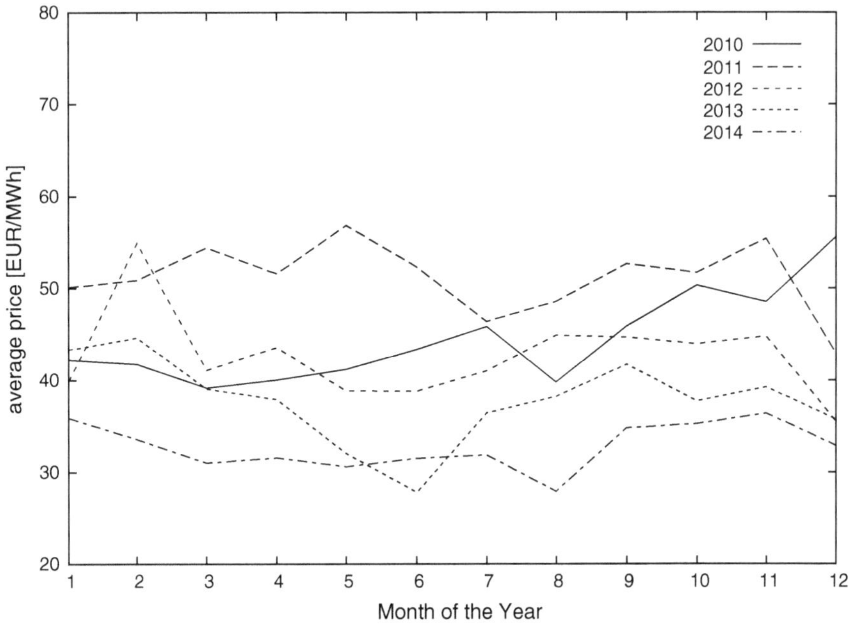

Fig. 3.7 Development of monthly average prices for electricity for Germany and Austria at the European Power exchange EPEX for 4 concurrent years from 2010–2014

the world market oil price. Countries with large renewable energy penetration rates have therefore generally higher tariffs. Another factor with impact on the production incentive tariffs is the wind resource. The higher the available onshore wind resource, the more dominant wind is and the less production incentives are required.

Given all the factors contributing to increased competition, it is not possible to point out a single reason why energy prices dropped as they did. One can also not exclude the possibility that energy saving investments world wide was planned during the 100–150$ oil peak price period, but the effect of such would only become noticeable some years later as oil storage systems fill up, because of oversupply.

At the end of the day, low prices will cause a reduced renewable capacity growth rate except in regions, where the onshore wind resource is good. These areas are likely to expand stronger, because the price of wind turbines will drop in periods of too high production capacity.

The lesson is that we will continue to experience a wavy pattern in the renewable energy growth and energy prices, because the system cannot reach an equilibrium state. Energy is believed to be the most important future business area, but the market changes and the expectations change as well in a much faster rate than the return of investment.

3.6.3 The Impact of Employment

The "German solar project" has had impact on a scale, which is similar to the Danish wind development in the years of 1980–1995. Without a similar *socio-political* support in Denmark for wind generated electricity, the world would not have seen the same level of wind generation world wide. Possibly there would have been 15–20 years delay of wind developments. In fact, the Danish evolution of wind turbines happened not only because of the pioneers doing the work, but also because of the Energy and Environmental Minister Svend Auken from the Danish Labor party. He was a strong believer of wind as the future energy source. He started his minister career as minister for employment after the 1973 oil crisis at a time, where the unemployment was at a record high. The unemployment experience is a likely reason why he was fighting for support to the Danish wind power development. It was a way to secure employment at the Danish wind turbine manufactures and all related services and to be able to export products and services to the rest of the world. By being an "example-country", Denmark was to push other countries to follow in their footsteps. It took in fact several decades to widely acknowledge his efforts towards a really sustainable and environmentally friendly growth of the country's economy.

Looking at the history of Svend Auken and referring back to the German Labor Government led by chancellor Gerhard Schröder with a record high unemployment of 11% and almost 5mio people in 2005, one could argue that this very similar pattern has been a catalyst for the German solar development.

If this analysis is correct we can conclude that the two main sources of renewable energy today developed in that speed both because of labor governments in Central Europe working with focus on increasing employment to secure the welfare states. The environmental impact may in fact have been a secondary priority and perhaps it was not even understood that both initiatives would have the capabilities to change the worlds energy supply.

The lesson from this analysis is that without a very long-sighted vision neither wind energy nor PV would have developed to a world-wide aggregated scale of Tera Watts. The Danish and German consumers have paid a socialized price of electrical power, which made the boom of wind energy and PV generated energy happen. Part of the investment came back in the form of increased export and high employment rates.

3.6.4 From Tariff Based to Market Based Settlement

The practical handling of the German solar generation follows the same pattern as was practiced for wind power throughout 15 years, while the capacity was growing to 38 GW with a yearly growth rate of about 2000–2500 MW. There is a strong geographical dispersion in the form of many small units, where the generation is not visibly increasing at specific points, but rather gradually increasing in an overall manor over periods of approximately one year.

Infrequent reading of the generation was possible for a long time, because the generators did not receive market prices, but instead fixed tariffs from the renewable support schemes. Therefore, the most cost efficient settlement approach was chosen, which was monthly readings for wind generation on all wind turbines and measuring of a number of reference sites in real-time that were then up-scaled to the total capacity. A so-called top-down approach was chosen (see Sect. 3.4.2).

Nevertheless, the lack of hourly or 15 minute generation data causes a high level of insecurity of what the actual generation from the renewable energy is, which becomes an issue with increasing penetration beyond 20% of consumption. Another issue is that simplified methods for estimating the actual generation were developed as well as simplified methods of forecasting were practiced, because the high dispersion level balances errors quite well and it was technically easier and cheaper to define an estimate of the wind generation than measuring over 20.000 turbines. Doing so meant that the deficit between the true and the estimated energy was considered inherent in the demand.

This practise continued in Germany until 2012, where direct marketing was launched in an attempt by the government to bring renewable energy closer to the market with economic incentives to balance or at least to contribute to balancing.

This change caused that each turbine formally qualifies for a personal price for its energy, which is related to the market price and how concurrent the individual turbine's generation is to the total wind generation. In this way the system encourages even more geographical spread.

The same financial instruments apply for PV as for wind except that the level of incentive is higher for solar than wind. The main difference is that most solar capacity is installed behind the meter in Germany (see Table 3.3 in Sect. 3.6). There is yet to develop a direct marketing strategy for the "behind-the-meter systems".

In Germany PV generation needs therefore to be seen as two separate systems. There is a "behind-the-meter" fraction, which is of the order 75% of the capacity and the remainder, where units are large enough to be direct marketed. The smallest systems of the second type are approximately 100 kw capacity. However, most of the direct marketed units, which make up at the beginning of 2015 to approximately 8 GW are in the MW-class, the largest plant being 80 MW.

The main disadvantage of pure fixed market invariant pricing structure is the risk of a too correlated generation pattern. This can potentially happen for wind and solar, if certain regions provide a superior resource. Consequently, the integration costs increase both with respect to transmission and balancing. With a market based pricing one would expect the investor to consider the risks more carefully. Thus, the aggregated generation will become smoother and reduce price volatility as well as ease the balancing. In addition one could expect existing transmission capacity to be utilized better.

The challenge for Germany has been that many wind farms and PV systems have been built at locations, where they would not have been built on pure market terms. Therefore, even modest reductions in the incentive payment can make projects unable to become profitable. The transition to market terms has been introduced as a difference to the average plant of the same type. That means each turbine receives the difference between the fixed price and average market price [1]. Effectively this means that a plant with a different generation pattern concurrent with a high correlation to demand receives a higher payment than those turbines with the opposite characteristics.

The original fixed prices co-exist with direct marketing. While the wind capacity migrated within 1 year with 80% to the market based structure, the solar capacity only slowly migrates to the direct marketing.

3.6.5 Why Not Working Under Pure Liberal Conditions?

There is a benefit of the high dispersion of renewable energy capacity in terms of reduced need of transmission capacity, which is non-trivial to estimate. In densely populated areas it is extremely difficult to get planning permission for increased transmission. A time horizon of 10 years is likely. For Germany, it would not have been possible to achieve today's penetration with renewable energy more concentrated to where the best resources are. The energy generation would have become even more concurrent. Thus, lower market prices of renewable energy, more price volatility and also more curtailment would have been the result of a less dispersed installation pattern. The socializing of renewable energy seems therefore not as waste but a pre-requisite to increase penetration and ease integration.

Formally, any plant is allowed to operate on market terms, but wind and solar generation cannot achieve a sufficiently high market price at high penetration levels, because all solar and wind generate power always in competition. This is not so for scheduled plant, which have periods, where they have no competition from solar and wind.

Thus, even a wind plant in a high wind resource would not be able to achieve a very high market price, because most of its generation would be correlated with other generation.

On top, there is a risk that a scheduled plant would try to cause a double loss for non-scheduled generation. This can be done, because the scheduled plant can predict the wind and solar generation and sell their energy at extremely low prices during a renewable energy peak. Consequently, the non-scheduled plant would not receive the contract, but would be forced to pay for the imbalance while generating power or alternatively stop generating power.

A solar or wind plant does not have similar ways to "cheat" a scheduled plant. This skewness in market conditions and competition is a benefit for the consumer, because the average market prices are pushed downwards due to increased competition. For this reason one can argue that the socialized production incentive for renewable energy serve as an insurance against high average prices.

The scheduled plant can act as price maker in all hours where the renewable generation does not exceed the demand. This is in competition with other scheduled units. Hence, the market price is under pressure also at lower penetration levels.

3.7 Decentral Versus Centralized Marketing of Renewable Energy

It is believed that centralized marketing of renewable energy is not efficient, because the transmission system operators naturally have insider information about the market participants, which nobody else has. Consequently, unbundling the renewable energy into independent companies is required to be able to bring the energy intelligently into the market.

Germany has chosen a strategy to reduce the financial risk for such trading companies to encourage them to take on parts of the renewable energy into pools and manage the trading, because they have the infrastructure to deal with the individual plant owners and capabilities to act as price makers on behalf of the owners of renewable energy.

At present only Spain has reverted back in 2013 from a decentral handling to central handling of their renewable energy, because of a miscalculation of their support scheme in the decentralisation process that made this solution too expensive for a government under economic pressure [13].

It should be noted that this is a decision that apparently was taken on the basis of using Spanish forecasting suppliers only. The market did not exist long enough to attract foreign competition or there was no willingness to utilize foreign technology. There are pros and cons on the decentral versus central handling of Renewables. One could argue that the forecasting process is easier and more accurate, if one IT solution can be used for a larger and more dispersed pool. In Germany and Denmark, the market did not develop this way and it is common practise that forecast providers serve competing traders with customized IT solutions.

The risk is in this way spread and competition ensures constant development in the forecasting process. However, practical experience says that it is not cost-efficient either to continue to maintain of the order 30–40 different IT solutions. A well-functioning market should not require more than 10–20 companies that are specialized in the services related to the sales and balancing of renewable energy. In countries like Germany this means that the market will have to consolidate with decreasing incentives and support to the decentralised marketing to ensure cost-effectiveness.

3.7.1 Data Provision Requirements for Grid Balancing

A number of jurisdictions operate today with a certain level of obligations via the grid code to provide data from the wind farm and PV plant as well as from meteorological instrumentation at or near the plant site. These data are used to keep control of the current weather conditions and the current in-feed of electricity into the grid.

Table 3.4 shows a number of areas and their obligations towards provision of data from RES.

It can be seen in Table 3.4 that the main difference in the obligations for providing data to the system operator (column 3 and 4) are between the most mature countries Denmark and Germany and the rest of the world. The reason being that these countries have a long history in the development of wind power with small projects of only a couple of wind turbines at a time. Because it is always more difficult to introduce new obligations in an established market, and the cost of data collection and transmission is relatively high for single turbines in comparison to large projects with 20 or more wind turbines, Denmark and Germany have been building their tools upon up-scaling methodologies in the real-time environment.

Neither Denmark nor Germany had obligations on the renewable energy generators to supply data from SCADA systems in the years of centralized balancing. When direct marketing via private balance responsible parties was introduced, this pattern changed to a large extend. However, most old turbines still are only measured on

Table 3.4 Comparison of obligations in various countries with significant amounts of wind power by variable generation units to provide observational data from their units to the system operators at the beginning of 2015

Country	Inst. Capacity (MW)	Wind power SCADA	Connection point	Meteorol. data at wind farm	Part of grid code
Denmark	4,845	No	Yes	No	Partially
Germany	36,643	No	Yes/partially	No	Partially
Spain	22,987	Yes	Yes	Met. Mast	Yes
Ireland	2,272	Yes	Yes	Met. Mast	Yes
Greece	1,980	Yes	Yes	Windfarm	Yes
Portugal	4,914	Yes	Yes	n/a	Yes
Sweden	5,425	Yes	Yes	Windfarm	No
UK	12,440	Yes	Yes	Windfarm	Yes
Canada/AL	1,450	Yes	Yes	Met. Mast	Yes
USA/OR	3,153	yes	No	Met. Mast	Yes
USA/TX	14,098	Yes	No	Met. Mast	Yes
USA/CO	2,593	Yes	Yes	Windfarm	Yes
				Met. Mast	
USA/CA	5,917	Yes	Yes	Met. Mast	Yes
USA/AZ	238	Yes	Yes	Met. Mast	Yes
India	22,645	Yes	Yes	No	Yes
China	114,763	Yes	Yes	No	Yes

grid point level and their real output often has to be estimated from the total in-feed at grid point level.

3.7.1.1 Data Exchange in Case of Outages

Outages are a phenomena that is not much discussed in public, because renewable energy sources are deemed to produce, as soon as there is sufficient wind for wind turbines or radiation for solar panels to produce power. As penetration increases along with increased competition, efficiency becomes a necessity and outages get more attention.

There are three types of outages:

- Turbine maintenance
- Turbine failure
- Curtailment

The first outage types can be considered scheduled and could easily be reported to all parties involved in the management of the power unit. The second and third types are usually non-scheduled and of indefinite length of time.

One of the issues that are often forgotten in all types of outage is the fact that this information is highly relevant for the forecaster of the energy source, whether it be wind power or PV. If this information is not available, forecasting tools and models are trained with wrong data and performance decreases. It is often not considered, when forecast performance is measured that any non-reported outage information is a problem for forecast training, even though quality checking of measurements can detect some of the outages of a wind turbine or a PV plant.

Practical experience shows that with increasing penetration, it is not only owner or operator that are responsible for what appears to be outages. It appears that outages are also caused by the DSOs, which becomes a challenges for the exchange of data as well as compensations.

In most centralized forecasting systems, there is full control of the level of curtailment, as it is done by the system operator who has the obligation to log, report and compensate the owners of the power plants. In Germany, there was no knowledge of the level of curtailment, while there was a so-called centralized forecasting system at the TSOs, because TSOs had no means of controlling wind plants, as all wind turbines were connected to the low voltage or medium voltage grid. Therefore, any curtailment on low- or medium voltage level was hidden in the system, also because total aggregated generation was estimated from a small subset of the individual units, covering approx. 5% of the installed capacity. DSOs are also not obliged to justify and report each single curtailment of a turbine or plant and reimburse the owners. Reporting is also reduced, if the curtailment is less than 15 h over one calendar year. It is in fact the owners of power plants that need to apply for compensation at the DSOs, if their plant has been in an area, where curtailment is reported (see paragr. 13 EnWG, 2014 [4] and paragr. 14 EEG, 2014 [2]).

The real issues with curtailment is that the curtailment information is not flowing to the responsible parties as it should, so that appropriate actions can be taken. It shall be noted that large wind farms with a SCADA system can provide information about curtailment, as the software in today's wind turbines and inverters for larger solar systems are capable of receiving instructions to reduce generation and hence also log this information. Theoretically, this information could flow further in the chain and reach the balance responsible party for the generated power. However, most SCADA reports reach in the best case the trader too late.

The level of curtailment is likely to increase with penetration level in any jurisdiction. Therefore, it is important to consider strategies for the market or the system operators to ensure that curtailment is considered as a market instrument that requires transparency on all levels to not cause unnecessary costs.

From the forecasters perspective the solution is to quality check the data. If there is a likelihood that there was curtailment, then such data shall be flagged to prevent a forecast bias.

3.7.2 Lessons Learned

The fact that renewable energy capacity is built over many years opens for a pitfall in the integration process. Stakeholders get used to not having any obligations and be autonomous, if such obligations do not exist at the time of construction. As the renewable energy capacity increases there is a serious difficulty in getting the mature capacity accepting new conditions. The downside of too strong obligations on the renewable energy generators is the risk of that few or nobody will build significant renewable energy capacity. However, it should be possible for sufficient large plant sizes to require a certain level of online data. Alternatively, one could tie part of the production incentive to provision of valid and regular real-time data. This would probably be a very efficient financial instrument to maintain the data quality.

Nevertheless, as a system matures, any changes take longer time for acceptance and for the implementation. In these phases there is an ongoing loss from running a less optimal system. Germany has paid a high price over 2–3 years to migrate the capacity from centralized handling to direct marketing. There is reason to expect that this was an investment in lower future energy prices. What was special in Germany was that 33 GW wind power and 8 GW PV was brought onto market-like conditions over a period of four years. Previous Danish experience brought 3.0 GW of wind power on market-terms, but over a period of 15 years. That was in the same pace as the fixed tariffs period ended. The majority of the German capacity was still under fixed price tariffs. Consequently, much stronger financial incentives were required to facilitate the transition. The Danish turbines were so to speak given up by the TSO the moment negative prices were introduced. The TSO had by the end of the support period encouraged the turbine owners to seek a private balance responsible party.

We have also learnt from Germany that centralized forecasting has made the process of handling individual wind turbines and solar plants less "blind" compared to the direct marketing process. However, because wind farm sizes vary from 100 kw to 400 MW all process must be automatic and robust. It is not economically feasible to use manual processes in the real-time operation of small units. Therefore, IT solutions must be able to automatically handle all data errors and generate alerts for repeated errors. Such IT solutions are nevertheless not an "off-the-shelf-product", which means that traders need to reach a certain size before direct marketing can be economic feasible, even in a well functioning market.

The German government wanted to ensure that anybody can become a trader of wind and solar energy when designing the direct marketing bonus for Renewables and had to realise after only 6 months that the premium was too high and had to adjust much more than expected in an adjustment to a at the time new version of the law [1]. The market had been growing too fast, because of too high premium and consequently a consolidation of the market had to take place shortly after. Overall, a lot of resources are lost, if this happens.

3.8 An Outlook on the "Direct Marketing Strategy" in Germany

Direct marketing in Germany started in 2012. After 2 years, 85% of all electricity produced by wind energy and 15% of all electricity produced by PV is sold through the direct marketing principles. In this process, the handling of these 2 renewable energy sources in the market has been improving despite the challenges with respect to data exchange with the DSOs and increasing curtailment as well as partial curtailment due to noise and light restrictions.

Handling of renewable energy is a new business area and appears to be important for trading companies, because the renewable energy penetration will increase world wide. Without such a capability and market share it is unlikely that a trading company can grow in the future. This is also the reason why the trading companies are under pressure to increase efficiency, if they want to survive after the market premium has been set down in 2014 to 16.67% alternative: to 1/6 of what it was at the beginning of 2012, when direct marketing started (see EEG, 2014).

The real benefits of bringing variable generation units into the market will only be really visible, when the market has become stable and market rules are settled. In the first 2 years in Germany since the beginning of the direct marketing, rules have been changed 2 times. This creates insecurity amongst the traders as well as the owners and the TSOs. The consequence of such insecurity is that contracts have a short binding and investments in infrastructure are postponed. It is only when the binding period of contracts reach more than one calendar year that trading companies can focus more on optimizing the handling of the energy sources than on increasing capacity.

Due to the market coupling and strongly interconnected systems in Europe, it is not possible with the capacity level of year 2015 to generate more electrical power from renewable energy than there is consumed. This is so day and night and also in holiday times. Consequently, negative prices should not occur, but they do. The peculiar pattern can be caused by long-term contracted nuclear power plant delivering power regardless of the market price. Another reason may be that there are no or too few renewable energy price makers.

To conclude, bringing renewable energy sources into a liberalised market structure takes time and has a cost for the society. Like all pioneer work, there are costs that can be avoided once there is enough experience available and mistakes have been analysed.

3.8.1 The Balancing Challenges on the Day-Ahead Horizon

Direct marketing of renewable energy is often looked at per unit, because a unit may drop out of a pool the following month while new may enter a pool. The forecast shall ideally be as accurate as possible per plant, but at the end of the day it is the pool imbalance that is generating the income and causing the balance costs.

A forecast for a mixed pool consisting of one or more large offshore plant, a portion of solar and a larger amount of onshore wind can appear quite noisy, because different units ramp at different times. Because some shorter time scales are not accurately predictable the result is that the day-ahead forecast may anti-correlate with reality. Such anti-correlation is costly to physically balance in an intra-day market. Urgency, shortage of volume and short contract length all contribute to a higher loss from the balancing process.

The best way to avoid volatile forecasts is to utilize a middle percentile rather than an average of individual deterministic forecasts. Although averages of multiple forecasts are less volatile than the individual forecast, outlier forecasts still cause some low amplitude noise, although the noise reduces with the number of individual uncorrelated forecast.

The central percentiles of multiple forecasts are invariant to outlier forecasts and contain therefore less noise. Consequently, they are useful on the day-ahead horizon to facilitate intra-day balancing. For a renewable energy price maker it is therefore essential to use percentiles to push the price up while still facilitating intra-day trading. There are periods, where a favorable forecast is a P50 forecast, but also periods where a P40 or P60 are preferable or even the average of P40 and P60.

Another major benefit of using the smooth pool percentile forecasts is that expectations to asymmetry with respect to balance costs can be taken account for. Such expectations can be built from the penetration percentage of renewable energy, because high and low penetration impact the intra-day market volume in a skew manor. The recipe for the trader is therefore to use day ahead bids with central percentiles with the skewness in balancing costs in mind while suppressing noise in the schedule. It is often mentioned that speculation in balancing reserve is not allowed, but it is certainly market logic to deliberately plan the schedule for cheapest possible balancing by forecasting the volume of positive and negative regulating capacity.

3.8.2 The Balancing Challenges on the Intra-day Horizon

Frequent short-term forecasts based on online data add value not only because they are more accurate than the day-ahead horizon. Nevertheless, deployment of short-term forecasts should take place with care, because they increase the trading volume and thereby the loss compared to the spot market price. This situation can occur, if the short-term forecast errors have opposite sign of the day-ahead error. This statistically occurs for up to 50% likelihood, if the error of the day-ahead is small. That means,

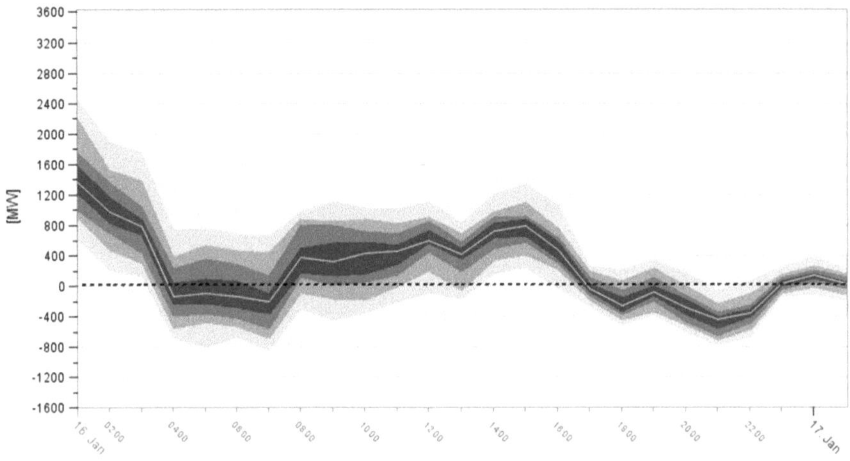

Fig. 3.8 Schematic depiction of the change of sign in forecast error and in uncertainty spread for different forecast horizons, starting with 144 h in 6 h intervals, up to the point in time when the forecast is valid. (2011/09/20 at 3:00UTC). The zero axis (*black dotted line*) is the measurement and the white line is the so called optimum forecast, the blue shaded areas are percentiles

if the error of the day-ahead forecast is small, then there is no reason to trade an expected error on the intra-day market with a loss [44].

In Fig. 3.8 the change in sign of the forecast error and the change in uncertainty spread is demonstrated for different forecast horizons, starting 144 h back in time.

A particular capability of frequent short-term forecasts is that certain forecasts are accurate further ahead than others. The forecast following the start of a multi-hour ramp is the best point in time to correct for a day-ahead phase error. The sign and magnitude of the error will hold for multiple hours. This means a multi-hour error can be balanced by one party with one startup and one stop, which is more favorable than trading the same error hour after hour with a fixed look ahead risking a higher loss. It is therefore the major pitfall in deployment of short-term forecasts to receive forecasts only once per hour. A proper forecasting technique can detect the timing of ramps early and contribute significantly to a reduced balancing cost. It is feasible to even use semi-automated algorithms for trading, because it is possible to compute a safe decision determined by simple algebra. In that way short term forecasts are deployed when time is right and otherwise not [44].

3.8.3 Summary and Outlook of the German Direct Marketing Experience

It is certain that the energy system will become more dependent on weather over time. This would even apply without increased renewable energy capacity and ignorance of all emission targets. Nevertheless, it is the experience from Central Europe that the market does in general not work pro-active. A possible explanation for that is that renewable energy is by the traditional energy companies not looked at as a business opportunity, but rather as an inconvenient competition factor, which is dependent on political support.

Clearly such competition, which is additionally a complexity factor to the market is not welcomed by an established industry. Nevertheless, the renewable energy industry has developed and it would not be possible to stop the evolution in the democratic part of the world, because such an attempt would not meet public support. The increased fear regarding nuclear energy, dependency on gas from Russia and environmental aspects will continue, because people think on behalf of their children and grand children. Even a serious instability issue in central Europe will not be able to stop or delay RES projects, not even if the event will trigger an accident on a nuclear plant. Neither will terror be able to influence decisions made to increase renewable energy sources.

Nevertheless, we do experience *let us wait and see* attitudes everywhere. Perhaps it is considered more economic to await a market consolidation, because it is more expensive to be pioneer and it might be possible to buy technology and step on board at a later stage. This is a likely scenario in which data and knowhow gained in a SME is suddenly deployed by large market players and well known from the mobile phone industry.

In the past a lot of effort has been spent on the unbundling processes to prevent that the trading department does not ask the DSO department to down-regulate a plant to increase profit of the trading department with a grid security explanation. There is reason to fear this, because the choice of turbines to curtail may not be objective. The DSO departments can potentially help the trading department by selecting a competitor's wind farm for curtailment, whenever the sister company is in balance. More serious is the opposite, if wind farms belonging to the sister company are curtailed while the sister's pool is generating too much power. In this way the wind farm is compensated for the loss and the balance costs are kept low.

It is technically not (yet) possible to provide evidence of such actions, but the abuse may at some stage reach a level, where potential whistle blowers decide for justice. Meanwhile it can happen that consumers pay the balancing costs and compensation for curtailment over tariffs. Regulators need to be aware of potential abuse and need to design the system to prevent abuse. One route is to split the total generation into shares as discussed in "The SuperGrid Study" [10]. Such a strategy would lead to a high level of competition and a centralized IT infrastructure at lower costs than today, where every TSO and every BRP develops their own infrastructure. The direct marketing approach chosen in Germany has mainly boosted the economy and so far made the energy process more expensive for the consumer. The native price of energy went downwards, while the costs of the management processing went up.

One way or the other, a direct marketing system is a catalyst for an increase in the number of court cases. The turnover is huge, contracts are in strong competition often made in a hurry with less care and disputes occur because of hard competition, non-compliance to system requirements and risk of settlement errors with economic impact.

A well planned unbundling of the renewable energy into a suitable company structure would save costs. Such a solution would also eliminate the waste caused by every direct marketing company running their own IT implementation, which is too simple to handle the task. Possibly up to 50 expensive but still too primitive interfaces were developed, which forecasters had to comply to. Direct marketing in Germany has therefore caused a significant layer of costs not related to the physical handling of energy. This layer will one way or the other be financed over a combination of tariffs and investor capital. Larger pool sizes with a high degree of online data will over time cause a strong consolidation.

On the other hand the imbalance of the largest pool is also likely to be correlated with the total balance cost and therefore be unfavorable. A few similar sized pools with experience will be more even in competition. The most likely outcome is that there will be a few pools with sizes of the order 10 GW once the market is consolidated.

After four years of experience with the German direct marketing project we still see a high level of inefficiency, although the market is competitive. On the one hand parties can see that the earnings are not as expected, but they are on the other hand not in a position to work with a sufficiently long horizon in mind. This is partially due to the immediate risk of loosing capacity and partially due to the complexity of a mixed pool of different generation sources.

However, if direct marketing companies are not concerned about negative prices and try to socialize costs as much as the TSOs had to at the outset of wind power becoming a significant supplier of energy with penetration rates of more than 10–15%, then there could be arguments to stop the incentivisation for direct marketing. Potentially, this can cause all capacity being back in centralized handling by the TSOs within a period of two months, as it has happened in Spain in 2012 [13].

3.9 Reserve Forecasting

Ensemble forecast theory says that there exists a possible relationship between the spread (or dispersion) of an ensemble and the skill of the forecast. If this relationship can be established, forecasting the predictability or skill of the forecasting system becomes another useful utility when using ensemble forecasting (e.g. [19, 21]). This utility in fact means that the forecast error of a previous forecast can be forecasted to a large extend. Pahlow et al. [51] and Möhrlen and Jørgensen [11] demonstrated the operational use of a multi-scheme ensemble to forecast the forecast error of a day-ahead forecast in order to allocate reserves more dynamically.

In market situations, the ratio between the day-ahead forecast error and expected accuracy of 2–3 hour forecast is often around unity. Therefore, it is often not economic to trade on the basis of a short-term forecast. At other times the day-ahead forecast seems so high that there is no doubt that it is beneficial to trade on the basis of the short-term forecast. In between there are possibly a third of the events, where it is worth to consider the expected system reserve volume. For those parties that have the possibility to allocate reserves dynamically or to buy reserves cheaper than using own reserves on a day-ahead or short-term market, ensemble forecasts can be used to do so.

Suppose a strong front sweeps concurrently over some large offshore farms, then there is reason to expect increased uncertainty and therefore also imbalance. As an example, in the North Sea there are now clusters of wind farms of up to 2000 MW over small areas compared to a low pressure system moving over the North Sea from the British Isles. At the coast of England as well as along the coasts of The Netherlands, Germany and Denmark, the amount of offshore wind projects is increasing in size and volume. Although the individual project would not cause balancing issues for any of the TSOs, the total volume on some of the power strings is increasing and coming to a size where short lasting strong ramps can cause issues to system security. Today, these wind farm clusters are only connected to land, in the future they may also be connected to each other into the common European grid. When the ENTSO-E network plan for 2020 and 2030 is coming to life, there will be more possibility for balancing, but also larger volumes, which will require more efficient and dynamic balancing handling.

For traders, but also TSOs, the large ramps are reason enough to begin forecasting reserve. Peaks in the reserve can be expected during ramps, but some ramps are more critical than others. The target for reserve forecasting is therefore to compute the risk of exceptionally high reserve deployment due to some special pattern in the weather such as the position of a front in the proximity of high wind power capacity. What is also relevant is the potential overlap between a weather generated peak and scheduled ramp of the load. At the 2014 international wind integration workshop in Berlin, there have been some first discussions and publications dealing with reserve forecasting (http://www.windintegrationworkshop.org/berlin2014/) in operational environments. Therefore, it can be expected that this topic is going to get more attention in the near future.

3.9.1 Risk Evaluation of Wind Energy and PV Generation

The core input to the risk evaluation for wind energy and PV is the ramp rate uncertainty of this aggregated non-scheduled generation. This is computed from the time derivative of the aggregated forecast for each ensemble member. The prices are likely to peak when the standard deviation of this vector approaches the amount of secondary reserve.

$$PPI \cong \sqrt{\frac{1}{member} \cdot \sum_{m=1}^{member} \left(\frac{\Delta P_m}{\Delta t} \frac{\Delta \bar{P}}{\Delta t} \right)^2} \tag{3.1}$$

where PPI is the peak price indicator, P is the power forecast, $\bar{P}$ is the mean of the power forecast, t is time and m is member.

The formula is developed to reflect fast changes. This is because the slower changes of larger amount are less likely to become expensive. The result is an objective prediction of a very uncertain parameter. The higher the value, the more reason there is to trade fully on the basis of a short-term forecast. Because of the high uncertainty of the ramp rate we choose to calculate the uncertainty with influence from all members. This is opposite to use of a central percentile, where the outliers are suppressed.

It is worth noting that it is possible to compute this risk on the short horizon as well as days in advance. As a minimum the trader can then be prepared for the risk.

3.9.2 Additional Opportunities for RES in a Market Environment

Renewable energy units can as minimum provide one sided reserve. Consequently, we must expect to see more and more skewness in the volume and therefore asymmetric prices for regulation. Renewable capacity that does not qualify for production incentives is more likely to be handled by traders by withholding some power and offer this power in the intra-day market with the intention to receive more than the spot-market price.

Direct participation as regulating reserve is a possibility too, but only in conjunction with uncorrelated generation in a larger pool with high predictability. Possibly, resources are spent better trying to only balance oneself than participate in a centrally managed reserve.

3.10 Changing Forecast Requirements

In the previous sections we described how percentile forecasts and short-term forecasts add value for different tasks. We also identified that the short-term forecast deployment must be taken on the basis of uncertainty information to not increase the trading volume and thereby the loss (see also [44]). The important and basic ingredients for any trading recipe is to use forecasts of the aggregate of a pool of units or sites from an ensemble, which addresses all the uncertainty factors relevant for the intra-day and day-ahead horizons.

The ensemble members must have a certain degree of independence, otherwise the percentiles do not smooth the uncertain ramps. It is not feasible to smooth all weather related ramps with a simple filter, because some of the ramps have a high degree of predictability. Such ramps can still exist in the percentile based forecast. The calculation of percentiles can be characterized as an inherent filtering process, which preserves the trusted ramps in the data, if the ensemble covers the real uncertainty of the parameters that are drivers of the generation. One can say that the

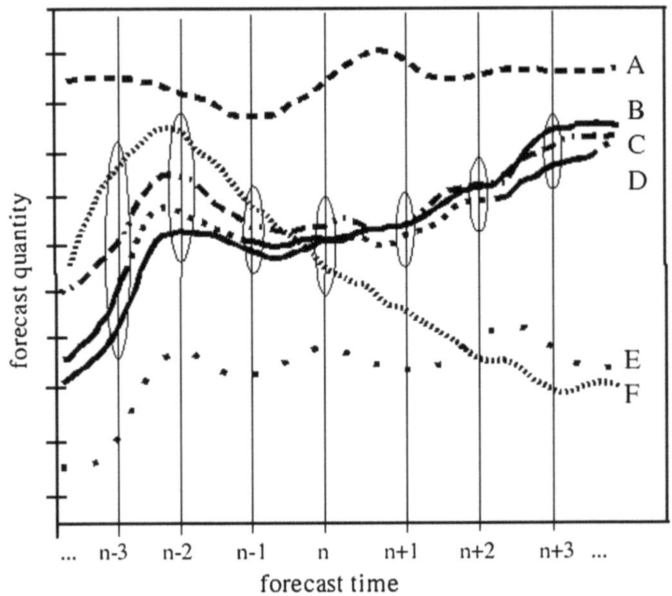

Fig. 3.9 Schematic of the way ensemble members can change position throughout a forecast and how certain ensemble members cluster together

percentile calculation is a democratic process, because all ensemble members get a weight, but it is the central majority of the spectrum, which has the most influence. The "extremes" have no or little influence on the final decision. Note, however that in contrast to human behaviour in politics, weather is characterised by chaotic pattern [41], which means that after say 12 hours, it can happen that the extreme member has become moderate and the other way around for some of the previously moderate members. The ensemble members can change from cluster to cluster any time. That is the core of ensemble forecasting and why an ensemble as a whole is of value to any forecasting process. As long as the ensemble members preserve their status inside the ensemble spread in time, high predictability can be assumed. The moment the distribution pattern of the members relative to each other changes it is equivalent to the loss of predictability.

Figure 3.9 shows schematically such a so-called clustering of ensemble members over time. In mathematical terms, this is expressed by using the ratio between the standard deviation increase and the covariance of the ensemble vectors forward in time. The typical pattern is that the covariance stays just around 1.0 for some hours. A sudden "blow-up" in the weather conditions and changed predictability can either be seen in increased ensemble spread or reduced covariance. These two parameters in fact indicate the level of predictability in time and space.

Figure 3.10 shows schematically, how the interpretation of predictability can be mathematically indexed by three classical distribution patterns:

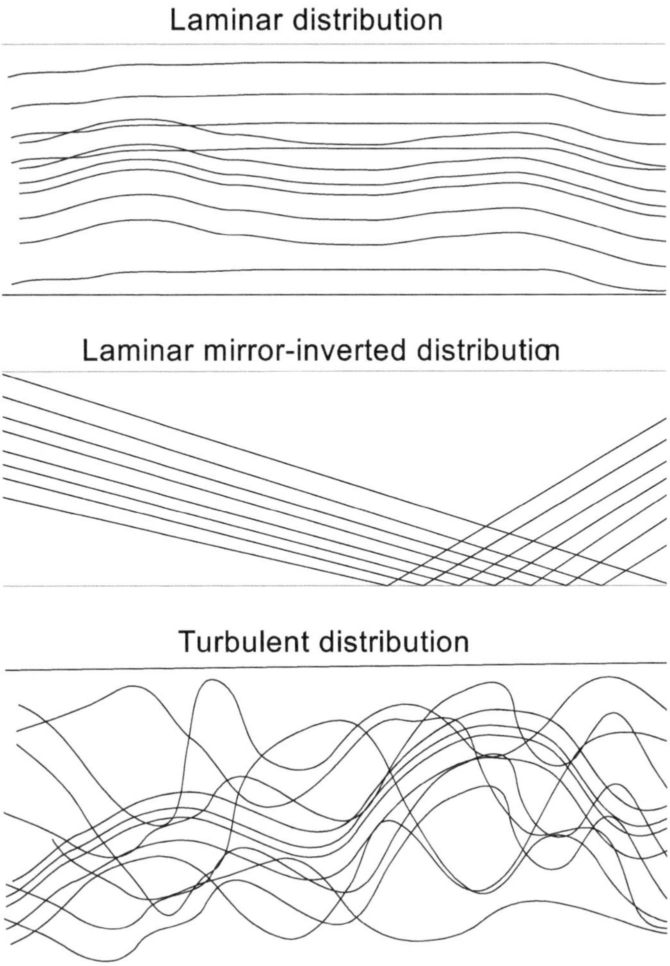

Fig. 3.10 Schematic of the three different ensemble member distributions indicating (1) high predictability, (2) reversed mirrored effects and (3) loss of predictability (chaotic distribution pattern)

- laminar distribution over time or space indicating high predictability
- laminar mirror-inverted distribution over time or space indicating unchanged predictability
- turbulent distribution over time or space leading to low predictability

In comparison to politics, a member in a multi-scheme ensemble does not belong to a party and does therefore not need to comply to the official view of a specific party. The ensemble members are solely selected as representatives of weather development. They are though bound by physical laws and they do not breach these laws, but instead use different, equally valid and correct methods to solve specific processes. Their different interpretations of the laws of physics keep them individual and independent from each other in their "behaviour". Moreover, a member can never become influenced by other members. This prevents so-called "in-breeding", which has been observed in statistically generated ensembles such as Monte-Carlo ensembles or ensembles that are solely built upon perturbations of initial conditions. This phenomena is especially known in the Ensemble Kalman Filter techniques [31].

Deterministic weather forecast does not improve as much as the collective ensemble. Nevertheless, deterministic forecasting may lead to better ensemble members. The only argument to utilize the deterministic approach is computational price. There are many parties who can deliver a deterministic forecast and this lowers the price. It is also possible to justify with simple MAE scores in MW that the deterministic approach is competitive on the day-ahead horizon, but as soon as the costs and the potential to reduce costs by short-term forecast is included, then the deterministic forecast looses out to the ensemble based approach.

The German TSOs have by law been forced to buy forecasts from different suppliers and combine these to a meta forecast. They shall use a minimum of three suppliers to prevent that the market can predict and speculate against what the TSO will base their decisions on. In this context it shall be noted that there has been a forecast publication obligation on each TSO. A closer look on the obligation however reveals the forecast which is published covers a sub-area and does therefore not reveal what is bid into the market.

In Germany not only a few forecast vendors are used to create a mixed meta forecast, but up to 8 suppliers, where each of them may already smooth the forecast with a multiple of weather input. This is the expensive way to obtain a smooth forecast. Expensive both in terms of forecast cost and in terms of forecast value, because the method does not filter ramps of low predictability and preserves extremes. Instead, this method tends to filter out all larger ramps and misses extreme events. Especially, the latter can be dangerous to the system operator or BRP once a critical penetration level has been installed and the errors can increase beyond available reserve.

3.10.1 Balancing Challenges of Noisy Forecasts

Inhomogeneous renewable energy capacity causes ramps in the forecasts, even if the weather forecast itself is rather smooth. Such ramps tend to cause two imbalances referred to as double punishment except, if the forecast has no phase error. If the noise could be balanced exclusively by primary reserve, there would be no reason to worry, but the amplitude and wave length enforces the use of additional reserve. Only in the case where the number of similar sized large units is high and sufficiently spatially dispersed it would be feasible to not deploy additional reserve.

However, the operator can without ensemble forecasts and short-term forecast by plant not know the cause of a developing system imbalance. Hence, the operator is forced to react early on ramps, because of the potential risk that the ramp continues. Ramps are therefore inconvenient in the forecast as well as in reality. If they are present in the forecast, then they are latent in the schedule and out of a sudden cause imbalance. Hence, total operating costs are lower for the smoother forecast with the same MAE as the more noisy forecast. This is the background for why the German TSOs smooth more than they are enforced to do by law.

3.10.2 Practical Implementation Considerations of Ensembles

A valid question to ask is whether ensemble forecasting is a prerequisite for handling RES everywhere in the world. Is RES really going to become such a challenge for the power system?

For wind power the answer is a clear "yes". We have not seen a location with a sufficiently good wind potential, where a deterministic forecast will do a sustained good job.

The wind power forecast error occurs, because high wind speed is a sign of a rapid changing weather and consequently lower predictability. Normally, wind farms are built, where the wind resource is best. In those cases low predictability is evident. However, also in locations with fairly poor wind resources and strong diurnal cycle, we observe low predictability at times. A location in India outside the Monsoon season may only produce 10–12% of the rated capacity, and one would expect some systematic diurnal cycle, but also such a location is sustained difficult to predict for.

This is because there is no momentum in the air and mesoscale weather develops with a short lifetime. Thus, all weather on the day-ahead horizon is a result of the NWP model's equations and not the initial conditions. This is so, because there is often no dominant initial movement, which is strong enough to last for more than a few hours.

So the experience is that wind power forecasting is not an out-of-the-box process. There is more or less predictability from location to location, but there is mostly variable uncertainty and therefore reason to assume that ensemble forecasts are required in the forecast process.

Core factors for forecast performance are:

- the availability of meteorological measurements in upstream direction
- the complexity of local as well as non-local terrain
- the spatial dispersion level of generating capacity

To provide another example, we observe significantly higher forecast errors on the day-ahead horizon in the western part of North America than anywhere in Europe. The Rocky Mountains play some role, but the key factor is most likely the difference in aircraft traffic in the Pacific versus the Atlantic ocean. The geographical size combined with the strengths of sea surface temperature gradients and currents play also a role for the forecast error and hence performance.

A forecast user can choose between buying an ensemble end-product or building the ensemble from externally generated ensemble forecasts from one or multiple providers.

The first option is simple to deploy. It won't be the cheapest, but the one with lowest total cost. The second option is rarely deployed, but should be feasible for parties with GW scale capacity. Most existing forecast providers would, though, prefer to deliver a blended forecast instead of the native forecasts. They would argue that they tune on some statistical parameters that level out large forecast errors. The problem for the end-user with a number of such forecasts is that every one of them filters out the extremes and leaves the user with a number of smoothed out forecasts and no knowledge about possible extreme values.

For the end-user such a strategy seldom is of benefit. The objective is rather to compute a central percentile from independent forecasts inclusive their noise. Thus, the forecast supplier's target should then be low absolute bias calculated with a window corresponding to double the width of a typical ramp. This target would not encourage the individual forecaster to smooth the forecast and therefore contribute positive to the users own ensemble (see also discussion in Sects. 3.2.1.3 and 3.2.3).

There may be commercial reasons to prefer the 2nd option or the hybrid, which is to ask the ensemble forecast provider to provide the native members instead of the end-products. Such commercial reasons could be to protect the know-how in cost optimization.

3.10.3 Summary and Outlook on the Forecasting Developments

Since the financial crisis in 2008, it can be concluded that economic fear of not doing well on the short horizon blocks required long-term strategic decisions. To establish strategic decisions has been difficult over the past 10 years and only been enhanced after the finance crisis. One reason was the continuously changing laws and market terms in the liberalisation phase of power markets inclusive the unbundling of power generating companies from system operation. The other reason was a growing Renewables industry with prospects for new employment that has driven more and more established companies to get into related, but new areas outside the traditional business. One of such area has been the power forecasting for variable generation units from wind and PV. Besides traditional companies going into this area, some start-ups have developed along side over the past 5 years. This development let to approximately 10–20 forecast providers delivering more or less identical deterministic forecasts into markets, where RES had become a mature energy source. Most of the new providers came without much prior experience in the particular business area, especially without prior experience in the meteorological part of the process.

In the case of Germany we have seen that competition has led to low prices and the possibility for TSOs and BRPs to purchase forecasts from multiple providers forming small and large poor-man ensembles. However, we have also seen that these types of forecasts suppress extremes, which are the driving factors of volatility and high prices on the power exchanges.

On the other hand, the poor-man's ensemble may also just be the predecessor of the next generation ensemble forecasting to be used that is more suitable and efficient for the power industry, just like Monte-Carlo simulations have been the predecessor of ensemble techniques using perturbations of initial conditions and the multi-scheme ensemble technique. We have seen from the examples in Europe, mainly Denmark and Germany, that being a pioneer requires courage and backup from the society, as a large portion of the pioneering costs has to be carried by the society or by some larger organisations that can afford to pay part of the development costs. The advantage of the success is that others will follow and development will continue.

Therefore, it is rather certain that the world wide expansion of renewable energy can be considered an overall success and it will continue to be that. And, forecasting is a key instrument for the continuous success of renewable energy integration into the power industry. We have also discussed the benefits and drawbacks of various approaches of the ensemble technique as a crucial part of the forecasting tasks already at hand as well as to come in the near future.

Unfortunately, when you most need predictability, that's usually when the atmosphere is most unpredictable. This comment made by McElroy from the American National Weather Service is expressing the situation nicely. A certain amount of predictability is always required to operate the electrical transmission grid in a safe manner. We nevertheless experience regularly issues that are directly connected to a lack of predictability of the generating units.

In the energy markets this lack of predictability is reflected in price volatilities, which are often triggered by uncommon weather phenomena that are difficult to predict, and/or some other abnormal conditions related to transmission or/and sudden failure of scheduled generation plant. These are moments, where the market participants experience the highest losses and profits. Therefore, there is an underlying need of forecasting tools that increase predictability and reduce risks.

This is what ensemble forecasts have been developed for since the early 1990's. In fact, this has been the starting point of scientific demonstrations proofing that ensemble forecasting is the state-of-the-art methodology to find the instabilities in the atmosphere that cause most forecast uncertainty and error.

Today, Ensemble forecasts occupy the largest fraction of the computing resources in two of the largest meteorological institutions NCEP and ECMWF. This is a clear sign of that the topic is prioritized and that there is no better way to manage uncertainty with increasing amounts of Renewables entering the electrical grids in the years to come. We have for that reason been discussing various applications of ensemble forecasting in the context of variable generation from wind and PV and how to use ensemble forecasting in a market, where discrete values have to be entered into the market-tools at the end of the day. Dynamic reserve allocation was one of the example application that has got recent attention. It is only possible to carry out with the help of ensemble forecasts, because it is necessary to predict the expected forecast error.

The future requires more of such applications to be developed and deployed in the power industry in order to manage the increasing volatility of energy prices that follows increasing penetration of wind power and PV, respectively.

3.11 Conclusions

In our review we have presented the evolution of three now rather mature technologies. By deployment of ensemble forecasting in the direct marketing of solar and wind power the amount of RES will continue to grow. The capacity growth rate will still be determined by political targets, because the return of any investment in RES will be 10–15 years and the best resources have already been utilized. The complexity of deployment of RES will increase from coupling of markets, more variable weather generation, variable consumption, transmission constraints and the average age of RES increases.

It is inevitable that the RES will continue to create more employment and reduce imports of energy into the western world. Without RES the world economy would be in a very different condition than we experience today. Possibly the consciousness about the environment would have been triggered the hard way in a later stage. Fortunately, we are in the middle of a controlled transition between the fossil fuel age and the market maturity of RES.

In our review we argued that such a massive transition could not happen in a liberal way without incentives. Perhaps an economist would postulate that there exist regions in the world where RES increases on pure market terms. However, this expansion is today only possible, because mature technologies are available where Labor Governments in Denmark led the way forward towards that technology development in wind energy and similarly in Germany for the development of PV. Without these long-term strategic decisions, the technology would not have been there and the emerging RES markets would have all followed the oil path for another decade or two until the wind energy and PV technology would have been developed to the level of today.

We have shown that financial incentives can boost a given development and too high financial incentive can cause unsustainable development. Therefore, it is important to start a transition process early with modest speed to get experience instead of burning off resources. The German direct marketing project was such an example. Denmark started slower and earlier, but reached the same fraction of energy on market terms. It is therefore still important that there is political will to take long-term decisions and risks. Without such initiatives one can expect long term unemployment, because existing technologies will otherwise be deployed at lower costs. Germany has long-term experience with this topic from subsidising their coal industry. In today's world most development is in fact short sighted on cost optimization. This is a useful and required process for existing technologies, but unhealthy for technologies in the immature phases.

Without the transition to RES, we would have reason to expect higher unemployment rates than today in the western world and more growth in the oil producing regions, where deserts would be converted to something artificial that would otherwise exist elsewhere.

We are in the fortunate position that synergies from wind, PV, direct marketing and ensemble predictions can be deployed in the years to come. It is though worthwhile to note that the cost of RES can only be kept low, if this evolution will be based on automated IT processes, which are utilizing existing resources and technologies at a high efficiency level and more advanced than presently used in the industry. These advances in technology also allow for a paradigm shift from "keep it simple" to "deal with complexity" to enhance results. Some companies are more than halfway through this process and their lead will probably increase and implies a market that is to be consolidated.

Acknowledgements The authors want to thank their valued customers, collaborators and partners for the inspiration to many of the topics presented in this chapter as well as the information provided to develop appropriate algorithms that can withstand in operational environments and in that way add value to the operation of the grid with growing amounts of renewable energies.

References

1. Act on granting priority to renewable energy sources. Renewable Energy Sources Act—EEG (2012) Consolidated (non-binding) version of the Act in the version applicable as at 1 Jan 2012, https://www.clearingstelle eeg.de/files/node/8/EEG_2012_Englische_Version.pdf
2. Gesetz für den Ausbau erneuerbarer Energien—EEG (2014), http://www.erneuerbare-energien.de
3. Press release (2014), http://www.marketcoupling.com/market-info-and-press/news/news-archive/date/2014-1. Accessed 01 Feb 2014
4. Gesetz ber die Elektrizitäts- und Gasversorgung (Energiewirtschaftsgesetz-EnWG), dated 7. Juli 2005, last modified on 21. Juli 2014, http://www.bmwi.de/DE/Service/gesetze,did=22154.html
5. Web page content (2015) Grid Control Cooperation, https://www.regelleistung.net/ip/action/static/gcc. Release 15.2
6. C. Möhrlen, Uncertainty in wind energy forecasting, Ph.D. dissertation, University College Cork, Ireland, DP2004 MHR (2004), http://library.ucc.ie/record=b1501384~S0
7. J.U. Jørgensen, C. Möhrlen, Increasing the competition on reserve for balancing wind power with the help of ensemble forecasts, in *Proceedings of 10th International Workshop on Large-Scale Integration of Wind Power into Power Systems as well as on Transmission Networks for Offshore Wind Power Plants, Aarhus, Denmark* (2011), http://download.weprog.com/public_paper_WIW11_032_joergensen_et_al.pdf, ISBN: 978-3-98 13870-3-2. Accessed Nov 2011

8. C. Möhrlen, J.U. Jørgensen, Forecasting wind power in high wind penetration markets using multi-scheme ensemble prediction methods, in *Proceedings of German Wind Energy Conference DEWEK, Bremen* (2006), http://download.weprog.com/mseps-dewek-2006.pdf. Accessed Nov 2006
9. C. Möhrlen, J.U. Jørgensen, A new algorithm for Upscaling and Short-term forecasting of wind power using Ensemble forecasts, in *Proceedings of 8th International Workshop on Large-Scale Integration of Wind Power* (2009), http://www.weprog.com/files/weprog_windintegration_2009_p54_paper.pdf, ISBN: 978-3-9813870-1-8. Accessed Nov 2009
10. C. Möhrlen, J.U. Jørgensen, Using Ensembles for Large-scale Forecasting of Wind Power in a European SuperGrid context, in *Proceedings of the German Wind Energy Conference DEWEK, Bremen* (2010), http://download.weprog.com/moehrlen_dewek2010_s10_p4.pdf. Accessed Oct 2010
11. C. Möhrlen, J.U. Jørgensen, Reserve forecasting for enhanced Renewable Energy management, in *Proceedings of 12th International Workshop on Large-Scale Integration of Wind Power into Power Systems as well as on Transmission Networks for Offshore Wind Farms, Berlin* (2014), http://download.weprog.com/Paper_WIW14-1035_moehrlen_joergensen_online.pdf, ISBN: 978-3-98-13870-9-4. Accessed Nov 2014
12. K. Sattler, H. Feddersen, *An European Flood Forecasting System EFFS: Treatment of Uncertainties in the Prediction of Heavy Rainfall using Different Ensemble Approaches with DMI-HIRLAM* (Scientific Report 03-07 of the Danish Meteorological Institute, 2003). ISSN Nr: 0905-3263 (printed), ISSN Nr: 1399-1949 (online), ISBN-Nr: 87-7478-480-3
13. Renewables International Magazine Online (2014) Spanish feed-in tariffs a wrapup, http://www.renewablesinternational.net/spanish-feed-in-tariffs-a-wrapup/150/537/71424/
14. TIGGE—the THORPEX Interactive Grand Global Ensemble (2015), http://tigge.ecmwf.int. Accessed ECMWF, 2015
15. R.L. Harrison, Introduction to monte carlo simulation. Proc. Proc. AIP Conf. **1204**, 1721 (2010). doi:10.1063/1.3295638, PMCID: PMC2924739, NIHMSID: NIHMS219206
16. S. Alessandrini, S. Sperati, P. Pinson, A comparison between the ECMWF and COSMO ensemble prediction systems applied to short-term wind power forecasting on real data. Appl. Energy **107**, 271–280 (2013)
17. C. Brankovic, T.N. Palmer, F. Molteni, S. Tibaldi, U. Cubasch, Extended-range predictions with ECMWF models: time-lagged ensemble forecasting. Q. J. R. Meteorol. Soc. **116**, 867–912 (1990)
18. R. Buizza, T.N. Palmer, The singular vector structure of the atmosphere global circulation. J. Atmos. Sci. **52**, 1434–1456 (1995)
19. R. Buizza, Potential forecast skill of ensemble prediction, and spread and skill distributions of the ECMWF ensemble prediction system. Mon. Weather Rev. **125**, 99119 (1997)
20. R. Buizza, P.L. Houtekamer, G. Pellerin, Z. Toth, Y. Zhu, M. Wei, A comparison of the ECMWF, MSC, and NCEP global ensemble prediction systems. Mon. Weather Rev. **133**, 10761097 (2005)
21. K.K.W. Cheung, A review of ensemble forecasting techniques with a focus on tropical cyclone forecasting. Meteorol. Appl. **8**, 315332 (2001). doi:10.1017/S1350482701003073
22. L. Delle Monache, F.A. Eckel, D.L. Rife, B. Nagarajan, K. Searight, Probabilistic weather prediction with an analog ensemble. Mon. Weather Rev. **141**, 34983516 (2013). doi:10.1175/MWR-D-12-00281.1
23. E.S. Epstein, Stochastic dynamic prediction. Tellus **6**, 739759 (1969)
24. G. Evensen, Sequential data assimilation with a nonlinear quasigeostrophic model using monte carlo methods to forecast error statistics. J. Geophys. Res. **99**(C5), 10143–10162 (1994)
25. I.-L. Frogner, T. Iversen, High-resolution limited-area ensemble predictions based on low-resolution targeted singular vectors. Quart. J. R. Meteor. Soc. **128**, 13211341 (2002)
26. A. Schierenbeck, D. Gräber, S. Semmig, A. Weber, Ein distanzbasiertes Hochrechnungsverfahren für die Einspeisung von Photovoltaik. Energiewirtschaftliche Tagesfragen, 60, Heft 12, 2010

27. R. Hagedorn, L.A. Smith, Communicating the value of probabilistic forecasts with weather roulette. Meteorol. Appl. **16**(2), 143–155 (2009)
28. T.M. Hamill, Interpretation of rank histograms for verifying ensemble forecasts. Mon. Weather Rev. **129**, 550–560 (2001)
29. D. Heizenreder, S. Trepte, M. Denhard, SRNWP-PEPS: a regional multi-model ensemble in Europe. Eur. Forecast. Newsl. **11**, 29–35 (2006)
30. P.L. Houtekamer, The construction of optimal perturbations. Mon. Weather Rev. **123**, 28882898 (1995)
31. P.L. Houtekamer, J. Derome, H. Ritchie, H.L. Mitchell, A system simulation approach to ensemble prediction. Mon. Weather Rev. **124**, 1225–1242 (1996a)
32. P.L. Houtekamer, L. Lefaivre, J. Derome, The RPN ensemble prediction system, in *Proceedings of ECMWF Seminar on Predictability, Reading, United Kingdom*, Vol. II (ECMWF, Shinfield Park, Reading, Berkshire RG2 9AX, United Kingdom, 1996b), pp. 121–146
33. P.L. Houtekamer, L. Herschel, H.L. Mitchell, Data assimilation using an ensemble kalman filter technique. Mont. Weather Rev. **126**(3), 796–811 (1998)
34. P.L. Houtekamer, H.L. Mitchell, Ensemble Kalman filtering. Q. J. R. Meteorol. Soc. **131**, 32693289 (2005). doi:10.1256/qj.05.135
35. P.L. Houtekamer, H.L. Mitchell, A sequential Ensemble Kalman Filter for atmospheric data assimilation. Mon. Weather Rev. **129**, 123–137 (2007)
36. P.L. Houtekamer, H.L. Mitchell, X. Deng, Model error representation in an operational ensemble Kalman filter. Mon. Weather Rev. **137**, 2126–2143 (2008)
37. C. von Junk, L. Bremen, M. Kühn, S. Späth, D. Heinemann, Comparison of postprocessing methods for the calibration of 100-m wind ensemble forecasts at off- and onshore sites. J. Appl. Meteorol. Climatol. **53**, 950969 (2014)
38. R.W. Katz, M. Ehrendorfer, Bayesian approach to decision making using ensemble weather forecasts. Weather Forecast. **21**, 220231 (2006)
39. S. Lang, C. Möhrlen, J. Jørgensen, B. Gallachir, E. McKeogh, Application of a multi-scheme ensemble prediction system for wind power forecasting in Ireland and comparison with validation results from Denmark, in *Proceedings of European Wind Energy Conference, Greece*, 2006
40. Y.-H. Lee, Loss Functions in Time series forecasting, Unicersity White Paper, University of Califoria, Dept. of Economics (2007)
41. E.N. Lorenz, Energy and numerical weather prediction. Tellus **12**, 364–373 (1960)
42. C.E. Leith, Theoretcal skill of monte carlo forecasts. Mon. Weather Rev. **102**, 409–418 (1974)
43. Z. Meng, F. Zhang, Tests of an ensemble kalman filter for mesoscale and regional-scale data assimilation. Part II: imperfect model experiments. MWR **135**, 1403–1423 (2007). doi:10.1175/MWR3352.1
44. C. Möhrlen, M. Pahlow, J.U. Jørgensen, Untersuchung verschiedener Handelsstrategien für Wind- und Solarenergie unter Berücksichtigung der EEG 2012 Novellierung, Zeitschrift f. Energiewirtschaft, No. 1/2012, **36**(1), 9–25 (2012)
45. H.L. Mitchell, P.L. Houtekamer, Ensemble Kalman filter configurations and their performance with the logistic map. Mon. Weather. Rev. **137**, 43254343 (2009)
46. F. Molteni, R. Buizza, T.N. Palmer, T. Petroliagis, The ECMWF ensemble system: methodology and validation. Q. J. R. Meteorol. Soc. **122**, 73–119 (1996)
47. F. Molteni, R. Buizza, C. Marsigli, A. Montani, F. Nerozzi, T. Paccagnella, A strategy for high-resolution ensemble prediction. I: Definition of representative members and global-model experiments. Q. J. R Meteorol. Soc. **127**, 20692094 (2001)
48. A.H. Murphy, E.S. Epstein, A note in probability forecasts and 'Hedging'. J. Appl. Meteorol. **6**, 1002–1004 (1967)
49. A.H. Murphy, The value of climatological, categorical and probabilistic forecasts in the cost-loss situation. Mon. Weather Rev. **105**, 803816 (1977)
50. A.H. Murphy, M. Ehrendorfer, On the relationship between the accuracy and value of forecasts in the cost-loss ratio situation. Weather Forecast. **2**, 243251 (1987)

51. M. Pahlow, C. Möhrlen, J.U. Jørgensen, Application of cost functions for large-scale integration of wind power using a multi-scheme ensemble prediction technique, in *Optimization Advances in Electric Power Systems*, ed. by Edgardo D. Castronuovo (NOVA Publisher NY, 2008), pp. 151–180. ISBN: 978-1-60692-613-0
52. T.N. Palmer, F. Molteni, R. Mureau, R. Buizza, P. Chapelet, J. Tribbia, *Ensemble Prediction, ECMWF Seminar proceedings, Validation of Models over Europe*, vol. 1 (ECMWF, Shinfield Park, Reading, UK, 1993)
53. T.N. Palmer, J. Barkmeijer, R. Buizza, Y. Petroliagis, The ECMWF ensemble prediction system. Meteorol. Appl. **4**, 301304 (1997)
54. W.S. Parker, Predicting weather and climate: uncertainty, ensembles and probability. Stud. Hist. Philos. Mod. Phys. **41**, 263272 (2010)
55. P. Pinson, Adaptive calibration of (u, v)-wind ensemble forecasts. Q. J. R. Meteorol. Soc. **138**(666), 1273–1284 (2012)
56. A.E. Raftery et al., Using bayesian model averaging to calibrate forecast ensembles. Mon. Weather Rev. **133**, 1155–1174 (2005)
57. N. Schuhen, T. Thorarinsdottir, T. Gneiting, Ensemble model output statistics for wind vectors. Mon. Weather Rev. **140**, 3204–3219 (2012)
58. D. Shepard, A two-dimensional interpolation function for irregularly-spaced data, in *Proceedings of the 1968 23rd ACM National Conference* (ACM, New York, 1968), pp. 517–524. doi:10.1145/800186.810616
59. D.J. Stensrud, J.W. Bao, T.T. Warner, Using initial condition and model physics perturbations in short-range ensemble simulations of mesoscale convective systems. Mon. Weather Rev. **128**, 2077–2107 (2000)
60. D.J. Stensrud, H.E. Brooks, J. Du, S. Tracton, E. Rogers, Using ensembles for short-range forecasting. Mon. Weather Rev. **127**, 433446 (1999)
61. Z. Toth, E. Kalnay, Ensemble forecasting at NMC: the generation of perturbations. Bull. Am. Meteorol. Soc. **74**, 2317–2330 (1993)
62. Z. Toth, E. Kalnay, Ensemble forecasting at NCEP and the breeding method. Mon. Weather Rev. **125**, 32973319 (1997)
63. T. Thorarinsdottir, T. Gneiting, Probabilistic forecasts of wind speed: ensemble model output statistics by using heteroscedastic censored regression. J. R. Stat. Soc. **173A**, 371–388 (2010)
64. R.L. Winkler, A.H. Murphy, Decision analysis, in *Probability, Statistics and Decision Making in the Atmospheric Sciences*, ed. by A.H. Murphy, R.W. Katz (Westview Press, Boulder, Colorado, 1985), pp. 493524
65. D.S. Wilks, A skill score based on economic value for probability forecasts. Meteorol. Appl. **8**, 209219 (2001)
66. H. Zhang, Y. Pu, Beating the uncertainties: ensemble forecasting and ensemble-based data assimilation in modern numerical weather prediction. Adv. Meteorol. **2010**, Article ID 432160, 10 (2010). doi:10.1155/2010/432160

Chapter 4
Wind and Solar Forecasting

John W. Zack

4.1 Nature of the Wind and Solar Forecasting Problem

The non-dispatchable variability of wind and solar power production presents a substantial challenge to electric grid operators who are assigned the task of balancing demand and generation at each moment at the lowest possible cost while maintaining ultra high system reliability. The economic factor is the key component of this problem since one can always adequately manage variability and maintain reliability if cost is not an issue.

The challenge becomes increasingly difficult as the percentage of system generation capacity supplied by wind and solar generation assets increases. A number of tools or approaches are potentially available to the system operator to assist in meeting this challenge. However, many require substantial long-term planning and significant implementation costs. These include a shift to more flexible (i.e. quicker response) generation assets, demand response programs, more accommodative market structures and the planning of the geographic diversity of wind and solar assets. However, one of the most cost effective and easily implemented tools to assist in the management of the non-dispatchable variability of wind and solar power production is the short-term forecasting of the production. This can provide system operators with the lead time and event visibility to make more economical decisions while maintaining the required ultra-high levels of reliability. The use and associated issues of wind power forecasts in grid management was recently summarized in a paper by Ahlstrom et al. [1], an analogous overview of the use and issues of solar power forecasts was provided in an article by Tuohy et al. [11].

There are four key components to a forecasting solution that should each be optimized to provide maximum value to the end user: (1) high quality and

J.W. Zack (✉)
AWS Truepower, 463 New Karner Rd, Albany 12205, NY, USA
e-mail: jzack@awstruepower.com

P. Du et al. (eds.), *Integration of Large-Scale Renewable Energy into Bulk Power Systems*, Power Electronics and Power Systems,
DOI 10.1007/978-3-319-55581-2_4

representative measurement data for input into the forecasting procedure, (2) skillful forecasting techniques or models, (3) meaningful assessment of forecast performance to provide users with confidence in using the forecast information for decision-making; and (4) effective and efficient communication of the critical forecast information to the user. This combination of components can be expressed in the acronym SMAC: Sense, Model, Assess and Communicate. A compromise in any of these components can result in less than the maximum value being obtained from a forecast for the application. The following subsections will address each of these four components of an optimal forecasting solution.

This chapter is organized into six sections. This overview of the nature of the wind and solar forecasting problem is the first section. The second section discusses the input data requirements and opportunities to improve forecast performance by acquiring additional input data. The third section presents an overview of the methods that are typically used in producing wind and solar forecasts and how they are typically applied. The fourth section discusses the issues associated with the evaluation of forecast performance, the metrics that commonly used to evaluate performance and provides a summary of the current state-of-the-art forecast performance. The fifth section provides an overview of the various types of forecast products that are frequently used. The concluding section provides some insight into the value of forecasting that has been diagnosed in a range of operating scenarios.

4.2 Sense: Gathering and Ingestion of Predictive Information

There are three fundamental types of data used by a forecast system: (1) quasi real-time data used as input in each prediction cycle of the forecast models, (2) historical data used for the training of the statistical components of the forecast system and (3) static data that describes the attributes of the generation facility (location of generation units, generating capacity, type of generation hardware etc.) and its physical environment. The historical data is also used to evaluate the performance of the forecasts generated by an integrated forecast system as well as by its individual components.

Ultimately all of the predictive information for each forecast scenario is provided to a forecast system via the quasi-real-time input data. From the broadest perspective, there are an enormous number of data types and data elements that are ingested by a forecast system. The input data includes the vast array of global atmospheric sensor measurements that are used by the various types of atmospheric prediction systems as well as the meteorological and generation data from the forecast target facility and nearby locations. However, different segments of the input data pool are employed by each component of the forecast system. For example, the physics-based atmospheric models utilize a multitude of data types from regional and global domains to specify their three-dimensional initial states. On the other hand, predictions from time series models are typically based only on recent data from the forecast site and perhaps nearby off-site locations.

4.2.1 Area of Influence

An important concept is the space-time envelope of data influence that determines which data locations and variables have a significant impact on a forecast for a particular or typical scenario. This has important implications for how to improve forecast accuracy. This concept is illustrated in Fig. 4.1. The images depict the fraction of the variance of an NWP forecast of the 80-m wind speed (the predictand) explained (i.e. the R^2 parameter) by variations in a "measurement" (predictor) variable (also the 80-m wind speed) at regional locations around the forecast site (the white box) for a 1-h (left panel) and 3-h (right panel) forecast. The purple and blue colors represent very low values of R^2, which indicate that there is essentially no relationship between changes in the predictor variable (simulated sensor measurement) and changes in the forecast value. That is. the forecast is not sensitive to measurements at these locations. In contrast the yellow and red shading depict relatively high values of R^2, which indicate a stronger relationship between variations in the measurement and variations in the resulting 1-h or 3-h forecast. Thus, the area covered by the green to red colors can be considered to the be the forecast sensitivity region. Measurements in this region have an impact on the forecast but measurements in the purple and blue regions do not.

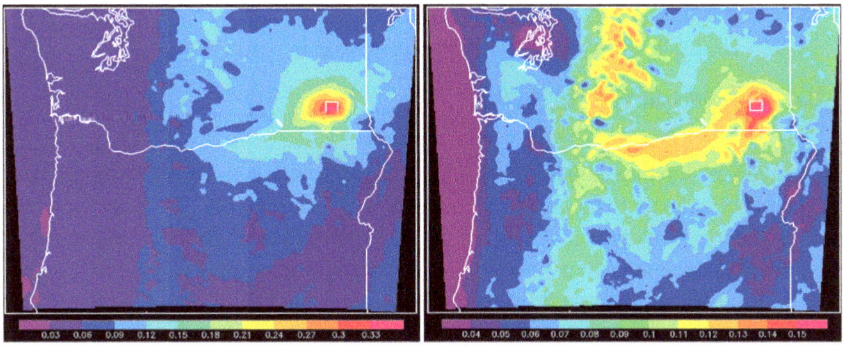

Fig. 4.1 An example of the increase in domain of forecast sensitivity as the look-ahead time for an 80-m wind speed forecast increases for a forecast site in southeastern Washington based on simulations from a Numerical Weather Prediction (NWP) model. The forecast target area is represented by the *white box*. The *color shading* represents the R^2 values of the relationship between the 80-m wind speed values at each location (the predictors) at the forecast issue time and the forecasted value on the target box (the predictand) for 1 h (*left*) and 3 h later (*right*)

A comparison of the left and right panels of Fig. 4.1 indicates that the area of forecast sensitivity increases by a large amount from a 1-h to a 3-h forecast. The implication is that measurements must be made over a much larger area to improve a 3-h forecast than to improve a 1-h forecast. Of course this expansion in the area of forecast sensitivity continues beyond 3 h and becomes very large as the look-ahead period expands to 1 day and then multiple days. Ultimately, the area of forecast sensitivity for a long range forecast (e.g. 10 days or more) is the entire world.

There are several points to note about the concept of forecast sensitivity. The first point is that the shapes of the areas of sensitivity are generally quite complex and not symmetric around the forecast target location. They are typically skewed to the prevailing regional upstream direction, which is typically to the west for the location depicted in Fig. 4.1. The patterns are also strongly influenced by the terrain as shown by channeling of the high sensitivity along the Columbia River Valley and the appearance of high sensitivity along the Cascade Mountains. Second, the charts in Fig. 4.1 represent an average (or climatological) sensitivity over a sample of many cases. The area and magnitude of the sensitivity is flow-dependent and therefore varies within a set of forecast cases. In some situations (e.g. weak and slow moving atmospheric features) the area of sensitivity expands slowly while in others (e.g. rapidly moving atmospheric features) it grows very rapidly. Third, the charts in Fig. 4.1 only depict the sensitivity of the forecast to a single predictor variable, which in this case is the same type of variable (i.e. the 80-m wind speed) as the predictand. Of course, the forecast of 80-m wind speed over the white box is, in general, sensitive to variations in a wide range of atmospheric variables (pressure, temperature etc.) and at locations in the vertical as well as the horizontal. Thus, there are many possible measurements to which the forecast can be sensitive. A fourth point is that the variables in their respective sensitivity regions are intercorrelated to some degree. Thus, it is generally not beneficial to measure each variable at each sensitive location.

4.2.2 Observation Targeting

The concept of forecast sensitivity can be used to provide guidance for the optimum placement of sensors to improve the performance of forecasts. The basic objective is to formulate a method that will enable the identification of the combination of variables and locations that provide the greatest positive impact on forecast performance at a specified cost level. Intuitively, one might think that this would be a very challenging task and it is.

A number of issues must be considered in order to get value from this approach. The first factor is that the area of forecast sensitivity expands rapidly with look-ahead time. This means that measurements must be made over a bigger region (i.e. more measurement locations) to get the same impact on forecast performance. Thus, the most value from a specified number of sensors (and therefore at a given cost level) is obtained for shorter look-ahead periods. This is visually depicted in the comparison of the left and right panels of Fig. 4.1. A big issue is the flow dependence of the sensitivity.

A number of approaches have been formulated and employed to address this objective. The simplest approach is a subjective approach in which one looks at the prevailing upstream direction and the average speed and estimates the direction and distance at which a measurement should be made. A second approach is to infer the spatial time-lagged correlation pattern around the forecast site from the available measurements. One can then identify locations with an inferred high time-lagged correlation that are currently not measurement points. These could be candidates for the deployment of additional sensors. A third approach is to employ an NWP-based ensemble sensitivity analysis [9] to construct spatial patterns of the sensitivity of forecasts for the target variable to variations in predictor variables at other locations at prior times. Experiments with the application of this approach for short-term wind power forecasting have been conducted for Tehachapi Pass and the Columbia River Basin in the United States [10, 13–15].

4.2.3 Historical Data

A wind and solar forecast system typically uses historical data as a training sample for statistical models. The amount, quality and representativeness of this data can have a substantial impact on forecast performance.

4.3 Model: Mapping Predictive Information to a Forecast

There are generally three broad classes of prediction techniques that are employed in state-of-the art wind and solar forecast systems: (1) physics-based meteorological techniques, (2) statistical methods applied to meteorological variables, and (3) power output models, which are statistical or physics-based relationships between meteorological variables and electrical power generation. The following subsections discuss each of these techniques.

4.3.1 Physics-Based Techniques

Physics-based atmospheric forecasting techniques are based on broadly applicable physical principles. A key attribute is that they do not require a training sample to generate the prediction equations. Therefore, skillful predictions can be made even in situations in which there is no historical data available from the forecast target entity. This also means that their range of predictions is not limited to what has been observed in a historical sample. Thus, they are theoretically capable of predicting events with little or no historical precedent.

More than one type of physics-based model is often used in the forecasting of wind and solar power production. However, the most prominent and sophisticated type is known as Numerical Weather Prediction (NWP). The NWP approach is described in the next subsection. Other types of physics-based approaches are generally based on highly simplified physics-based models that account for a small subset of physical processes that can be assumed to be the dominant factors for a specific application or time scale. The most widely-used approach of this type in renewable energy forecasting are the cloud advection models used for short-term solar forecasting. These are based on the concept that the motion of existing features (advection is the term used by meteorologists) is the dominant physical process. The non-NWP physics-based models are discussed in the second subsection of this section. Although NWP is the most dominant type of physics-based prediction technique used in the forecasting of wind and solar power production, it is not the only physics-based technique. Another class of physics-based technique can broadly be referred to as feature detection and tracking approaches.

4.3.1.1 Numerical Weather Prediction (NWP)

The NWP approach [8, 2, 4] is based upon the application of the fundamental physics principles of conservation of mass, momentum and energy and the equation of state for moist air to the atmosphere. These principles are formulated as a set of differential equations, which are then "solved" by numerical methods such as finite difference approximations or spectral techniques. The most basic set of NWP model equations accounts for processes in an atmosphere during which the energy content of a parcel of air does not change in time ("adiabatic") on the scale of the NWP grid. The form of the equations is well known from basic physics and there is very little uncertainty in their formulation.

Additional terms are then added to the NWP model equations to account for processes that change the energy content of an air parcel. These are processes such as long and short wave radiative transfer, water phase changes, motions (i.e. turbulence and moist convection) that occur on scales smaller than the NWP grid and the fluxes of heat, moisture and momentum from the underlying surface of the earth. The mathematical formulation of these processes is heavily rooted on basic physics principles but they also have a significant empirical component. The empirical component is mostly related to processes (such as the formation of a rain drop or the impact of a turbulent eddy) that occur on too small of a scale to explicitly model on the NWP grid. Therefore, the bulk effect of these processes are modeled through statistical relationships with grid-scale variables. While an effort is made to develop sets of relationships that are universally applicable, they often have some location or atmospheric regime dependence. As a result, these relationships often introduce biases (systematic errors) into the NWP forecasts.

NWP prediction systems are actually composed of two major components: (1) a NWP model; and (2) a data assimilation system. The NWP model contains the physics-based prediction equations and operates on a specified initial atmospheric state for the model's domain to generate time-dependent predictions.

The function of the data assimilation system is to create an accurate representation of the initial state (i.e. the starting point for the NWP forecast process) from a large and diverse set of atmospheric sensor data as well as previous NWP forecasts. The primary issues faced by the data assimilation system are (1) measured data is not available for every model variable at every model grid point, (2) the available atmospheric sensor data has a wide range of spatial and temporal coverage and resolution, variables measured and characteristic measurement errors, (3) atmospheric variables are tightly coupled via the physics-based equations and therefore a modification made to the value of one variable must be accompanied by appropriate modifications to other variables in order to maintain a physically realistic state. The most typical approach used by data assimilation systems is to start with the three-dimensional state provided by a forecast from the previous cycle of the same model (referred to as a warm start). This is sometimes referred to as the "first guess" or "background state". The available sensor data are then used to update this first guess. These updates or corrections to the first guess essentially represent model forecast errors. The key issue to how to spread the influence of a measurement from the point of measurement to nearby model grid points. The spreading of the influence typically uses some estimate of the spatial model error covariance. This essential provides the information about how much connection there is between the model error diagnosed at the measurement point and the error at nearby grid points where there is no measurement data (hence errors can't be explicitly computed). The spatial error covariance patterns are typically flow-dependent.

It is difficult to estimate these covariance patterns although techniques based on ensembles of NWP simulations have been developed to estimate the flow dependent variations in the spatial error covariance. However, in practice, climatological spatial error covariance patterns are typically compiled from a historical set of NWP forecasts and these are used in the process of spreading the influence of measurement data when updating the first guess state. The improvement in the specification of the spatial covariance is a key area in which NWP forecasts are likely to be improved in the future.

The forecast errors and the associated uncertainty of NWP model predictions are associated with three primary factors: (1) resolution of the grid and the numerical methods used to solve the equations, (2) uncertainty in statistical relationships and approximations of physics-based principles in the formulation of the NWP model physics, and (3) uncertainty in the specification of the initial state due to data sparseness, unrepresentativeness or sensor error and unrepresentative spatial error covariance used in the initialization process. These issues produce a combination of systematic and random errors. As will be discussed later, the presence of systematic errors provides an opportunity to improve NWP forecasts via statistical techniques that can diagnose and partially correct these errors.

In addition to the correction of systematic errors via statistical postprocessing, the uncertainty associated with NWP predictions is typically also addressed through the use of the ensemble concept. The basic concept of a forecast ensemble is that a set of forecasts are created for a given forecast period by varying the input data or model parameters within their range of uncertainty. This yields a range of forecasts. There are two fundamental types of NWP ensembles: (1) those generated by perturbing the initial state and (2) those generated by varying the formulation of the model physics. A given ensemble may contain members of only one type of a combination of both types. It seems reasonable to expect that a combination of both types would more accurately represent the forecast uncertainty. However, in practice, some scenarios are more sensitive to the initial state and others are more sensitive to the model physics.

If all the significant sources of input uncertainty are appropriately represented in the formulation of an ensemble then the forecast ensemble should accurately represent the uncertainty in the forecast. However, there are so many individual sources of input uncertainty in an NWP system that it is virtually impossible to have a comprehensive representation of all the significant sources. Hence, the output from a typical NWP ensemble will underestimate the forecast uncertainty. This is typically addressed by statistically calibrating the forecast distribution produced by the ensemble through the use of historical forecast ensemble data and the associated outcomes so that it more accurately represents the true probability distribution of the forecast.

One of the most significant attributes of NWP models is that they supply physically-consistent predictions of virtually the entire set of meteorological variables. Thus, a single NWP run can be used for virtually all types of atmospheric forecast applications. An example of a high resolution NWP forecast of wind speed and solar irradiance over the island of Oahu is shown in Fig. 4.2.

NWP prediction systems can be applied to a forecast problem on many different space and time scales with a wide variety of model configurations. NWS systems are typically configured to operate in one of three modes: (1) global, (2) regional or (3) rapid update. Global NWP systems have a forecast domain that encompasses the entire world and they are typically used to generate forecast with look-ahead periods that extend to 1 to 2 weeks. However, they are now also periodically run to 30 days or longer to provide guidance for monthly or seasonal forecasts. Global NWP models are typically run on a 6-h cycle although some NWP centers use a 12-h cycle. Current examples of global NWP models that are widely used for wind and solar forecast applications are the Global Forecast System (GFS) operated by the US National Weather Service, the Global Deterministic Prediction System (GDPS) run by Environment Canada, the global system run by the European Centre for Medium Range Weather Forecasts (ECMWF) and the global prediction system of the United Kingdom's Meteorological Office.

Regional NWP operates on a limited area domain such as a continent or part of a continent. Current examples of regional NWP models are the North American Mesoscale (NAM) model operated by the US National Weather Service and the Regional Deterministic Prediction System (RDPS) run by Environment Canada.

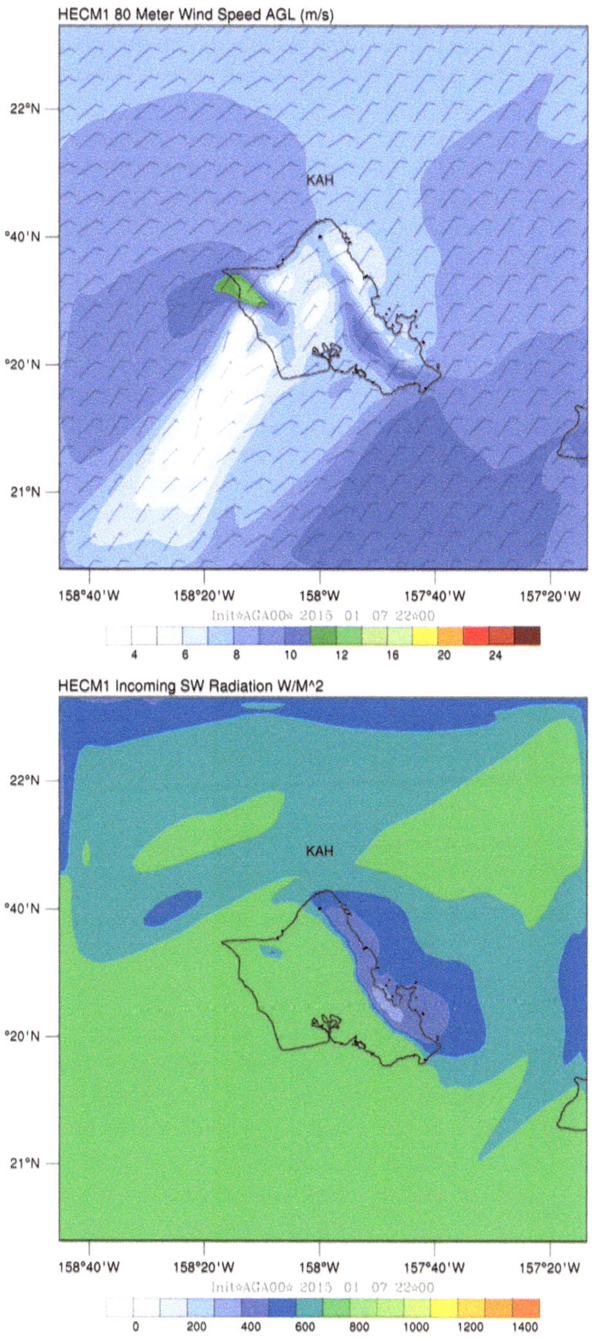

Fig. 4.2 An example of a forecast of wind speed and global horizontal solar irradiance produced by a high resolution (horizontal grid spacing of 3 km) NWP model

Many other countries also operate regional NWP models over the region surrounding their area of interest. The high density of countries in continental Europe results in the availability of output from many overlapping regional NWP model domains for many locations. This provides a substantial regional NWP ensemble that is not available in other parts of the world.

Rapid update NWP models are similar to regional NWP systems but they are operated on a shorter update cycle. These are typically run on 1-h or 2-h cycles and frequently cover smaller domain with a higher resolution grid than the standard regional models. Their primary objective is to frequently assimilate the latest sensor data and make a very short term forecast. A forecast duration of 15–18 h is often used. Current examples of rapid update models are the Rapid Refresh (RAP) and High Resolution Rapid Refresh (HRRR) models run by the US National Weather Service. The RAP is run hourly on a 13 km grid while the HRRR employs a 3 km grid for its hourly update cycle. These forecasts are available within an hour after the initialization time.

Wind and solar forecast producers often also run their own in-house NWP models to supplement the data from the government center NWP systems. There are three primary motivations for this: (1) to downscale the government NWP forecast to a higher resolution to more effectively represent the local terrain, land-water boundaries and other surface features; (2) to assimilate data into the initial state from sources not available to government NWP systems; (3) to customize the model physics formulations for wind or solar forecasts for a specific region. The most popular model used for this purpose is the Weather Research and Forecasting (WRF) model.[1] It is an open source community model maintained by the National Center for Atmospheric Research (NCAR) and the National Oceanographic and Atmospheric Administration (NOAA) with major ongoing contribution from all sectors of the atmospheric science community including academia, government and the private sector.

4.3.2 Statistical Approaches

Statistical techniques are applied to a very wide range of prediction problems and the forecasting of atmospheric variables is one of them. Many text books have been published that supply an overview of statistical prediction techniques, the details of specific methods and the application of statistical methods to atmospheric prediction problems. A number of these traditional methods have been employed for short-term wind and solar power prediction. However, recent years have seen the application of some of the newest and most advanced statistical prediction techniques. This section provides a high level overview of the most widely used methods in solar and wind power prediction and how they are applied.

---

[1] http://www.wrf-model.org/index.php.

4.3.2.1 Methods

This section provides an overview of the statistical methods that have been frequently applied in the statistical components of wind and solar forecast systems. It is intended to indicate the type of tools that are typically applied in such systems and not to provide an exhaustive list of all of the possibilities.

Multiple Linear Regression (MLR)

MLR is one of the most widely-used statistical prediction method for wind and solar forecasting. As a predictive tool, the multiple linear regression is used to explain the relationship between one continuous dependent variable (the predictand) from two or more independent variables (the predictors). In linear regression, the relationships are modeled using linear predictor functions whose unknown model parameters are estimated from the data.

Artificial Neural Networks (ANNs)

Artificial Neural Networks are a family of statistical learning models inspired by biological neural networks (e.g. the central nervous systems of animals). They are widely used to estimate or approximate functions that can depend on a large number of inputs that are generally unknown. ANNs are generally presented as systems of interconnected "neurons" which send messages to each other. The connections have numeric weights that can be tuned based on experience, making neural nets adaptive to inputs and capable of learning.

Although ANNs can be very effective predictive tools, in practice they suffer from some issues that often limit their performance. ANNs tend to work best when they are trained on large and representative datasets that have low noise levels. In these situations they will often outperform simpler methods such as MLR by a substantial margin. However, this is often not the case in wind and solar applications. The training datasets are often relatively small and frequently have a high noise level. In addition, the predictive relationships in a training sample may not be stationary since they often have a seasonal or weather-regime based component. In these cases the ANN often does not perform better than simpler techniques.

Support Vector Regression (SVR)

The SVR method is rooted in the **support vector machines** (**SVM**) concept [3], which originated as a tool for classification problems. SVMs are supervised learning models with associated learning algorithms that analyze data and recognize patterns. Given a set of training examples, each marked for belonging to one of two categories, an SVM training algorithm builds a model that assigns new examples into one category or the other, making it a non-probabilistic linear classifier.

A version of SVM for regression was proposed in 1996 by Drucker et al. [6]. The regression version of SVM is the one that is most commonly used for renewable energy forecasting applications. Instead of minimizing the observed training error, Support Vector Regression (SVR) attempts to minimize the generalization error bound so as to achieve generalized performance. The idea of SVR is based on the computation of a linear regression function in a high dimensional feature space where the input data are mapped via a nonlinear function. The model produced by support vector classification depends only on a subset of the training data, because the cost function for building the model does not care about training points that lie beyond the margin. Analogously, the model produced by SVR depends only on a subset of the training data, because the cost function for building the model ignores any training data close to the model prediction.

Decision-Tree Regression

There are numerous statistical prediction methods that have been constructed upon the decision-tree concept. One of these, which is widely used in wind and solar forecasting applications, is known as Random Forests. Random Forests is an ensemble learning method for classification, regression and other tasks, that operates by constructing a multitude of decision trees at training time and outputting the class that is the mode of the classes (classification) or mean prediction (regression) of the individual trees. Random Forests correct for decision trees' habit of overfitting to their training set.

Other Methods

Other statistical methods have also been adopted in the wind and solar forecasting domain. One of the newest statistical prediction techniques used in wind and solar forecasting is the gradient boosted model (GBM) concept. Analog Ensemble (AE) is another unique statistical approach that has recently been applied to wind and solar forecasting. One of the first formulations of this concept for renewable energy forecasting applications was by Delle Monache [5].

4.3.2.2 Applications

Statistical techniques are typically applied to the wind and solar resource prediction problem in a number of ways. There are four prominent types of applications. Three of them are described in the following subsections. The fourth (the power output model) is discussed in Section 4.3.3.

Time Series Models

The objective of time series prediction techniques is to use the predictive information contained in the values, trends or patterns of the recent history of the forecast variable or related variables.

The simplest time series method is the "persistence" model, which employs the most recent value of the forecast variable as the prediction for all of the forecast intervals. It is a forecast of "no change". However, in practice, operational intra-hour or multi-hour forecasts typically employ time series data from 5 to 10 intervals before the forecast production time as predictors in the statistical forecast procedure. The use of autoregressive and moving average approaches are typical for this purpose. However, the application of advanced machine learning approaches (such as Artificial Neural Networks or decision-tree methods) to construct forecasts from time series data is rapidly becoming more common. In some cases, time series data from sensors at offsite locations (for example, at "upstream" sites) are used as supplementary predictors.

Model Output Statistics (MOS)

The fundamental concept of Model Output Statistics is to reduce the magnitude of systematic errors ("biases") in the forecasts from an underlying prediction model and/or generate predictions of variables that are not explicitly produced by the underlying model. The term "MOS" originated in a paper by Glahn and Lowry [7] in a description of their method to predict a set of meteorological variables at airport measurement sites using output variables from early NWP models as predictors in a screening multiple linear regression model.

As a result of its origins, the term MOS is often associated with the application of multiple linear regression methods to the output of NWP models. This approach is still widely used in general meteorological forecasting and in the prediction of wind and solar power production. However, the concept of MOS should be viewed in a much broader perspective. First, a wide variety of statistical techniques can be employed for the purpose of MOS. Second, the approach can be applied to many types of prediction models, not only NWP systems.

There are a number of factors that should be considered in the application of MOS to the prediction of wind and solar power production. An initial issue is whether to use the MOS approach to directly predict the power production from a facility or to employ MOS to improve upon the NWP prediction of the meteorological variables and subsequently employ an explicit power output model to create the power forecasts.

A second fundamental decision is the type of statistical method that is to be employed. Multiple linear Regression is the most basic method that is typically employed. However, a number of advanced machine learning approaches have been used in recent years in the application of MOS for wind and solar prediction applications. It may seem logical to simply use one of the more sophisticated methods. However, they do not always yield better performance than linear regression and typically are much more computationally intensive. Experience has indicated that the benefits of the advanced methods are more frequently realized when large high quality training samples can be used.

A third issue is the representativeness of the data in the training sample. The most significant version of this issue is the case in which the underlying model that supplies the predictors to the MOS procedure has changed during the period for which data is available for use in the training sample. This often is the result of upgrades, such as an increase in model resolution or changes to the formulation of the model physics of an NWP system. These type of changes can significantly change the patterns of systematic errors in the model forecasts. Therefore, the use of data from before the change in the underlying model may not represent the error patterns that occur after the change. If one chooses to use only data from the period after the model change, the training sample size will be limited. Since upgrades to government-run NWP systems routinely occur, this effectively eliminates the possibility of using a long training sample. There are two approaches that are widely used to mitigate this limitation. First, many of the changes to the underlying model are minor and have little impact on the systematic error patterns of particular variables. Therefore, data from before and after the model change can be used with little or no negative impact on MOS performance. However, a considerable amount of caution must be used when applying this approach. The impact of a model change typically varies substantially among model variables and even between geographic areas for a specific variable. A second approach is to use predictors from an in-house model that is controlled by the user. This enables the user to generate a historical training dataset produced by a model with an unchanging configuration.

A fourth issue is the training sample strategy. There are an almost infinite number of possible training sample configurations. The training sample strategies are frequently classified into three broad categories: (1) static, (2) dynamic and (3) regime-based.

In the static approach, a single training sample is used to generate a fixed set of MOS equations. The training sample typically covers a long period of a year or more. The same set of equations are used for all subsequent forecast cycles.

In contrast, the dynamic approach typically uses a much smaller training sample that is frequently updated. A typical dynamic MOS configuration is a rolling 60-day sample that ends on the forecast cycle before the current forecast cycle. In this approach the oldest data element is deleted on each cycle and data from the most recent cycle is added. In this configuration, the composition of the training sample changes for each forecast cycle and a new set of MOS prediction equations are used for each cycle. There are several advantages to this approach. First, the training sample is typically more representative of the current error patterns in the model forecasts since it is drawn from the current season and often from the current weather regime. Second, the use of a short-training sample avoids many of the issues associated with impact of changes in the underlying model except immediately around the time of the model change. A disadvantage of this approach is that the MOS equations are less stable because they are derived from a short sample.

The regime-based MOS approach is based on the concept of defining groups of cases for which the underlying model has similar error patterns. The use of the word "regime" in this context is often misinterpreted to mean "weather regime" but the clustering should be based on "model error regimes" to have value. Certainly, in many cases the "error regimes" are correlated with "weather regimes" and this information may be useful in designing a regime-based MOS approach, but ultimately it is the error regimes that are critical to identify. There are also many possible implementation strategies for the regime-based concept itself. A commonly employed strategy is to define a fixed set of error regimes either subjectively or via an objective approach such as a clustering algorithm or principal components analysis (PCA). The training sample is then divided into "N" regimes (clusters) by one of these approaches and a separate set of MOS equations is derived for each regime by employing a statistical prediction method. The production of a forecast is then accomplished by assigning the current forecast scenario to the most appropriate regime and using the MOS equations from that regime to generate the prediction. The primary benefit of this approach is that more case-specific error regimes can be identified, which can yield more effective error correction. However, a substantial disadvantage is that the division of the training sample into several subsets results in the training sample being smaller for each regime. Since the regimes may not have equal size, some of the regimes may not have a statistically meaningful sample size. Another approach is to formulate a dynamic regime-based strategy in which there is no pre-defined set of regimes. Instead, a custom-regime is created for each forecast scenario by selecting historical cases that are the best matches for the current case. This is essentially the approach employed by the Analog Ensemble method [5]. In general, it is difficult to employ many of the advanced machine learning methods with the regime based approach because of the small sample sizes that typically occur in the subdivided samples. However, the regime-based approach is essentially doing explicitly what many of the advanced machine learning techniques are implicitly doing.

An example of the impact of MOS on an NWP forecast of wind speed at hub height for a wind generation facility in Hawaii is depicted in Fig. 4.3. The data depicted in this chart indicates that the application of a MOS algorithm to the raw NWP forecast data substantially reduces the wind speed forecast error for a large fraction of the forecast intervals in this example. This type of impact on forecast performance is fairly typical although there is a substantial amount of variability associated with weather regimes, the characteristics of the NWP model, the look-ahead period and other factors. It is interesting to note that the application of MOS to an NWP forecast can substantially change the temporal structure of the forecasted time series. In this example, the temporal correlation between the raw NWP forecast and the MOS-adjusted forecast is only 0.63.

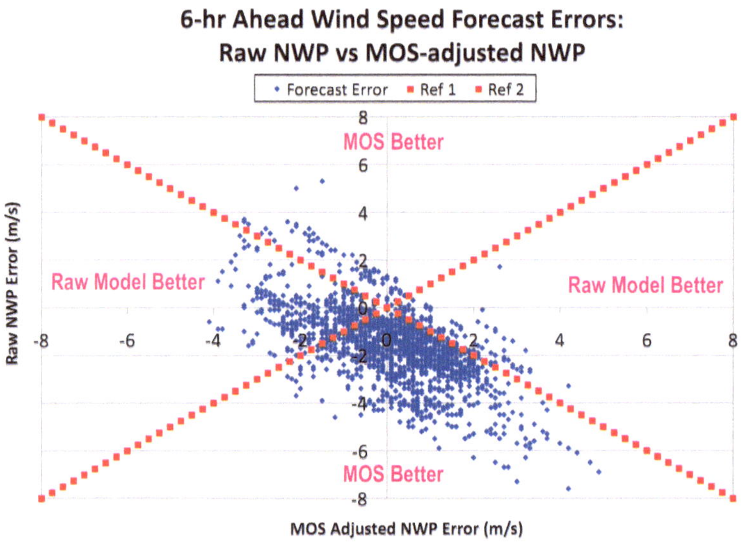

Fig. 4.3 Depiction of the relationship of errors (m/s) from a raw NWP and MOS-adjusted NWP 6-h ahead forecast of the 15-min average hub-height wind speed at a wind generation facility in Hawaii for the month of June 2015. The *blue* markers represent the forecast errors from each method for an individual 15-min period during the month. Reference *lines* that indicate where the magnitude of the errors from both methods is equal are shown in *red*

Ensemble Composite Models (ECM)

The objective of the ECM is to construct the best-performing composite forecast from an ensemble of individual forecasts. Although the term "ensemble" is frequently used to describe a set of NWP forecasts, an ensemble for a typical state of the art wind and solar forecast system is typically composed of forecasts produced by several types of methods including raw NWP forecasts, MOS-adjusted NWP forecasts, time series models, and feature detection and tracking models. The ECM is actually a form of MOS with the predictors coming from many prediction models rather than only one. Therefore, most of the same techniques that are employed for the single model MOS are applicable to the ECM. But there are some issues that distinguish the ECM application from the typical single model MOS application.

One issue in the ECM application is which members of a forecast ensemble are distinguishable or indistinguishable. The members are indistinguishable if the members are generated by using the identical forecast model and randomly perturbing (within some probabilistic limits) the same set of input data. In this case, there is no basis to distinguish one member of the ensemble from another and it is not useful to attempt to assign differential weighting to each of these members. However, the ensemble mean of an indistinguishable ensemble will typically produce a lower forecast error than an individual member. In contrast, the distinguishable members are produced through the use of characteristically different prediction methods (i.e. different models of different configurations of the same model) or input datasets (i.e. systematically omitting or adding specific input datasets). In this case, there is a basis for differential weighting of the members since it is possible some methods or input datasets may yield better forecasts under specific circumstances (e.g. regimes). Therefore, a typical approach is is to create a composite forecast from the members of a distinguishable ensemble through the application of one of the machine learning methods.

A second issue in the formulation of an ECM is whether to use all of the raw predictors from each member method as input into the ECM (a "super-MOS" approach) or to first create a forecast for the ultimate target variable (e.g. hub height wind speed or solar irradiance) from each method and then use only the target variable forecasts as input into the ECM. In general it is better to apply MOS to each individual model and then create a composite from the resulting forecasts since each model has it own unique error patterns, which may be difficult to distinguish in a training sample with forecasts from many other models.

A third consideration is the selection or filtering of member inputs into the ECM. It is tempting to use as many methods as could be available to the ECM. However, an indiscriminate application of that philosophy can be detrimental. The issue is based on two factors: (1) intercorrelation of errors and (2) historical sample size available to the ECM. The first factor is a reflection of the intuitive fact that the construction of a composite will have no benefit if the errors of all the methods are the same for each forecast period. In that case any composite of the methods will of course yield the same error as any individual method since all the errors are the same and there is no basis for distinguishing the performance of the methods. A less extreme and more typical occurrence is that the errors of individual methods are highly correlated. In this case, there will be minimal benefit in the construction of an ensemble. The point is that ensemble members with high error correlations to other members do not provide much value in the construction of the composite. However, the result can be worse than no impact. Indiscriminate use of highly correlated members in an ensemble composite can have detrimental effects. For example, in the case when some members have highly correlated errors and some don't, a simple equally weighted ensemble average will have the beneficial impact of less intercorrelated members diluted by the highly intercorrelated members. In effect, one is placing heavier weight on the forecast represented by the highly correlated members since it basically represents a multiple occurrence of the same forecast in the ensemble. On the other hand, poor performing uncorrelated forecasts will not be beneficial either.

The use of an appropriate statistical technique will serve to minimize the weighting of the highly correlated members and achieve an optimal blending of the high correlated and less correlated methods. This is where the training sample size can become an issue. The use of a large number of input methods along with an advanced statistical technique (with many adjustable parameters) can result in an overfitting issue if only a small training sample is available. The availability of a long representative training sample can minimize this issue but this can often be difficult to assemble because many of the input models will change periodically.

4.3.3 Power Production Models

Once a high quality prediction of the key meteorological variables are generated via a composite of the methods discussed in the previous sections, the meteorological data must be transformed to a prediction of wind or solar power production. This is accomplished with the use of a facility-scale power output model. The facility power output model represents the relationship between the meteorological variables and the power generation by a wind or solar generator but also implicitly or explicitly accounts for other non-meteorological effects.

Although the basic concept and objective is the same, a number of different strategies can be employed and there are notable differences between the modeling of wind and solar power production. This section provides an overview of the different strategies and the differences between wind and solar power production models.

The most fundamental option is whether to use an explicit or implicit power output model. The implicit approach predicts the power output at the MOS or ECM step by training the statistical model to go directly from predictions of meteorological variables to power production in the MOS or ECM process. In this case the equivalent of a power production power is implicit within the MOS and/or ECM equations. This simplifies the prediction process and also reduces the need for high quality meteorological data (in addition to power production and outage data) from the generation facility. The disadvantage of this approach is that it makes it more difficult to separate the component of the forecast error that is associated with the meteorological predictions from that associated with the power production model. This makes it more challenging to analyze the performance and refine the prediction system. Results indicate that the explicit approach generally yields better forecast performance for facilities that supply high quality meteorological data while the implicit approach may be as good or better when high quality meteorological data is not available.

If the explicit approach is pursued, then the modeling approach and the granularity of the model must be selected. There are two fundamental types of power output models: (1) physical and (2) statistical. The physics-based models attempt to simulate the behavior of the facility based on the physical layout and hardware specifications of the facility. These models use meteorological and operating data as input and the hardware and layout specifications are used along with the physics-based model equations to determine the response of the facility to the inputs. These models have considerable detail and the engineering processes are generally modeled quite well. However, the detail of input data required for these models to perform well is generally more than is typically available. This generally limits their performance in operational forecast production applications.

In most operational application, statistical power output models are employed because they provide better performance. These models are statistical relationships between measured meteorological data and actual power output. Any of the statistical methods previously described can be employed for this purpose.

However, the data may be noisy and in many cases simple models will perform as well or better than more sophisticated machine learning methods. The facility models can be constructed at different levels of granularity. For example, in the case of a wind generating facility, statistical relationships could be constructed for the output of each turbine or for the aggregated output of the facility. The aggregated approach is more typically employed because the data are often not available at higher granularity and even in cases where such data are available the impact of modeling with additional granularity on forecast performance is often minimal.

An example of a simple aggregated facility-scale wind power output model is depicted in Fig. 4.4. The horizontal axis is the hub-height wind speed measured by an anemometer on a meteorological tower located at a representative location on the prevailing upwind side of the facility. The vertical axis is the power output at the interconnection point. Each blue marker represents the average power production for a 1-hr period and the concurrent average hub height anemometer wind speed. The black line is a least squares polynomial fit to the data which represents a one-variable facility scale power output model. In most applications, a multi-variant approach is used with wind speed, wind direction and temperature as predictors.

An analogous example for a facility-scale solar power output model is shown in Fig. 4.5.

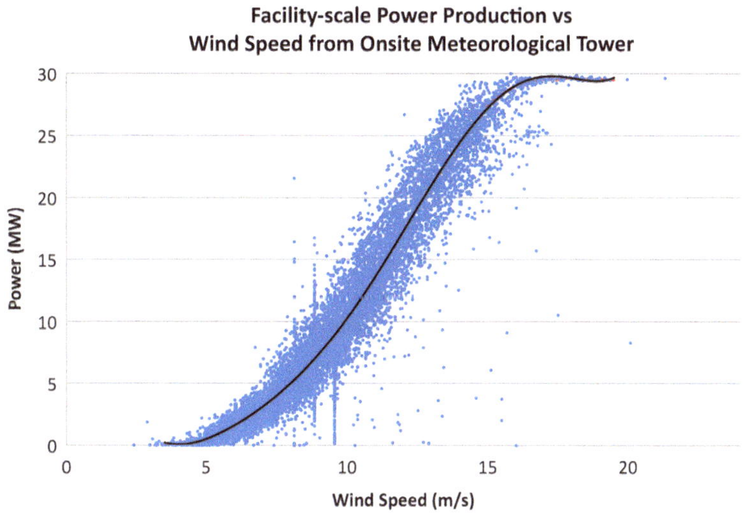

Fig. 4.4 A depiction of a simple statistical facility-scale wind power production model based on the measured plane-of-array solar irradiance (W/m^2) and the measured facility output. The *blue* markers indicate the 1-h average facility power output and the concurrent 1-h point measurement of the hub height wind speed

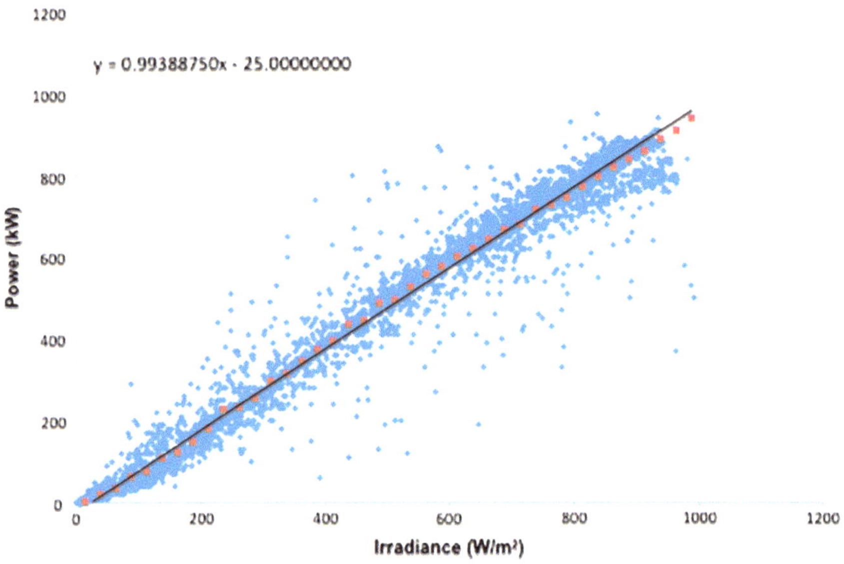

Fig. 4.5 A depiction of a simple statistical facility-scale solar power production model based on the measured plane-of-array solar irradiance (W/m^2) and the measured facility output. The *blue* markers indicate the 1-h average facility power output and the concurrent 1-h point measurement of the plane of array solar irradiance

4.3.4 Integrated Forecast System

A typical state-of-the-art wind and solar forecast is based on a multi-method ensemble approach.

The components and dataflow of such a system is depicted in Fig. 4.6. The top row of this chart depicts the primary forecast models, which ingest data from a variety of sources as discussed in the description of the individual methods. The second row represents MOS methods that mitigate the systematic prediction errors in the primary models. The single red object in the third denotes the ECM functionality in the system.

4.4 Assess: Evaluation of Forecast Performance

A key issue in the application of wind and solar forecasts to the integration of wind and solar generation resources into an electric grid is the expected level of forecast accuracy. A thorough understanding of the attributes of the errors of the power production forecasts being used for a renewable energy integration application provides essential guidance on how to obtain the most value form the forecasts and possibly even how to structure other grid management processes in order to

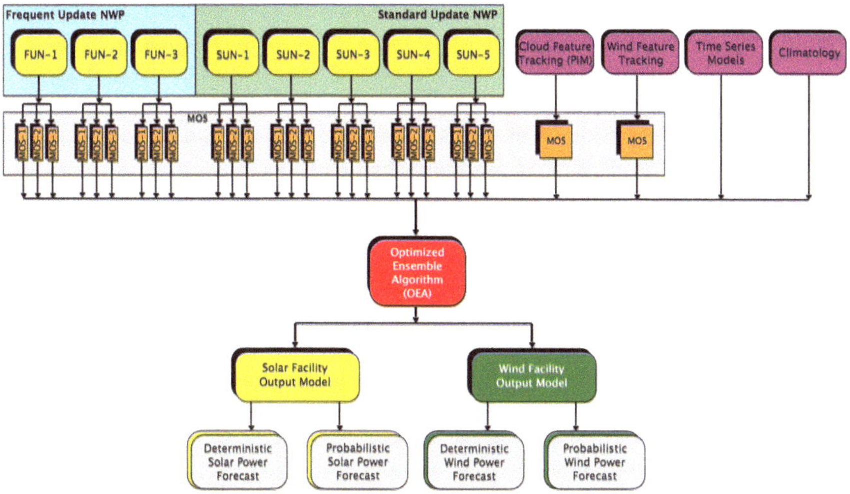

Fig. 4.6 A schematic depiction of the components and data flow of a typical multi-method state-of-the-art wind and solar forecast system

optimize the value provided by the forecast information. However, the attributes of forecast error patterns are typically quite complex, variable and dependent on the objectives and format of the forecast. This often poses a substantial barrier to the productive use of forecast error information.

The most widely employed approach to the assessment of forecast error by users is focused upon the evaluation of the typical error of deterministic forecasts. This is the most basic evaluation concept for the most basic type of forecast.

However, this very basic approach provides a very limited set of information about the forecast error attributes and therefore provides less opportunity to optimize the value obtained from the use of forecast information. This section provides an overview of the metrics and issues associated with the assessment of forecast performance. It is divided into two subsections. The first provides an overview of potentially useful metrics with a brief discussion of the specific information each provides. The second subsection presents a summary of the factors that impact forecast performance and the typical performance levels that are currently achieved by state-of-the-art forecast systems with respect to the most commonly used metrics.

4.4.1 Forecast Performance Metrics

Forecast accuracy can be measured with a wide variety of metrics, which provide different information about the error characteristics of the forecasts. A recent project funded by the US Department of Energy compiled and analyzed a wide range of forecast metrics for the purpose of evaluating solar forecasts. The results of this study are presented in Zhang et al. [17] and provide a very useful reference for forecast users. Most of these metrics are also appropriate for the evaluation of wind power forecasts or predictions of other meteorological variables.

The most appropriate metric for a specific application depends on the characteristics of the application. In principle, the best metric is the one that evaluates the way in which the application is sensitive to forecast error. However, in many applications, the sensitivity to forecast error is not quantitatively known or even the qualitative aspect of it may be somewhat fuzzy. Hence, in many cases, generic widely-used metrics are employed to assess forecast performance. The use of standard widely used metrics facilitates comparison of forecast performance but may not provide the most relevant information about the value of the forecast for a specific application.

It is easy for a user to become overwhelmed with the wide range of complex evaluation concepts and metrics. This reinforces the tendency to employ only basic widely used metrics. It is suggested that the formulation of an evaluation approach for a specific application should consider three factors: (1) quantitative (if known) or qualitative relationships of the sensitivity of the user's application to forecast error, (2) the transparency of the metric to the forecast user and (3) the ability to compare performance to that obtained by other users, which implies the use of at least some commonly used basic metrics.

4.4.1.1 Deterministic Forecasts

As noted previously there are a wide variety of metrics that have been formulated for the evaluation of deterministic forecasts. The most widely used are the mean error (ME) also known as the bias, the Mean Absolute Error (MAE) and the Root Mean Square Error (RMSE). These metrics are commonly used and defined in many publications. The ME provides information about the tendency of the forecasts to be higher or lower than the actual values. A well-calibrated forecast should have a bias near zero over a long sample. However, the ME is frequently observed to be appreciably non-zero for shorter samples and this often indicates a period of atypical conditions. The MAE is obtained by computing the average of the absolute values of the forecast error and thus is a measure of the average magnitude of the forecast error. The RMSE is also a measure of the typical magnitude of the forecast error but it is based on the average of the squares of the forecast errors. A square root operation is applied to the resulting average so the units are consistent with those of the underlying variable. However, the largest errors are weighted more heavily in this process. Thus, RMSE is perhaps a more useful metric when the user has greater sensitivity to large errors than small or average errors. However, one needs to understand the application's sensitivity to forecast error. Many user's think their application is more sensitive to large forecast errors but in fact it is not. Many binary decision-making applications are in this category. Once the error is sufficiently large to cause a wrong decision to be made, it does not matter how much larger the error actually is.

When ME, MAE and RMSE are used to evaluate wind and solar power production forecasts, the values are typically expressed as a percentage of a reference value (often referred to as "normalized") so that forecast performance for facilities or aggregates of different sizes can be compared. The choice of the reference value can have a large impact on the metric values and also one's impression of forecast performance. The two fundamental choices for a reference value are the "average production" and the "generation capacity". An immediate issue is that the "average production" will vary among evaluation periods so the reference value will not be constant when intercomparing periods. The generation capacity is typically more invariant but of course the capacity occasionally does change over time as facilities are expanded or new facilities are added to a regional aggregate. The more significant issue is which approach yields a better representation of the sensitivity of the broadest set of integration applications to forecast error. An example of the monthly MAE for one year of regional wind power production forecasts using each of the two normalization approaches is shown in Fig. 4.7. The two approaches yield very different perspectives on forecast performance. The black columns in this chart represent the capacity factor (the % of installed capacity that was actually produced in each month). The capacity factor peaks in the late spring and early summer (i.e. April–July) and is a minimum in the autumn and winter (October–February). The red columns depict the MAE as a percentage of the capacity. When normalized by the capacity, the monthly MAE ranges from 5 to 12%. The capacity-normalized MAE peaks in the months with the highest capacity factors and the lowest MAE values generally occur in the months with the lowest capacity factors. In contrast the production-normalized MAE values (blue bars) range from 25 to 61%. This is a result of monthly average capacity factors that range from 12 to 35%, which

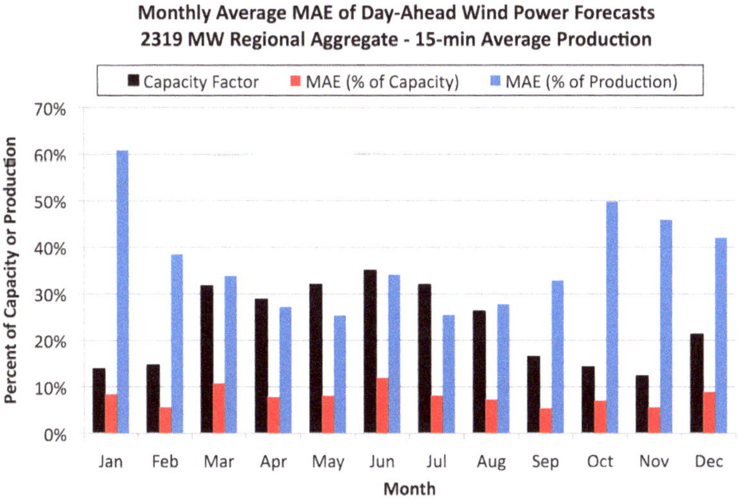

Fig. 4.7 An example of the monthly MAE for system-wide wind power forecasts over a 1-year period

essentially multiplies the capacity-normalized values by factors in the three to six range. In addition, the yearly pattern is substantially different, the production-normalized MAE has its largest values in the autumn and winter (the lowest production season) and its smallest values in the spring and early summer (the highest production period). As previously proposed, the question is which MAE perspective is most relevant to the integration application. A system operator typically has to balance unscheduled generation or demand with an equivalent amount of energy from another source. Thus, the relative error is not a significant factor. For example if a 100 MW wind facility is forecasted to generate 2 MW but it actually generates 1 MW, this is a 100% error relative to the actual production but the system only has to adapt to a shortage of 1 MW. From a capacity-normalized perspective this is only a 1% error, which is a very good forecast. A comparison is often made to the evaluation of electric demand ("load") forecasts, which are almost always expressed as a percentage of the average or actual load. However, the variability of the system-load is much less than that of wind or solar generation so the use of the production-normalization is more acceptable. It is analogous to a situation in which the wind generation only deviates by a modest percentage from its capacity. In reality, the most rigorous and appropriate normalization factor is a measure of the variability. That is, how big is the error relative to the variability of the forecast variable. The variability metric could be the standard deviation or the absolute range of variability.

The normalization of solar power forecasts has a second issue because the effective capacity of a solar facility varies with the time of day. Obviously, the capacity is zero at night since a facility can't generate any power unless there is an extreme astronomical anomaly. A typical approach is to use the installed capacity of the facility or aggregate but to only include the daylight period (i.e. times with significantly non-zero observed solar irradiance) in the calculations. This approach tends to understate the solar power forecast errors relative to those of wind since the capacity may be much lower during a substantial portion of the day (depending on the solar tracking characteristics of a facility)

As noted, the ME, MAE and RMSE, regardless of how they are normalized, are measures of the "typical" error of the forecasts and they do not provide much information about the distribution of forecast errors, which may be quite important to many integration applications.

Zhang et al. [17] proposed a number of additional metrics for the evaluation of deterministic forecasts that are not as widely used in the renewable energy forecasting community. Many of these provide more information about the full error distribution than the basic ME, MAE and RMSE metrics. One of the most straightforward of these additional metrics is the 95th percentile of the forecast errors (i.e. 5% of the errors are larger than this value). This is based on the fact that power system operators typically rely on reserves to manage the anticipated and unanticipated variability in generation and load. These reserves are usually referred to as "operating reserves" and are used to manage variability on the timescales of minutes to multiple hours. High penetration of wind and solar generation can require additional operating reserves. More accurate wind and solar forecasts can benefit the system by reducing the amount of these additional operating reserves. Zhang et al note that that the use of the 95th percentile of forecast errors is a generally accepted method in the power industry for load and other variability forecasts to determine the amount of operating reserves needed. Therefore, the 95th percentile of power forecast errors can serve as an approximation of the amount of reserves that need to be procured to accommodate increasing penetrations of wind and solar generation. Thus, improvements to the forecasting process which achieve a lower value of this metric (i.e. reduction in the size of the largest forecast errors) might be expected to translate into lower costs of operating reserves. This is an example of a metric that attempts to model an application's sensitivity to forecast error.

4.4.1.2 Probabilistic Forecasts

Probabilistic forecasts require a different set of evaluation concepts and metrics than deterministic forecasts since they do not attempt to make explicit prediction of the exact value of the forecast variable. They embody both predictive and uncertainty information and should be evaluated from that perspective.

There are three key concepts that should be considered in the evaluation of probabilistic forecasts: (1) reliability, (2) sharpness and (3) resolution. The best probability forecast will exhibit good performance for all three of these attributes.

Reliability refers to the relationship between the forecasted probability and the frequency of the actual outcome. The most reliable forecast will result in an outcome frequency that matches the forecasted probability. That is, an outcome that has a probability of 70% will happen 70% of the time. The most widely employed method to assess the reliability of probabilistic forecasts is a chart commonly referred to as a "reliability diagram". An example of a reliability diagram for probabilistic wind ramp rate forecasts in Texas is shown in Fig. 4.8. The horizontal axis represents bins of forecasted probabilities. In this case there are ten bins with each bin having a width of 10% (i.e. 0–10%, 10–20%, etc.) and labeled with the midpoint value. The vertical axis represents the observed outcome frequency for each bin. The red and blue lines depict two different forecasts. The black line denotes a perfectly reliable forecast. The forecast that most closely follows the black line is the most reliable forecast. In this case, both forecasts are fairly reliable but the forecast represented by the red line follows the black line more closely and hence can be considered to be more reliable.

A more quantitative approach to measuring the reliability is through the use of the reliability score (RS):

$$RS = \frac{1}{N} \sum_{k=1}^{K} n_k (f_k - \bar{o}_k)^2$$

where N is the number of forecasts, K is the number of forecasted probability bins (10 in the example shown in Fig. 4.8), n is the number of forecasts in bin k, f is the forecasted probability of bin k (values on the horizontal axis of Fig. 4.8) and o-bar

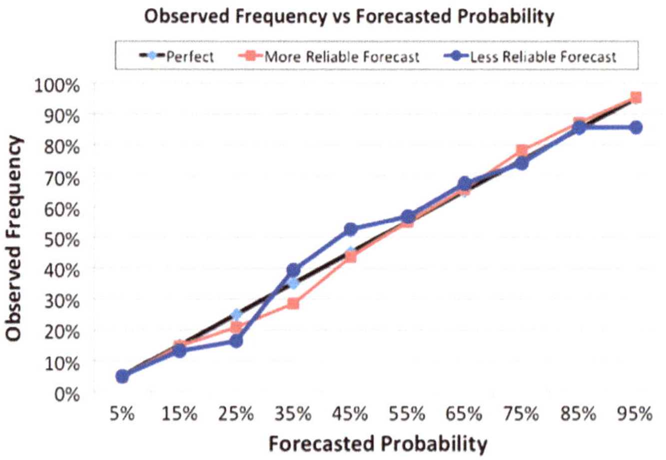

Fig. 4.8 An example of a "reliability chart", which depicts the forecasted probability on the *horizontal axis* and the corresponding observed frequency of the outcomes on the *vertical axis*

is the observed frequency of bin k. The RS essentially measures the sum of the squared deviations from the black line in Fig. 4.8. Since the RS is a measure of the deviation from perfection, lower values indicate a more reliable forecast. In the case of Fig. 4.8, the red forecast has an RS value of 0.008 and the blue forecast has an RS of 0.025 which indicates the red forecasts is more reliable than the blue forecast as indicated by a visual inspection of the reliability diagram.

However, reliability is only one of the key attributes of a probabilistic forecast. After all, a long-term climatology of a variable such as wind or solar power production will be reliable (unless there is rapid climate change). However, such forecast will not perform well with respect to the other two key attributes: sharpness and resolution. Sharpness refers to the tendency to forecast probabilities near 100% or 0%, or from another perspective it can be viewed as the width of a probability distribution. Obviously, such a probabilistic forecast is desirable since there is less uncertainty but only if it is reliable. Thus, it is not useful to compromise on reliability to obtain a sharper forecast (Fig. 4.9).

Resolution refers to the ability to discriminate among situations with characteristically different probability distributions. A climatological forecast has no resolution since every forecast has the same probability distribution (Fig. 4.10).

4.4.2 State-of-the-Art Performance Benchmarks

Two of the most commonly asked questions by existing or potential users of wind and solar power production forecasts are: (1) What are appropriate expectations for the accuracy of forecasts for my application? and (2) Why is the performance of the

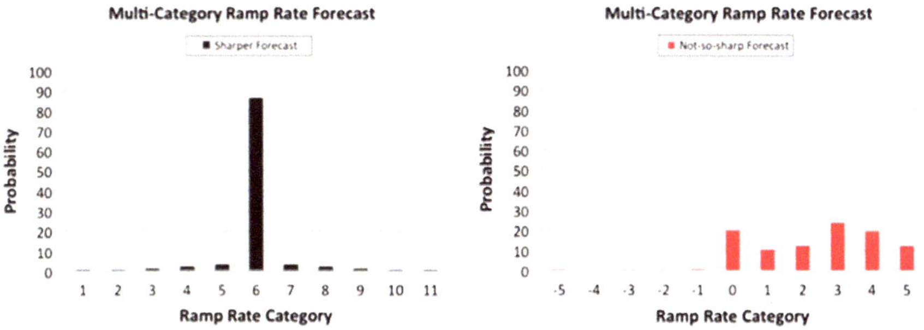

Fig. 4.9 An example of a sharp (*left*) and not-so-sharp (*right*) multi-category wind ramp rate forecast. The horizontal axis is the ramp rate bin and the vertical axis is the probability that the actual ramp rate will be in that bin

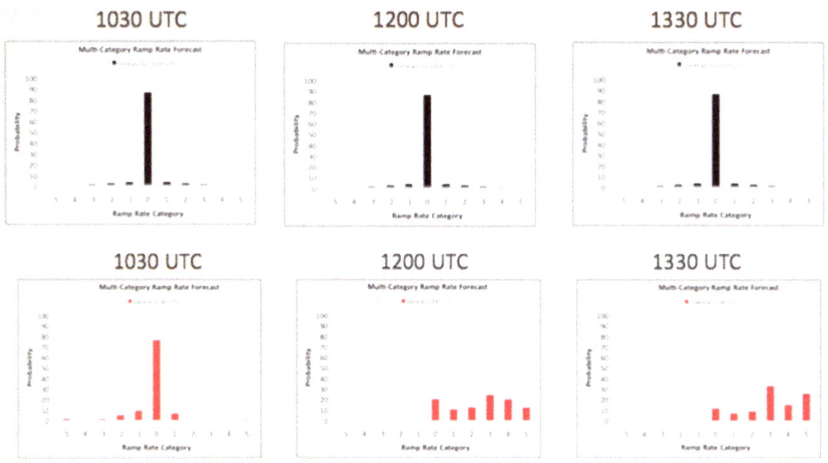

Fig. 4.10 Two multi-category wind ramp forecasts issued at three consecutive times that are 90 min apart. The forecast depicted by the *top row* of charts (*black columns*) has no resolution because each forecast has the same probability distribution. The forecast depicted by the *bottom row* of charts (*red columns*) has substantial resolution because the forecasted probability distribution changes substantially in time

forecasts that I use significantly different from those reported by others? These are two very difficult questions to precisely address because there are a wide variety of factors that impact forecast performance. There have been several recent attempts to develop methods to provide baseline performance benchmarks for state-of-the-art wind and solar power forecasts [16] but each attempt has raised a number of issues that are difficult to address. There are many ways to analyze the causes of variations in forecast performance but it generally comes down to five primary factors: (1) the amount of variability in the forecast parameter, (2) the spatial and temporal scales of the variability, (3) the look-ahead period, (4) the quality of the input data to the forecast process and (5) the skill of the forecast methods in forecasting resource variability at the target sites. The relative importance of these factors vary among forecast target entities as well as among evaluation periods for a specific forecast target entity.

The many factors that drive variations in forecast performance make it difficult to compare performance among applications. Therefore, it is difficult to provide a simple answer to the question of what level of forecast performance should expected for a specific application. One useful approach to provide guidance to address this issue, is to compile performance results for a specific look-ahead period by the size (capacity) of the forecast target entity. The size of the forecast target entity is one of the primary factors that control the amount of production variability and hence forecast difficulty. Smaller target entities tend to be more variable and therefore have larger forecast errors.

The annual MAE (as a % of capacity) for recent (2013–2014) day-ahead wind power forecasts for a range of forecast target entities is shown in Fig. 4.11. An analogous chart for the annual MAE of solar power forecasts is shown in Fig. 4.12. Only the daylight hours are included in the solar power MAE data. It is clear that although there is a substantial amount of variability the annual MAE has a substantial dependence on the size of the forecast target entity. Day-ahead forecasts for individual facilities (target entities smaller than about 500 MW) have MAE values mostly between 11% and 18%. Hence, the rule of thumb of about 15% is often used for those desiring a single number. The MAE for day-ahead solar power forecasts has a similar pattern with similar magnitudes. However, one must keep in mind that this is for the entire daylight part of the day and that the magnitude of the MAE near the time of peak solar irradiance (~noon) is somewhat larger. The wind MAE does not have as prominent of a diurnal cycle although there is certainly some tendency to have higher values at the times of the day when the wind speeds are higher.

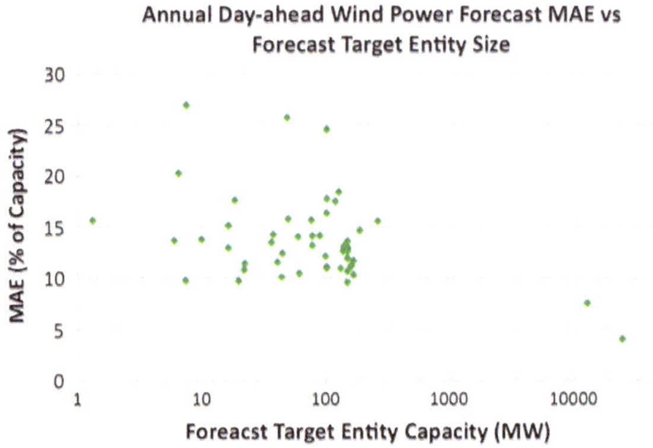

Fig. 4.11 The relationship of day-ahead wind power production forecast MAE (% of capacity) to the capacity of the forecast target entity. Each *green* marker represents the MAE for an individual forecast target entity over a 1-year period

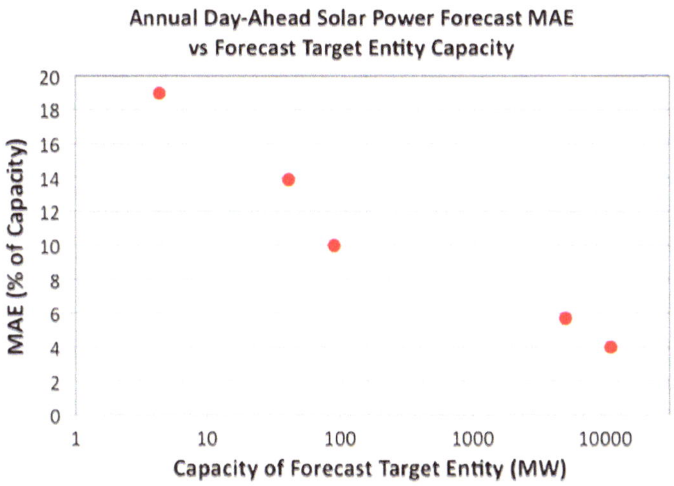

Fig. 4.12 The relationship of day-ahead solar power production forecast MAE (% of capacity) to the capacity of the forecast target entity. Each *red* marker represents the MAE for an individual forecast target entity over a 1-year period

4.5 Communicate: Inform the User for Decision-Making

The final component of the process that facilitates the optimization of value from wind and solar forecasting for grid integration applications is the effective and efficient communication of the forecast information to the end user. This communication is a two-way information exchange between the forecast service provider and the end user. Since the end users are involved in the application of the forecast service on the daily base, they have a first-hand knowledge about the forecast performance, and a better understanding of how the end user applications are sensitive to the forecast errors. The end user's feedback on the performance will provide guidance for improvements to the forecast service. This feedback is critical since it communicates not only a level of expected performance from the end user but also the circumstances in which the end user sees more value for a particular type of improvement. In this case, the performance metrics tied to particular events or weather conditions are preferred during the communication. The end user's applications could also vary from one user to another and change over the time, which requires different performance or features for the forecast service. This should be factored into the development or improvement of the forecast service.

The end user, if owning or operating generation resources, should play an important role in closing the gap in communication between generation resources and the forecast service provider. One of these roles is to ensure that consistently good quality telemetered data is provided from the resources to the service providers.

The job of the service provider is to deliver the best forecast service at a reasonable cost to the end users so that their business needs are met. To understand the end user's business need is prerequisite to the success of such a forecast service. The service provider is responsible for communicating advantages and disadvantages of the different options for the improvement to the service so that the end user can decide which option best serves the user's application. As users gain more experiences with the forecast service, they can learn from the forecast service provider about how the meteorological conditions drive the change in the power generated by wind or solar generation resources. This qualitative information complements the numerical values of the forecast, and helps the end user to improve the situational awareness by quantifying the uncertainties of the forecast services in real-time to some extent. The service provider also needs to help the end users to understand the limit of the technologies, so that the end user can look at alternatives to hedge against the large forecast errors which may be encountered. A good communication between two parties also requires dedicated resources and staffs to address issues on each side of the forecast service relationship.

4.6 Conclusions

The increasing penetration of variable renewable generation resources into the power grid poses a great challenge in managing the variability and uncertainties resulting from these resources. It has been widely recognized that wind and solar power forecast is one of the most cost effective and easily implemented tools to assist in the management of the variability and uncertainties. While wind and solar power forecasting is evolving and gradually being integrated into the control room, the fundamental principles to guide how a forecasting solution should be optimized in order to provide maximum value to the end user are applicable worldwide and remain the same for different end user applications. It essentially consists of four components, which are Sense, Model, Assess and Communicate. The combination of four components can be expressed in the acronym SMAC. This chapter discusses in depth the concept, significance, and connectivity of these four components. First, high quality and representative measurement data are a critical input into a state-of-the-art forecasting procedure. Second, skillful forecasting techniques or models are essential for mapping the complex relationship between the measurement data and the predicted power generation. Third, meaningful assessment of forecast performance will enable users to be more confident in using the forecast information for decision-making and to track the performance of the forecast to provide guidance for further improvement. Finally, the effective and efficient communication between the service provider and the user is essential to optimally customize the forecast content and format to the application and identify where improvements should be made to increase the value of the forecast service in a cost-effective manner.

References

1. M. Ahlstrom, et al., Knowledge is power. IEEE Power Energy Mag **11**, 45–52 (2013)
2. J. Coffier, *Fundamentals of Numerical Weather Prediction*, reprint edition (Cambridge University Press, 2012), 368 pp. ISBN-10: 110700103X, ISBN-13: 978-1107001039
3. C. Cortes, V. Vapnik, Support-vector networks. Mach. Learn. **20**(3), 273 (1995)
4. A. Decaria, G. van Knowe, *A First Course in Atmospheric Numerical Modeling* (Sundog Publishing, 2013)
5. L. Delle Monache, F. Eckel, D.L. Rife, B. Nagarajan, K. Searight. Probabilistic weather prediction with an analog ensemble. Mon. Wea. Rev. **141**, 3498–3516 (2013). doi:10.1175/MWR-D-12-00281.1
6. H. Drucker, C.J.C. Burges, L. Kaufman, A.J. Smola, V.N. Vapnik, Support vector regression machines, in *Advances in Neural Information Processing Systems 9, NIPS 1996*, (MIT Press, 1997), pp. 155–161
7. H.R. Glahn, D.A. Lowry, The use of model output statistics (MOS) in objective weather forecasting. J. Appl. Meteor **11**, 1203–1211 (1972)
8. T.N. Krishnamurti, L. Bounoua, *An Introduction to Numerical Weather Prediction Techniques* (CRC Press, 1995)
9. E.J. Natenberg, J. Zack, S. Young, R. Torn, J. Manobianco, C. Kamath, Application of ensemble sensitivity analysis to observational targeting for wind power forecasting, in *14th Symposium on Integrated Observing and Assimilation Systems for the Atmosphere, Oceans, and Land Surface (IOAS-AOLS), AMS Annual Meeting, Atlanta* (2010)
10. E.J. Natenberg, J. Zack, S. Young, J. Manobianco, C. Kamath, A new approach using targeted observations to improve short term wind power forecasts, in *AWEA WindPower 2010 Conference and Exhibition, Dallas, TX* (2010)
11. A. Tuohy, J. Zack, S. Haupt, J. Sharp, M. Ahlstrom, S. Dise, E. Grimit, C. Mohrlen, M. Lange, M. Casado, J. Black, M. Marquis, C. Collier, Solar forecasting: methods, challenges, and performance. IEEE Power Energy Mag **13**(6), 50–59 (2015). doi:10.1109/MPE.2015.2461351
12. S. Vannitsem, Dynamical properties of MOS forecasts: analysis of the ECMWF operational forecasting system. Wea. Forecasting **23**, 1032–1043 (2008)
13. J. Zack, E.J. Natenberg, S. Young, G.V. Knowe, K. Waight, J. Manobianco, C. Kamath, Application of ensemble sensitivity analysis to observation targeting for short term wind speed forecasting in the Tehachapi region winter season, LLNL Technical Report LLNL-TR-460956 (2010)
14. J. Zack, E.J. Natenberg, S. Young, G.V. Knowe, K. Waight, J. Manobianco, C. Kamath, Application of ensemble sensitivity analysis to observation targeting for short term wind speed forecasting in the Washington-Oregon region, LLNL Technical Report LLNL-TR-458086 (2010)
15. J. Zack, E.J. Natenberg, S. Young, J. Manobianco, C. Kamath, Application of ensemble sensitivity analysis to observational targeting for short term wind speed forecasting, LLNL Technical Report LLNL-TR-424442 (2010)
16. J. Zhang, B.-M. Hodge, S. Lu, F.H. Hamann, B. Lehman, J. Simmons, E. Campos, V. Banunarayanan, Baseline and target values for PV forecasts: toward improved solar power forecasting, in *Conference Paper NREL/CP-5D00-63876, Presented at the IEEE Power and Energy Society General Meeting, Denver, Colorado, 26–30 July 2015* (2015). http://www.nrel.gov/research/publications.html
17. J. Zhang, Anthony Florita, Bri-Mathias Hodge, Lu Siyuan, Hendrik F. Hamann, Venkat Banunarayanan, Anna M. Brockway, A suite of metrics for assessing the performance of solar power forecasting. Sol. Energy **111**, 157–175 (2015)

Chapter 5
Reserve Estimation in Renewable Integration Studies

Brady Stoll, Rishabh Jain, Carlo Brancucci Martinez-Anido, Eduardo Ibanez, Anthony Florita and Bri-Mathias Hodge

In 2011, the International Energy Agency (IEA) reported that electricity from renewable sources accounted for 19.3% of the total world energy generation [1]. This share is expected to grow to 46% by 2040 [2], and renewable generation in some markets, such as in California, will reach 50% by 2030 [3]. While hydropower is currently the largest renewable generation source, the share of variable renewable technologies (such as wind, and solar) is growing rapidly. These variable technologies have recently experienced significant cost reductions [4] and favorable policies [3, 5, 6], which indicate a likely acceleration of the penetration of variable generation. As variable renewable generation increases on the grid it becomes important to adequately capture the impact of these generation sources on bulk power system operations. Several renewable integration studies have analyzed these impacts using production cost models, which simulate the scheduling process of power system operations. Correctly representing reserves in these studies is one important facet of adequately capturing the implications of integrating these resources while maintaining grid economics and reliability. This topic of how to correctly handle reserves in grid integration studies is the primary topic of this chapter.

5.1 Need for Renewable Integration Studies

Variable renewable technologies are now emerging as cost competitive and sustainable power generation sources for the grid. Growing concerns regarding the emission of carbon and other pollutants have also influenced policies, which are

B. Stoll · R. Jain · C. Brancucci Martinez-Anido · A. Florita · B.-M. Hodge (✉)
NREL, Golden, CO, USA
e-mail: Bri.Mathias.Hodge@nrel.gov

E. Ibanez
GE Energy Consulting, Schenectady, NY 12306, USA

P. Du et al. (eds.), *Integration of Large-Scale Renewable Energy into Bulk Power Systems*, Power Electronics and Power Systems,
DOI 10.1007/978-3-319-55581-2_5

now placing greater restrictions on emissions and requiring renewable generation to be installed [6, 7]. Additionally, decreasing costs of solar and wind are leading to economic installations of these resources [8, 9]. These policies and cost reductions have led to tremendous growth in installations of both solar and wind in recent years, with expectations for this growth to continue.

As variable generation continues to increase, there are growing concerns for the ability to adequately incorporate these resources into the electric grid while maintaining grid economics and reliability. Renewable integration studies present means of analyzing the future needs of the electricity sector when penetrations of variable renewable generation increase from current levels. In some cases, a need to incorporate high penetrations of renewable generation may require significant changes to the transmission network, operational strategies, and/or policies. It is important to be able to adequately plan for these challenges in order to not disrupt future operations.

Studies of this nature have been performed on a large scale since 2010. Initial renewable integration studies focused on transmission expansion required for increases in wind generation [10], and have since began to focus on either wind and solar [11–13], or with recent policy requirements, solar alone [14]. These studies have also been performed for a wide range of system sizes, from the small island grids of Hawaii [15, 16] to the entire Eastern Interconnection [13] and the European Grid [17].

There are many challenges associated with renewable integration studies, in large part due to the modeling complications of studying such large and integrated systems with many constraints on their operation. An important aspect of these models is their representation of reserves. The rules for reserves provision vary widely throughout the United States, however all balancing areas must provide enough reserves to ensure grid reliability in the event of system outages, and to compensate for variability in renewable generation. Variability and uncertainty associated with renewable generation leads to additional reserve needs as changes in variable generation output could lead to stress on the electric grid. Regulation reserves can be held to compensate for some of these instances, and some regions are beginning to require specific reserve products to cover the variability associated with wind and solar [18]. Correctly modeling these ancillary services, in addition to the normal unit commitment and dispatch process, is important to understanding how renewable generation will impact grid operations.

5.2 Need for Modeling Reserves Correctly

In many modern power systems, electricity is being traded as a commodity in both day-ahead markets and spot markets. Auctions include energy, capacity, and reserves. System operators receive bids to reserve capacity for both anticipated and unforeseen variability and uncertainty in the system generation and loads. Costs of reserves depend on their capacity and type (typically based on their response time).

Over-estimation of resource needs will lead to an increase in system operating costs. On the other hand, under-estimation can cause violation of reserve regulations, system instability, and, in extreme situations, cascading failures.

These requirements influence the ability of real systems to meet dispatch needs and influence the commitment status of generators, as the need to provide reserves may require that a generator start or not shut down when it might otherwise. For example, on a very windy night wind generation may be able to meet all demand, however the need to provide reserves may require that a natural gas plant remain online and some of the wind energy would be curtailed so that the thermal plant could remain at its minimum generation level.

Just as it is important to recognize the challenges associated with holding reserves in real systems, it is equally important to model them correctly. Important aspects of reserve modeling include the total amount of reserves held, the ramping requirements for each reserve product, and the response times of these reserves. Holding too few reserves can lead to incorrect conclusions regarding system stability and operational impacts, where the model would underestimate the ability of a particular system with increased renewables to handle system stress. On the other hand, holding too many reserves could lead to an overestimation of total production costs, leading to concerns regarding the economic viability of increased renewables, even when these concerns may not be realistic.

Additionally, correctly modeling reserves helps identify the adequacy of existing operational practices and infrastructural capabilities for future growth in variable generation. Models with accurate reserve requirements can identify operational areas that anticipate significant increases in expenses and can initiate discussions into alternate technologies or operational practices to increase efficiency and reliability without significantly increasing investments.

In summary, without the correct modeling of reserves the results of a production cost model focusing on renewable integration would not be able to adequately predict the operations of the electricity grid in high penetration cases, nor would it be able to correctly identify the costs associated with commitment and dispatch. As such, the results from such inadequate models would not be suitable predictors of the ability of a system to further integrate renewables.

5.3 Challenges with Renewable Integration

The increasing share of solar and wind across the globe is magnifying operational challenges for power systems related to variable generation. These challenges are predominantly related to an increase in the system variability and uncertainty due to the addition of variable renewable generation.

Variability refers to the fluctuation of solar and wind generation or demand at different time scales, e.g., across seconds, minutes, hours, days, or seasons that may be forecasted but is not able to be controlled. One example of variability is the rising and setting of the sun on a perfectly clear day; the amount of sun will vary throughout the day, but in a generally predictable manner. Uncertainty refers to the predictability of both load and generation from wind or solar plants, and in particular the forecasting errors associated with predictions hours or days ahead of operation. Output from wind and solar generators is largely driven by weather patterns, which are difficult to forecast exactly even hours ahead, leading to some uncertainty in the actual output of wind and solar facilities at a given future time.

Both variability and uncertainty are known phenomena for power system operators, as load also has these qualities. One widely used strategy to mitigate variability and uncertainty is the use of operational reserves, which are also used to account for potential transmission and generator outages across the system. The most common types of reserves are contingency (spinning and non-spinning) and regulation reserves. Occasionally, systems use a load-following or flexibility reserve [19]. Each reserve type is maintained to counteract variability or uncertainty of a particular type at a particular timescale. In the presence of variable renewable generation, the variability and uncertainty are also becoming a greater function of supply. Therefore, reserves now need to account for changes in demand and supply, as well as equipment failures, for time scales from seconds to several minutes. Increasing penetrations of wind and solar will continue to increase the fraction of variability in supply, thus increasing the need for reserves.

Variability and uncertainty can be broadly classified into categories based on three time scales:

1. Immediate—impact the stability of the system in real time (order of seconds to minutes).
2. Operational—impact the balancing of the system over an average period of operation (minutes to days).
3. Long Term—maintain the reliability and adequacy of the resources over a period of months to years. System loads and generation may have seasonal variation. Therefore, the ability of system to meet the peak demand has traditionally been the chief concern. However, with increased levels of wind and solar other time periods (such as low load/high wind seasons) could also present planning issues, such as the need to secure more ramping capability.

These time scales influence the reserve products needed to ensure grid stability. In particular, the immediate and operational time periods are most impacted by increasing renewables in renewable integration studies. Long term variability and uncertainty are important considerations, but tend to influence capacity expansion models to a higher degree than production cost models, which assume system adequacy requirements have been met.

The degree of challenges associated with variability and uncertainty differs based on a system's size and location. The PJM independent system operator (ISO), for example, foresees a stronger impact on regulation [20]. Geographical diversity helps reduce the total uncertainty of solar photovoltaics (PV) and wind, as the forecast errors tend to be relatively uncorrelated at relevant timeframes and thus can reduce total uncertainty through aggregation. In PJM's case, optimizing the estimate for longer-term reserves is a greater challenge [20]. On the other side of the spectrum, islanded systems like the Hawaiian Islands or smaller countries must be more rigorous in short term and immediate reserve requirements due to a lack of geographic diversity.

These challenges can have many aspects, some technology specific (e.g. unavailability at night for PV or predictability of wind) while others are general (viz. scale of integration). The rest of this section elaborates the technology specific and common challenges of wind and PV.

5.3.1 Challenges with Wind Integration

Wind turbines are typically designed to be operated primarily between three parameters, (1) Cut-in speed, (2) Rated speed and (3) Cut-out speed. Usually for the region between rated and cut out the turbine regulates its aerodynamics to regulate the output. This means that for higher wind speeds, the turbines will adjust their aerodynamics to discard a portion of the possible wind energy and limit generation to rated capacity. If the wind speed exceeds the cut out speed, the turbines will stall by design. However, the control loop takes a few seconds to regulate this, and with the wind speed changing continuously, the output from the turbine can continue to change. Additionally, turbulence can lead to uncertainty in the output.

Variability in wind speeds leads to power output changes from wind turbines. This variation can be on time scales of minutes to seconds, and is forecastable with some accuracy. Even fast generation changes due to wind may not be completely unexpected, as wind power forecasts can provide times when large ramps are likely. Additionally, longer-term variability on the scale of minutes to hours may be more generally forecastable, but must still be accounted for in terms of generator output. This variability can lead to a higher need for load-following or flexibility reserves [21].

Wind gusts can also cause sudden spikes in wind turbine output. This introduces a need for additional ramp down reserves/alternatives, to offset for this potentially excess power. Sudden drops in wind speeds, on the other hand, lead to a net deficit in the supply. Utilities have to manage this using reserves or local storage to offset the gap, or active load management programs to curtail demand.

Wind turbines can also provide reserves themselves, albeit to a limited degree. Turbines generally do not provide contingency reserves, but can provide both up and down regulation reserves. Regulation reserves in the upwards direction can be provided by de-rating the turbine capacity, which allows for increases in output when regulation is required [22]. Wind turbines also have a small amount of inertia through their rotating blades that can be used to provide frequency response [23]. Additionally, down reserves may be provided by wind power facilities via curtailment. While not desirable, the opportunity cost of not providing energy could be partially met through reserves payments.

5.3.2 Challenges with Solar Integration

Challenges associated with solar integration largely relate to clouds and weather patterns. Solar power is unique in that it has a definite maximum profile that may be calculated for any location and point in time based on the position of the sun in the sky. This maximum limit is commonly referred to as clear-sky power, i.e., power in the absence of clouds. However, there is still a significant amount of uncertainty that may be associated with reductions from this maximum output. Solar output is typically reduced due to the presence of clouds or aerosols in the atmosphere that reduce the amount of incoming radiation that is able to reach the Earth's surface and thus solar PV panels or mirrors associated with concentrating solar power. While cloud cover may be forecasted with some degree of accuracy, the exact motion and thickness of clouds can be difficult to predict, and can greatly impact the output from solar facilities.

It is interesting to note that PV and wind do not tend to have coincidental peaks or dips. Several studies have shown that the forecast errors for PV and wind are anti-correlated [24, 25]. This anti-correlation helps to reduce the overall requirements when both technologies are installed. However, local geographic resources may not favor both wind and PV at the same time. Solar power facilities can provide reserves themselves largely through reduction in output via curtailment [26].

5.3.3 Centralized and Distributed Renewable Penetration

Conventional dispatchable generators are typically installed by utilities and their size usually influences the reserves that need to be carried by the system. Typically, contingency reserves are held to cover unexpected-but-credible events, such as the loss of the largest generator or the tripping of one or more transmission lines.

Unlike traditional generators, wind and PV have a more modular nature given that their basic building blocks are the turbine and the solar panel, respectively. This makes it possible to deploy these technologies in a centralized or distributed fashion. A wind or solar plant would be an example of the former, while rooftop solar is an example of the latter. This brings up two possible scenarios for scaling of renewable generation, which have different impacts on the reserves carried by a system:

1. Multi-MW-scale Centralized Generation: Chiefly utilities and large generation owner/operators will install the generation at this scale. Interconnection requirements for sensing and communications infrastructure makes the generation at these plants directly (or at least more easily) controllable by the system

operator. Wind and solar plants are capable of providing regulation, frequency, and secondary reserves in both the upward and downward directions, though upward reserves are still relatively rare from an operational perspective. With centralized generation, the grid could potentially lose a larger capacity at once due to lack of diversity in weather patterns, hence centralized generation may impose slightly higher reserve requirements than distributed generation.

2. Small-scale Distributed Generation: These generating units are mostly privately owned or by smaller investors (although utilities are starting to consider entering this space). The generation is typically deployed at a load site in the distribution system, which improves the geographical diversity of the resource and causes the variability in generation due to weather patterns to also be distributed. For instance, clouds are unlikely to start covering an entire region at exactly the same time over a broader geographic region. This may reduce the need for reserves, when compared to centralized generation. On the downside, operators currently often have little or no have visibility or control for these distributed generators and thus the resource may not be able to provide reserves and respond to the system under stress.

5.4 Existing Operational Practices for Reserves

Currently, the definition of reserve products and how they are provisioned varies greatly across electricity systems, even in terms of similar reserve products. While the rules and requirements vary between regions, the ultimate goal is to regulate the imbalance between load and generation in a reliable and economic way. The terminology used to refer to these reserves also varies between regions and countries. This section attempts to provide an overview of the different methodologies and classifications used by most system operators. For consistency, the definitions are presented with reference to the North American Electric Reliability Corporation's (NERC) classification of reserves [27].

Reserves are often classified into primary, secondary, and tertiary reserves, depending on their role and response rate. Primary reserves (in any category) are the reserves called first when the operator needs them and may even be deployed automatically. These are usually spinning reserves (SR) and have very fast start-up times. Secondary reserves are backup resources and/or usually have slower response times. The reserves in these categories may be non-spinning reserves or demand-side reserves. Tertiary reserves are even slower reserves and are called to restore faster reserves after they have been deployed, and often consist of non-spinning reserves.

5.4.1 Contingency Reserves

NERC has clearly laid out in its reliability standards that balancing authorities (BAs) must ensure that their system is immune to the single worst case contingency, in BAL-002, typically transmission line or generator failures. In addition, they should be able to restore the balance within 15 min of this contingency. In the United States, following these requirements from NERC, regional ISOs have established restrictions in their operating regions to adhere to the NERC requirements. Often the contingency reserve requirement is based on at least the single largest credible contingency in the system, and may be accompanied by secondary and tertiary reserves.

Contingency reserves typically consist of a mixture of spinning and non-spinning reserves. Sometimes, ISOs may opt to procure additional reserves by demand side response schemes, or cancelling scheduled energy sales.

5.4.2 Normal Condition Reserves

Regulation reserves are fast-acting reserves with a response time from seconds to a few minutes. They are required to avoid deviations from the nominal system frequency (typically 50 Hz or 60 Hz) due to second-to-second and minute-to-minute variations in load and generation. These reserves are often provided by generators with available headroom that follow automatic generation control (AGC) signals to correct area control error (ACE), i.e., energy imbalance.

Load following or flexibility reserves are slower than regulation reserves, i.e., they are deployed over longer periods of time. These reserves help the system manage the demand changes in sub-hourly periods. They are often referred to as flexibility reserves, because they provide the operator with some flexibility between the generation and demand for a short interval. Recent or proposed implementation of this type of reserves by the Midcontinent ISO or the California ISO have an emphasis on improving system economics, rather than reliability. Flexibility reserves in both ISOs seek the reduction in real-time energy spikes due to insufficient ramping, by committing additional resources than can respond to ramps in load, variable generation, and interties over minutes or hours.

5.4.3 Differences in Reserve Classification Terminologies

Different ISOs and countries have their own methods of classifying reserves. NERC notes many of these differences in [27]. Table 5.1 presents the classifications used by the ISOs within the U.S. as well as those from select countries.

Table 5.1 Comparison chart: reserve classification in different ISOs/countries chart adapted from Table 5.2 of [27]

	Frequency Regulation: Primary	Frequency Regulation: Secondary	Frequency Regulation: Tertiary
PJM	Frequency Response	Operating Reserves: Regulation	Operating Reserves: Spinning; Quick-start / 30-min Reserves
CAISO	Spinning Reserve	Operating Reserves: Regulation; Spinning	Operating Reserves: Contingency; Spinning; Non-Spinning / Replacement + Supplements
VDN, Germany	Primary	Secondary	Minutes / Hours + Emergency
RTE, France; Belgium	Frequency Containment Reserve	Automatic Frequency Restoration Reserve	Tertiary: Manual Frequency Restoration Reserve / Replacement Reserve
Spain, Netherlands,	Primary	Secondary	Tertiary
Great Britain	Frequency Reserve, Disturbance Reserve	Does not Exist	Operating: Regulating; Standing / Contingency: Fast-start; Warming; Hot standby
Australia	Contingency Reserve (fast/slow/delayed)	Regulating Service, and Network Loading Control	Short Term Capacity Reserve
New Zealand	Instantaneous Reserve Fast/Sustained Over-frequency	Frequency Regulating Reserve	Not named separately
Canada (NPCC)	10-minute Reserves: Synchronized	10-minute Reserves: Non-synchronized	30-minute Reserve / Inter-BA Reserve
ERCOT, NYISO	Regulation Reserve	10-min Reserve: Spinning; DSR	30-minute Reserve: Non Spinning; Spinning; DSR
CAISO	Frequency Response	Regulating Reserve: Reg Up; Reg Down	Tertiary: Spinning; Non-Spinning
ENTSO-E	Operational Reserves: Frequency Containment Reserve	Operational Reserves: Frequency Restoration Reserve	Operational Reserves: Frequency Restoration Reserve / Replacement Reserve

5.5 Reserve Estimation Methods by System Operators

The previous discussion highlights the diversity in definitions, and reasons why different usage profiles, generation fleets, and regulation goals lead to different practices. Table 5.2 presents a comparative chart between the reserve estimation practices by system operators. Note that, this diversity will also be reflected in their reserve implementation methods while conducting their renewable integration studies.

Table 5.2 Reserve practices of system operators

	Regulating Reserves	Operating Reserves	Contingency reserves	Comments
IESO, Ontario	Contracted service; offered by AGCs; Min limit of ±100 MW at all times. Overall minimum ramp rate requirement is 50 MW/min	3 Classes: (1) 10-min SR, (2) 10-min NSR, (3) 30-min OR; 10-min reserves based on SLC; SR >=25%; 30-min OR based on SLC +50% SeLC		
NY-ISO	Determined hourly; dispatched every 6 s through BA's energy system	10-min OR >= SLC; 50% <= unused SR; 30-min OR = 50% of 10-min OR		10-min SR—unused capacity or resource available via PHSU, or cancelling energy sales
MISO	300–500 MW maintained in each direction depending on the load and time of the day	Currently at 2000 MW, based on SLC. 50% = SR; other half = NSR + DSR		
HELCO	Islands operate autonomously; all imbalances presented as frequency unbalance errors; Requirements are hourly determined, based on system frequency, past hour wind generation, and forecasts. Only increased when generation is around mid-point of capacity			Up-regulation has increased, down-regulation has decreased with renewable integration
BPA	(actual regulation – 10-min average)	Imbalance = LFPS – LFES; Imbalance is used to determine reserves with 95% confidence		LFPS = (10-min Average – perfect schedule); LFES = (10-min Average – estimated schedule)
CAISO	Based on the projected worst 10-min ramp rate; 20-min Intertie schedule changes, 20 min of self-scheduled generation, and actual system demand variations are the 3 primary factors	5% of generation from hydro resources, plus 7% of generation from other resources, plus 100% of any Interruptible Imports, or the single largest contingency (if the latter is greater). Half must be spinning, half non-spinning		
ERCOT	Deployed every 4 s; Input data include historical 5-min net load (Load–Wind) changes, Installed capacity, Recent CPS1 performance, Historical Regulation deployments	Maintains a 10 min reserve ϵ [2300, 2800] MW. DSR <= 50% of this	Calculated hourly for all day of a month based on historical wind forecast errors and load forecast errors. It is a 30-min reserve	CR is mostly available cap on generators, gen with more under-freq tolerance or DC tie-line response (fully deployed < 15 s after under freq. event). Load following and NSR deployed as a practicable to minimize the use of 10 min reserves

(continued)

Table 5.2 (continued)

	Regulating Reserves	Operating Reserves	Contingency reserves	Comments
PJM	Determined in whole MW for the on-peak (0500–2359) and off-peak (0000–0459) periods of day; On-peak RR = 1% of day's forecast peak load; Off-peak RR = 1% of day's forecast valley load; may be adjusted in consistency with NERC control standards	max(Reliability First Corporation (RFC) imposed minimum requirement, SLC)	CR = Synchronized Reserves + quick-start reserves. SR >= 50% CR; DSR <= 25% of CR	
NPCC	10-min reserve should be at least equal to SLC. SR is based on the past performance of ACE to return to pre-contingency or zero error; 30-min reserve should be equal to at least 50% of SeLC; Inter-BA reserves are sometimes reserved and used for a DCS recovery or NPCC reportable event			
ENTSO-E	3000 MW; Need to be >= reference incident (the loss of at least the biggest generation/consumption in it or n − 1 failure)	Three methods specified: (1) Statistical analysis of open loop ACE, (2) based on ACE using Monte Carlo Simulations with 90% confidence, (3) Statistical analysis of Control block imbalances		pp. 17, 25–28
RTE, France	Method not specified; varied between 550–600 MW throughout 2015	Balancing Mechanism provides 1000 MW of rapid reserves (13-min) and 500 MW of complimentary reserves (30-min); Exact method not specified;		
ELIA, Belgium	AGC of the generators	Contracted 140 MW reserves. Contributions from all European TSOs for a loss of two 1500 MW Generating units within 15 min of the incident	Method not specified; contracted to producers for capacity injection, and grid users for interruption	

1. SR—Spinning Reserve
2. RR—Regulation Requirement
3. NSR—Non-Spinning Reserve
4. SLC—Single Largest Contingency
5. SeLC—Second Largest Contingency
6. OR—Operating Reserve
7. PHSU—Pumped Hydro Storage Units
8. DSR—Demand Side Response
9. LFES—Load Following Estimated Schedule
10. LFPS—Load Following Perfect Schedule
11. DCS—Disturbance Control Standard
12. CR—Contingency Reserves

5.6 Reserve Estimation in Renewable Integration Studies

The purpose of many renewable integraton studies is to understand the impact of high penetrations of variable generation on a given power system. Hourly (or sub-hourly) simulations of the system operation for a given period of time (generally one year) under potential future variable generation deployment scenarios are used to evaluate the effect of additional variability on the power system, and identify potential mitigation options. Studies often present different levels of renewable penetration (including a reference, or business-as-usual case) and reserve requirements are calculated for each scenario. This section summarizes the treatment of operating reserves in general in these types of studies, and then presents the treatment of reserves from several integration studies completed to date.

Contingency reserves are typically considered unaffected by increased renewables. By definition, these reserves are meant to assist the system in riding through credible losses of transmission lines and generators. Given the dispersed nature and small relative size of PV and wind installations, the largest contingency in the system normally does not change and, thus, contingency reserve requirements are unchanged. Regulation and other operating reserve requirements are typically increased with increasing renewable penetration to ensure reliable operation. The methods and level of confidence used for reserve estimation with forecasts varies among studies, and includes methods such as calculating a percentage of output or a standard deviation of historical forecast errors. Table 5.3 lists the methods used by several studies when calculating reserves.

Renewable integration studies with sub-hourly dispatch see much more variation between time steps than hourly dispatch, since much of the variability of these resources occurs at time-scales of minutes. As such, studies with hourly dispatch may not fully reflect the actual variability and reserve needs of a system. Additionally, many studies saw reduced net variability when high levels of penetrations from wind and PV were included, since studies accounted for the relationship between variability in load, PV, and wind, which frequently tended to be non-coincident. Lastly, studies pertaining to islanded systems often have more aggressive reserve requirement practices, given their lack of interconnection to neighboring areas for support and more limited geographical diversity in resources.

5.6.1 Contingency Reserves

These reserves are maintained to handle unexpected loss of large generation units and are typically based on the size of the largest generator, or set as a percentage of load. Almost unanimously, integration studies do not consider the influence of PV and wind on contingency reserves. The requirements, however, do vary between studies.

Table 5.3 Reserves requirements from several renewable integration studies

	Study		Contingency reserve requirements	Regulation reserve requirements	Flexibility reserve requirements
1	New England Wind Integration Study (NEWIS)—2010		10 min requirement = single largest credible contingency. Usually 50% (could be as low as 25%) is carried as spinning reserve, and the rest as 10-min NSR. 30-min operating reserve requirement = 50% of the second largest credible contingency		
2	Eastern Wind Integration Studies (EWITS)—2011		Function of conventional equipment, and the network. Defined to be 1.5 times the single largest contingency when no information about current practices is available for given region	Geometric sum of standard deviation from load and 3*standard deviation of wind variability	SR = (standard deviation of the expected error); NSR = 2* (standard deviation of the expected next hour wind generation forecast error)
3	Eastern Interconnection Planning Collaborative (EIPC)—2012		Operating reserves are estimated hourly based on the requirements respective of each reliability region, such as loss of the largest single generator, loss of largest single generator and half of second largest generator, or a percentage of peak demand		
4	Western Wind and Solar Integration Study (WWSIS)	WWSIS—Phase 1 (2012)	6% of the load; 50% SR, 50% NSR	1% of the peak load	Not estimated
		WWSIS—Phase 2 (2014)	3% of the load SR	Geometric sum of base requirement (1% of the load) and the contribution of the wind and PV to cover 95% of the 10-min forecast errors. NSR not modeled	Held to cover 70% of the 1 h forecast errors of wind and PV
5	Hawaiian Solar Integration Studies (HSIS)	Oahu (2012)	SR to cover the loss of the largest generating resource (185 MW—AES plant, Oahu)	Defined to cover 99.99% of the renewable variation. For other times, contingency reserves will be used. Down reserves >= 40 MW at all times	Defined to meet or exceed the renewable variation (defined by DP vs. P0 for a 10 min interval). Final SR should cover all the different time intervals within an hour
		Maui (2013)	Same as Oahu	Based on an algorithm which increases the amount of up-regulating reserves as a function of delivered wind power. The requirements are the converged solution from the iterative calculations, based on the amount of wind power	Same as Oahu

(continued)

Table 5.3 (continued)

	Study		Contingency reserve requirements	Regulation reserve requirements	Flexibility reserve requirements
				delivered to the system	
6	Idaho Power	(a) Wind Integration Studies (2013)	3% of the load +3% of the generation—50% is Spinning Reserve	Balancing reserves: confidence interval of 90% for wind and load forecast errors	
		(b) Solar Integration Studies (2014)	Similar to (1a) Wind Integration studies, except that the confidence level is 95%		
7	PJM Renewable Integration Studies (PRIS)—2014		Largest possible contingency; 50% is online generation; DSR <= 25%	Defined to cover 99.7% (3σ) of all 10-min variations of load, wind and PV. Wind and PV standard deviation approximated by curve fitting (quadratic) the points of standard deviation versus hourly production	Spinning reserve requirement is to meet the RFC and VACAR requirements. No additional reserves due to PV/wind considered
8	National Offshore Wind Energy Grid Interconnection Study (2014)		Contingency and regulation reserve estimation is same as in WWSIS-2; flexibility reserves were not considered		
9	Duke Energy—Solar Integration Studies for Carolinas (2014)		Considered a component of DA planning reserve; Excluded from the DA planning reserve calculations	(Actual net load—Real time schedule of the net load) using persistence models. Same procedure as Planning reserve estimation, except that the L10 limit of CPS2 is deducted from the RR requirement	DA PR estimates = (Actual net load − DA forecast of the net load) calculated hourly; Net load = (load − PV output); Requirements are grouped hourly for the month and ranked to truncate excess of a 95% confidence. Final requirement for a given hour = Average of max & min values of DA PR requirements from 10 Monte Carlo simulations
10	Eastern Renewable Grid Integration Study (ERGIS) (2015)		Same as WWSIS-2; Based on 10-min persistence forecasts with 95% confidence. Dynamic Reserve requirements for each region as a function of wind forecast		No flexibility reserves

(continued)

Table 5.3 (continued)

	Study	Contingency reserve requirements	Regulation reserve requirements	Flexibility reserve requirements
11	Carolina Offshore Wind Integration Case Study (COWICS)—2015	Sufficient to cover the single most severe contingency; responds within 10 min	Geometric sum of current reserve (L10) and 1.65 times (representing 90% confidence level) the standard deviation of 10-min wind dataReferred to as Frequency Reserve; based on the NERC's BAL-003-1—Attachment A	Geometric sum of current reserve (L10) and 1.65 times (representing 90% confidence level) the standard deviation of 10-min wind dataNone

Studies that calculated reserves based on the largest generator or two generators include EWITS, HSIS, COWIS, and EIPC. EWITS [10] defines the requirement to be 1.5 times the single largest contingency possible. These reserves are split equally between spinning and non-spinning units. HSIS, [15, 16], models reserves as the largest or two largest contingencies for Oahu and Maui, respectively. Similarly, Pacific Northwest National Laboratory (PNNL) [14] sets the requirements adherent to conventional Duke operating practice; 95% confidence interval of historical forecast errors. COWICS Phase 2 studies [28] also estimate their reserves based on the loss of the single largest generation unit, and are set with an expected response time within 10 min. EIPC's [29] estimates were hourly and sometimes also included additional requirement attributed to 50% of the second largest generator capacity or a percentage of peak demand. Note that, in this study, the supply reserves were specified based on the type of unit. PRIS [20] maintains reserves to suffice the largest possible contingency in the system: 50% of this is online generation (spinning reserve), and demand response reserve is set to a maximum of 25%.

The other common method for contingency reserves bases the requirement on a fixed percentage of load. The Idaho power studies reserve a capacity based on of 3% of load and 3% of generation, for both integration studies [30, 31], with half of the reserve modeled as spinning. WWSIS-1 [11] allocates generation equivalent to 6% of the load as 'Contingency Reserves'—half of which is spinning, and the other half is non-spinning. Similarly, WWSIS-2 [12] allocates 3% of the load for contingency reserves. Both ERGIS [13] and NOWEGIS [32] followed WWSIS Phase 2 and adopted its approach for modeling all reserves. They will therefore not be discussed separately, except for any points of difference.

5.6.2 Regulation Reserves

These reserves are expected to be available immediately to help manage the imbalance of the load and demand on a second-to-second basis. The studies we analyzed have a variety of methods for calculating regulation reserves. Some considered regulation as a fixed percentage of the load, while others took a statistical approach with different confidence levels to define the reserves. A few had a mixture of the two approaches.

In WWSIS-1 [11], regulation requirements were estimated to be 1% of the peak load, following the practice of the Western Electricity Coordination Council's Transmission Expansion Planning Policy committee. However, the WWSIS-2 approach [12] additionally considered the contribution from wind and PV to cover 95% of the historical 10-min forecast errors, by taking the geometric sum of the three components. The geometric sum results in domination of the load component for low VG scenarios and of PV and wind components for high penetration scenarios. ERGIS and NOWEGIS followed the same calculations.

In EWITS [10], reserve requirements for each region and scenario are calculated from hourly load and 10-min wind production data, resulting in an hourly profile which varies both with the load and wind. For production simulations, all the reserves are maintained as spinning reserves. Based on the net standard deviation with and without wind, it was concluded that the impact of wind would be larger for sub-hourly time scales. The regulation requirement was defined as the square root of 1% of load and the third standard deviation of wind forecast error to cover 99.7% (3σ) of the wind variability, presuming that the forecast errors follow a normal distribution. EWITS [10] states this as:

$$Reg_{Req} = \sqrt{(1\%\ of\ Hourly\ Load)^2 + \sigma_{ST}.(3*Hourly\ Wind)^2}$$

where, σ_{ST} is the function obtained by curve fitting the 10-min ahead wind generation forecast errors with respect to hourly MW production for the specific operating area and wind generation scenario. The requirement of additional spinning reserve was set to one standard deviation of the expected error. For supplemental (non-spinning) reserves, twice the standard deviation of the expected next hour wind generation forecast error was used.

PRIS [20] used similar methodology for their regulation requirements estimation. The corresponding σ functions for wind and PV were approximated by curve fitting (quadratic) the points of standard deviation, and hourly production data. PJM considered nine scenarios ranging from the 2% Business As Usual (BAU) to 30% renewable penetration with different combinations of PV, on-shore, and off-shore wind generation. Given the geographic diversity of PJM, the variability of solar and wind were greatly reduced by aggregation. The short term intermittency was regarded as a greater challenge in operational modelling.

For HSIS [16], given their nature of operation, Oahu's restrictions were more stringent at 99.99% of the historical forecast errors. Also, they followed an approach similar to the actual methods used in operation to estimate flexibility reserves, using 2 s intervals for 2 years as the basis data. The requirement should envelope 99.99% of all actual variability in wind and solar. It is assumed that the remaining 0.01% will be accommodated by the contingency reserves. Oahu also requires a minimum of 40 MW in down reserves at all times. Maui [16], on the other hand, estimates reserves by defining up-regulating reserves as a function of the delivered power from solar and wind generation. The final requirements are based on the converged solution from iterative calculations of the above function.

PNNL [14] estimated regulation reserves for the Duke Solar study based on real time forecast errors and the variability of PV and load at the minute time scale. The difference between the actual net load and the real time schedule of the net load is estimated using persistence models. The criterion used was similar to its approach for operating reserves, except that the L10 limit of CPS2 is deducted, instead of maintaining a 95% confidence level. Also, note that, the operating reserves used in the report refer to the combined regulation and load following reserves.

COWICS Phase 2 estimates reserves based on the L10 limit for no wind scenarios, which represents a confidence interval of 90% based on the historical data. For wind scenarios, the net reserve is a geometric sum of current reserve (L10) and 1.65 times (representing the 90% confidence level) the standard deviation of the 10-min wind data.

NEWIS [33] on the other hand, has a more dynamic approach for calculating reserves based on system conditions. It classifies its contingency reserves as a 10-min product (both spinning and non-spinning) and has a 30-min operational reserve product. Ten-minute requirements are based on the single largest contingency. Usually 50% of this, though potentially as little as 25%, is carried as spinning reserves, and the rest as non-spinning reserves. The 30-min operating reserve requirement is 50% of the second largest credible contingency. For the study, a simplistic approach was taken by combining the reserve requirements based on time of the day, and day of the week. The procurement was finally determined as [33]:

1. 0700-2300—Weekdays:
   a. Total 10 min reserve = 1500 MW (750 MW—Spinning, rest—non-spinning)
   b. 30-min Reserve—750 MW
   c. Total—2250 MW
2. 2300-0700 weekdays and all hours weekends
   a. Total 10 min reserve = 1300 MW; (650 MW—Spinning, rest—non-spinning)
   b. 30-min Reserve—650 MW
   c. Total—1950 MW

For their wind integration studies [30], Idaho Power considered a reserve requirement to provide for the 90% confidence interval of forecast errors as reserve capacity, combined with individual load and PV reserve components with a confidence of 90% added, to yield the total bidirectional balancing reserve requirement. It considers the incremental reserve capacities to be scheduled for dispatchable generators to account for uncertainty and variability from the forecast for the given hour. For solar integration studies [31] the same approach was used, except that a 95% confidence level is used.

For the solar integration study in the Carolinas region of Duke Solar, PNNL [14] estimates the requirements based on the deviation of the net load (difference of actual load and PV output) from the forecasted values. The requirements were then grouped hourly month-wise and a reserve requirement was determined based on the 95% confidence level of these deviations.

5.6.3 Flexibility Reserves

Terminologies and implementations for flexibility reserves vary significantly among different integration studies, and not all studies include this type of reserve product. For those that did, the dependence on PV and/or wind, and use of statistical analyses remains the focal point for all studies. Different studies use different implementations and levels of coverage, adherent to their respective operational practices. Islanded systems like Oahu and Maui were more stringent in defining their reserve requirements.

Some studies, like EIPC [29] have not distinguished between the regulating and operating reserve requirements. For spinning reserves, the requirement was based on the largest generator or largest generator and half of next largest generator, depending on the region.

Interestingly, in PRIS [20], PJM analyzed its 10-min load and wind profiles against hourly production simulations. Following that, it ruled out the need of augmenting additional flexibility reserves. It was concluded that given its large footprint and geographical diversity, solar and PV penetration didn't have any noticeable effect on the operating reserve requirement of PJM. Hence the reserves were estimated based on their current practice of spinning reserve requirements.

WWSIS-1 did not consider flexibility reserves, however, WWSIS-2 [12] estimated these reserves to cover 70% of the 1 h forecast errors of wind and PV. Note that WWSIS-2 assumed perfect load forecasts due to the lack of available data and, thus, load had no contribution to flexibility reserves. There was a single flexibility requirement held for the entire Western Interconnection. As mentioned previously, ERGIS [13] followed the same approach as WWSIS-2 for reserves.

5.7 Conclusion

Increasing variable renewable capacity worldwide has led to increasing desires for a better understanding of the ways in which these resources will impact the existing infrastructure in the future. In order to determine the impacts of variable renewable resources, many renewable integration studies have been performed to directly analyze how renewable generation will affect the bulk power system operations. In order to ensure grid stability, these studies must also incorporate the impacts of renewable energy on reserve requirements. These requirements are modeled in ways similar to how reserves are actually procured by independent system operators, however there are many differences between these systems and the models representing them. This chapter highlights the methods used for modeling required reserves, in addition to providing a description of the current policies for reserve provision. In many cases, statistical analysis was used to quantify the variability and uncertainty of forecasts in load and renewable generation particularly for regulation and flexibility reserves. Renewable penetrations did not typically impact contingency reserves in these models. It was also found that islanded systems are generally more conservative with their reserve requirements compared to large interconnected systems, likely due to the fact that they cannot count on support from neighboring regions. Many studies also identified that using sub-hourly scheduling practices can reduce the net reserve requirements for the system.

References

1. I.E.A. Deploying Renewables 2011: Best and Future Policy Practice. International Energy Agency (2011)
2. S. Henbest, E. Giannakopoulou, V. Cuming, *New Energy Outlook 2015* (Bloomberg New Energy Finance, 2015)
3. K. de Leon, D. Williams, M. Leno, SB-350 Clean Energy and Pollution Reduction Act of 2015. SB-350, 7 Oct 2015
4. N. Blair, K. Cory, M. Hand, L. Parkhill, B. Speer, T. Stehly, et al., *Annual Technology Baseline* (National Renewable Energy Laboratory, 2015)
5. California Air Resources Board. Article 5: California Cap on Greenhouse Gas Emisssions and Market-based Compliance Mechanisms to Allow for the use of Compliance Instruments Issues by Linked Jurisdictions [Internet]. Title 17, California Code of Regulations. Sections 95800 to 96023, 26 Jun 2014, http://www.arb.ca.gov/cc/capandtrade/ctlinkqc.pdf
6. N.C. Clean Energy Technology Center. Database of State Incentives for Renewables & Efficiency [Internet]. DSIRE, 2015, http://www.dsireusa.org. Accessed 20 Mar 2016
7. E.P.A. Carbon Pollution Emission Guidelines for Existing Stationary Sources: Electric Utility Generating Units. 40 CFR Part 60, 23 Oct 2015
8. R. Wiser, M. Bolinger, 2014 Wind Technologies Market Report [Internet]. Lawrence Berkeley National Laboratory, 2015, http://newscenter.lbl.gov/2015/08/10/study-finds-that-the-price-of-wind-energy-in-the-united-states-is-at-an-all-time-low-averaging-under-2-5%c2%a2kwh/. Accessed 20 Mar 2016

9. S. Lacey, Cheapest Solar Ever: Austin Energy Gets 1.2 Gigawatts of Solar Bids for Less Than 4 Cents [Internet] (2015), http://www.greentechmedia.com/articles/read/cheapest-solar-ever-austin-energy-gets-1.2-gigawatts-of-solar-bids-for-less. Accessed 20 Mar 2016
10. EnerNex Corporation. Eastern Wind Integration and Transmission Study (EWITS) (Revised) [Internet]. Report No.: NREL/SR-5500-47078 (National Renewable Energy Laboratory, Golden, Colorado, 2011), http://www.nrel.gov/docs/fy11osti/47078.pdf
11. GE Energy. Western Wind and Solar Integration Study [Internet]. Report No.: NREL/SR-550-47434 (National Renewable Energy Laboratory, Golden, Colorado, 2010), http://www.nrel.gov/docs/fy10osti/47434.pdf
12. D. Lew, G. Brinkman, E. Ibanez, A. Florita, M. Heaney, B.M. Hodge, et al., Western wind and solar integration study phase 2 (2013)
13. A. Townsend, D. Palchek, A. Bloom, Eastern Renewable Generation Integration Study (Forthcoming)
14. S. Lu, N. Samaan, D. Meng, F. Chassin, Y. Zhang, B. Vyakaranam, et al., Duke Energy Photovoltaic Integration Study: Carolinas Service Areas [Internet]. Pacific Northwest National Laboratory. Report No.: PNNL-23226, 2014, http://www.pnnl.gov/main/publications/external/technical_reports/PNNL-23226.pdf
15. G.E. Energy, Hawaii Solar Integration Study: Final Technical Report for Oahu, 2012
16. G.E. Energy, Hawaii Solar Integration Study: Final Technical Report for Maui, 2013
17. Transpower stromübertragungs gmbh, European Wind Integration Study. Report No.: TREN/07/FP6EN/S07.70123/038509, 2010
18. California I.S.O, Flexible Ramping Product, Revised Draft Final Proposal [Internet], 2015, https://www.caiso.com/Documents/RevisedDraftFinalProposal-FlexibleRampingProduct-2015.pdf
19. E. Ela, M. Milligan, B. Kirby, Operating Reserves and Variable Generation. A comprehensive review of current strategies, studies, and fundamental research on the impact that increased penetration of variable renewable generation has on power system operating reserves [Internet]. Report No.: NREL/TP-5500-51978, 2011, http://www.nrel.gov/docs/fy11osti/51978.pdf
20. GE Energy, PJM Renewable Integration Study [Internet] (2013), http://www.pjm.com/committees-and-groups/task-forces/irtf/pris.aspx
21. E. Ela, B. Kirby, E. Lannoye, M. Milligan, D. Flynn, B. Zavadil, et al, Evolution of operating reserve determination in wind power integration studies, in *2010 IEEE Power and Energy Society General Meeting* (2010), pp. 1–8
22. Y. Dvorkin, M.A. Ortega-Vazuqez, D.S. Kirschen, Wind generation as a reserve provider. IET Gener. Transm. Distrib. **9**(8), 1–9 (2015)
23. E. Muljadi, V. Gevorgian, M. Singh, S. Santoso, Understanding inertial and frequency response of wind power plants, in *2012 IEEE Power Electronics and Machines in Wind Applications (PEMWA)* (2012), pp. 1–8
24. D.A. Halamay, T.K.A. Brekken, A. Simmons, S. McArthur, Reserve requirement impacts of large-scale integration of wind, solar, and ocean wave power generation. IEEE Trans Sustain Energy. **2**(3), 321–328 (2011)
25. J. Zhang, B.M. Hodge, A. Florita, Joint probability distribution and correlation analysis of wind and solar power forecast errors in the western interconnection. J Energy Eng. **141**(1), B4014008 (2015)
26. E. Ela, *Variable Renewable Generation Can Provide Balancing Control to the Electric Power System* (National Renewable Energy Laboratory, Golden, CO, 2013)
27. NERC. NERC IVGTF Task 2.4 Report Operating Practices, Procedures, and Tools [Internet]. North American Electric Reliability Corporation, 2011, http://www.nerc.com/files/ivgtf2-4.pdf
28. Duke Energy Business Services. Carolina Offshore Wind Integration Case Study Phase 1 Final Technical Report [Internet], 2013, http://nctpc.org/nctpc/document/REF/2013-06-06/COWICS_Phase_1_Final_Report1%5B1%5D.pdf

29. Eastern Interconnection Planning Collaborative, Phase 2 Report: Interregional Transmission Development and Analysis for Three Stakeholder Scenarios. [Internet]. Report No.: DOE Award Project: DE-OE0000343, 2012, http://www.eipconline.com/uploads/20130103_Phase2Report_Part2_Final.pdf
30. Idaho Power, Wind Integration Study Report [Internet], 2013, https://www.idahopower.com/pdfs/AboutUs/PlanningForFuture/irp/2013/windIntegrationStudy.pdf
31. Idaho Power, Solar Integration Study Report [Internet], 2014, https://www.idahopower.com/pdfs/AboutUs/PlanningForFuture/solar/SolarIntegrationStudy.pdf
32. J.P. Daniel, S. Liu, E. Ibanez, K. Pennock, G. Reed, S. Hanes, National Offshore Wind Energy Grid Interconnection Study. Report No.: DOE Award No. EE-0005365, 2014
33. G.E. Energy, *New England Wind Integration Study* (ISO New England, 2010)

Chapter 6
Balancing Authority Cooperation Concepts to Reduce Variable Generation Integration Costs in the Western Interconnection: Consolidating Balancing Authorities and Sharing Balancing Reserves

N.A. Samaan, Y.V. Makarov, T.B. Nguyen and R. Diao

6.1 Introduction

In an electric power grid, demand (load) and supply (generation) must always be balanced. Significant imbalances could result in interconnection frequency deviations, transmission system violations, stability problems, etc., that ultimately could lead to widespread system blackouts. System load varies with time and is to a large extent not controllable. To provide an adequate balance of supply and demand, generation must be dispatched to follow load variations.

Challenges of maintaining the balance become more significant when an interconnected power grid is operated locally and separately by each individual balancing authority (BA), which is the case in most large power systems. For example, the U.S. Western Interconnection is a large integrated and interconnected power system. Organizationally, it is divided into 38 BAs. Within each BA, operators are responsible for maintaining the balance between load and generation within their territory, as well as for following interchange schedules among BAs. Working locally and separately, each BA must maintain balance with its own, sometimes limited, resources. A BA with limited balancing reserves and high wind and solar power penetration could encounter significant balancing problems. To maintain the balance, it may have to resort to more expensive resources. Sometimes, it may even run out of resources to maintain balance. Associated economic and reliability concerns may create hurdles to high levels of renewable penetration.

N.A. Samaan (✉) · Y.V. Makarov · T.B. Nguyen · R. Diao
Pacific Northwest National Laboratory, Richland, WA 99352, USA
e-mail: Nader.Samaan@pnnl.gov

P. Du et al. (eds.), *Integration of Large-Scale Renewable Energy into Bulk Power Systems*, Power Electronics and Power Systems,
DOI 10.1007/978-3-319-55581-2_6

In 2009, the Western Electricity Coordinating Council (WECC) formed the Variable Generation Subcommittee (VGS) to explore the issues surrounding the integration of variable generation (VG) resources such as wind and solar into the Western Interconnection in terms of grid marketing structure, operating, and planning. To help manage the increased variability and uncertainty that VG resources will bring to the power system, a number of approaches have been considered, ranging from technological to institutional. Most of these approaches are complementary and, therefore, can be developed together. Most of the approaches benefit power system reliability and reduce costs with or without variable renewable generation.

In the Western Wind and Solar Integration Study (WWSIS) Phase I funded by the U.S. Department of Energy [1], the National Renewable Energy Laboratory (NREL) concluded that, with high wind and solar penetration rates, managing the net load variability will become very expensive if it is not aggregated over large geographic areas. It was shown that if the system is operated in aggregate, the variability of a wind penetration up to 20% of annual energy can be managed with existing balancing resources. The WWSIS focuses on hourly production simulations, coupled with statistical analysis of 10 min data. The NERC Integration of Variable Generation Task Force (IVGTF) recognized the need of BAs cooperation with higher penetration level of VG.

The objective of this study is to demonstrate the benefits of BA consolidation through development and use of a detailed model and methodology. The study is working to determine the savings in production cost and reduction in balancing reserve requirements in the WECC system. The effect of transmission congestion on potential benefits is evaluated in addition. The analysis is being performed for two different scenarios of VG penetration: 11% (8% wind and 3% solar) and 33% (24% wind and 9% solar) of WECC projected energy demand in 2020.

6.1.1 Related Studies

Several approaches to BA cooperation have been adopted in the power industry or can be proposed for the future such as ACE diversity interchange, regulation and load-following sharing, VG-only BAs, dynamic scheduling, and BA consolidation options [2].

An Energy Imbalance Market (EIM) represents another form of cooperation among BAs. Currently, different versions of an EIM are under consideration in the WECC system [3–6]. The California Independent System Operator (ISO) has announced that, in October 2014, EIM went live with PacifCorp, with customers in six states. There are plans to incorporate Arizona Public Service Company (by October 1, 2016), Portland General Electric (by October 1, 2017), and other entities outside of California (NV Energy and Puget Sound Energy).

In the last few years, there have been several studies that investigate the benefits of WECC BA coordination using Production Cost Models (PCM), which used hourly and sub-hourly WECC PCM. These include:

- WECC/Energy and Environmental Economics (E3) EIM (hourly, WECC-wide) [5]
- WECC Variable Generation Subcommittee (VGS) Full BA Consolidation and Reserve Sharing (WECC-wide, hourly analysis) "presented in this chapter"
- E3 (CAISO-PAC) EIM benefits (hourly analysis) [4]
- WECC/VGS benefits of intra-hour scheduling (10 min analysis) [7]
- NREL/Public Utility Commission EIM (10 min analysis [3]
- PNNL/NWPP EIM (10 min analysis) [6]

These studies have been performed by E3, NREL, PNNL, Energy Exemplar, and WECC staff, and with participation from various Western Interconnection entities. The studies ranged from hourly PCM studies (the first three studies) to intra-hour simulations (the last three studies) and have looked at various methods for BA cooperation and markets including an EIM. Figure 6.1 shows how the models used in the different studies relate to each other.

6.1.2 Comparison Among Different Market Structures

Several potential market structures in WECC are compared with current market structure in Table 6.1. The first structure represents current practice or "business-as-usual" scenario with hourly transactions between BAs. The second

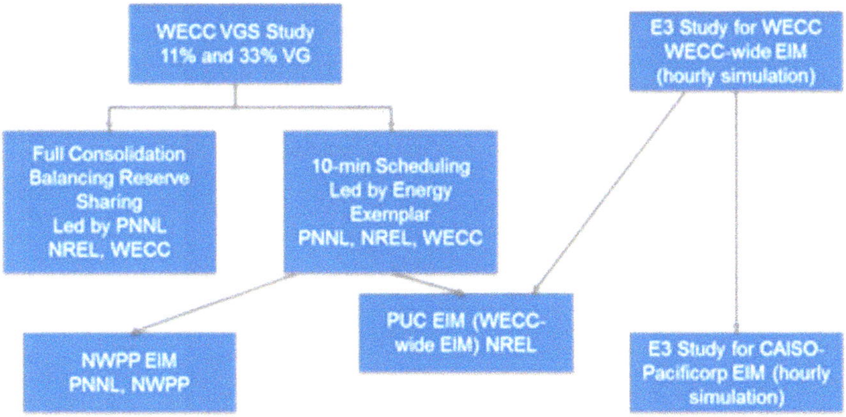

Fig. 6.1 Recent WECC balancing authorities coordination studies

structure (Centralized Market) represents a full BAs consolidation scenario. The third structure represents "business-as-usual" scenario with intra-hourly transactions between BAs which reflects FERC order 764 enabling 15-min transmission scheduling. Finally, the fourth structure represents the EIM market operation.

6.1.3 *What Is Balancing Authority Consolidation and What Benefits can it Provide?*

Problems with balancing the system can be mitigated through BA collaboration and consolidation. The total percentage of power variation can be reduced through coordination among BAs. Wind and solar generation are generally not correlated with each other over a wide geographic area. Thus, the average power variation is smaller than the lumped variation for balancing areas spread over a large territory. In addition, the spare resources in neighboring areas can be used to manage power variation through coordination. Working together, BAs can balance reduced power variance with shared resources. This benefits power system operation from both economic and reliability perspectives. The principle of savings achieved through BA consolidation is explained in this brief example (see Fig. 6.2).

Table 6.1 Basic differences among different current WECC operations versus three potential market structures

	Current hourly scheduling among WECC BAs	Centralized market	Intra-hour scheduling among WECC BAs	Energy imbalance market
Day-ahead forecasts	UC/ED at BA level with BA exchange	UC/ED at WECC level	UC/ED at BA level with BA exchange	UC/ED at BA level with BA exchange
Hour-ahead forecasts	UC/ED at BA level with FINAL BA exchange	UC/ED at WECC level	UC/ED at BA level with BA exchange	UC/ED at BA level with FINAL BA exchange
Intra-hour	ED to meet imbalance at BA level	ED to meet imbalance at WECC level	ED at BA level with FINAL exchange	ED to meet imbalance at WECC level
Regulation	At BA level	WECC level	At BA level	At BA level
Hurdle rate on BA-to-BA transactions	Yes (DA, HA)	No (DA, HA, RT)	Yes (DA, HA, RT)	Yes (DA, HA) No (RT)
Contingency and balancing reserves	Individual BA obligations	CBA obligation (lower requirements)	Individual BA obligation Lower load-following?	Individual BA obligations Lower load-following?

UC unit commitment; *ED* economic dispatch; *DA* day-ahead; *HA* hour-ahead; *RT* real time

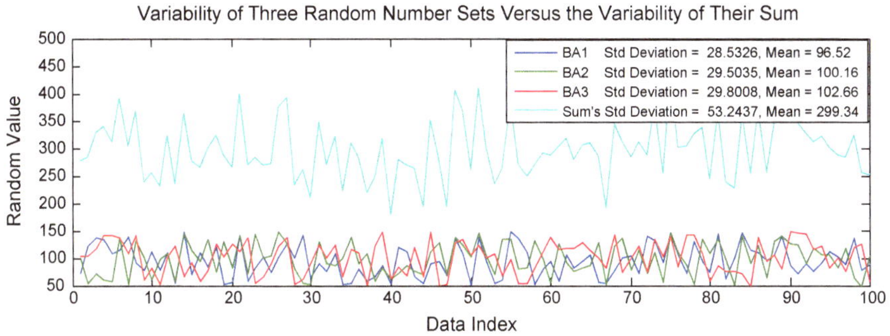

Fig. 6.2 Concept of variability reduction due to balancing authority consolidation

The figure shows the variability (measured as standard deviation) of random number sets versus the variability of their sum. One might expect that, just as the averages of each data set are additive, so would be the variability; however, that is not what we see. The fact that the peaks and valleys of each BA's values do not correlate perfectly results in the sum of the values having a smaller variability than simply adding the variability of each individual data set, which the reader can verify by using the standard deviation values that appear in the figure's legend. When BAs are combined, the resulting variability of the combined system is expected to be less than the sum of each individual system's variability. Of course, a major issue influencing potential benefits is the degree to which the wind and solar power will be correlated across CBAs.

The consolidation of individual BAs is the integration of two or more BAs into a single CBA. By consolidating BAs, one could effectively employ the diversity factor to minimize the collective system balancing requirements in terms of the required balancing capacity and energy, to reduce ramp rates, to effectively share operating reserves, and to provide more cost-effective and flexible UC and dispatch patterns. The CBA will provide the necessary infrastructure to facilitate the intrazonal transfers (formerly BA-to-BA schedules). These transfers are initiated by the allocation of wide area regulation, load-following and scheduling requirements. Transmission owners would provide Available Transfer Capability (ATC) information so intrazonal transfers would not exceed security constraints. Ultimately, transmission owners would provide loading and line ratings so that the CBA could calculate security constraints.

BA consolidation becomes one of the most effective solutions among the other BA cooperation measures because it eliminates the key factor limiting the positive effect of diversity—the existing BAs' structure and their responsibility to provide balance in limited parts of an interconnection. Balancing authority consolidation essentially can incorporate many other solutions (e.g., common ACE, common reserves, etc.), and it helps to provide more cost-effective and flexible UC and dispatch patterns. The potential effect of this solution for the WECC members is under investigation in this research; at the same time, it should be noted that this analysis concentrates on the benefits of CBA as they are related to balancing functions, and it does not consider other aspects of consolidation such as transmission system impacts, extreme imbalances, etc.

The authors are not advocating BA consolidation as the optimal way of BA cooperation, but rather is using a single dispatch model of the entire Western Interconnection as a proxy for maximum potential benefits. The objective of WECC BA consolidation analysis is to propose and evaluate key technical metrics to demonstrate the potential gains that would accrue by consolidating all the BAs within a large geographical area and by operating them as a single CBA.

6.1.4 Types of Benefits Realized Through Balancing Authority Consolidation

Sharing the variability of resources and loads across a broader region provides a natural aggregation impact that reduces the per-unit variability with a resulting reduction in required reserves. Similarly, the per-unit variability of wind and solar has an aggregation benefit that results from the different wind speeds or solar irradiance that occur at the same time in different locations. The net load (load minus wind and solar and any other VG) is what must be managed by the balance of the generation fleet (and responsive loads). Because the variability increases less than linearly as it relates to size, large areas that provide balancing require relatively smaller balancing reserves. In addition, BA cooperation will result in a linear addition of ramping capability from existing generators. Therefore, cooperating BAs have more ramping capability and less need, on a per-unit basis, for this capability than they would have needed in the absence of cooperation. This benefit, which follows the same rationale for creating reserve sharing pools to reduce contingency reserve requirements for individual BAs, makes it attractive for balancing over larger electrical footprints. Allowing sharing of variability at a less than hourly timescale among BAs can provide access to additional existing physical response capability.

The CBA with a much wider geographical boundary results in significantly lower scheduling, load-following and regulation requirements compared to the sum carried by the individual balancing areas. These savings could be realized immediately by reallocating regulation reserve requirements to the balancing areas and reducing operational burden.

Numerous benefits from the consolidation of individual BAs can be identified and evaluated:

- *Reducing regulation requirements in terms of their required upward and downward reserved capacity, actual use (energy), reduced impact on the regulation units (wear-and-tear, efficiency), and cost.* In this chapter, we will analyze the reductions in terms of regulating reserve capacity.
- *Reducing load-following requirements in terms of their required upward and downward capacity, actual use (energy), reduced impact on the regulation units*

(wear-and-tear, efficiency), and cost. In this chapter, we will analyze the reductions in terms of load-following capacity.

- *Reducing ramping requirements.* Fast and continuous ramps, especially when they are not expected, create problems for grid operators. The ramping capability of most thermal generators is limited, so to follow fast ramps, more generators need to be used, which means more online, but underutilized capacity.
- *Having a larger pool of generating resources available for dispatch.* This results in reducing total operation cost as BA consolidation allows for the access to more generation units for UC and economic dispatch.

6.1.5 Study Objective

The main objectives of this study are (1) to determine the saving in production and balancing costs resulting from WECC BA consolidation as an upper bound for BA cooperation benefits; (2) to evaluate the benefits of intra-hour (10 min) scheduling for the current WECC BA structure; and (3) to analyze two different scenarios of VG penetration (11 and 33% as percentage of WECC projected energy demand in 2020). The work is not implying or premeditating a specific method of BA consolidation or cooperation used to achieve the benefits (there are multiple ways to benefit from the geographic diversity factor). The effect of transmission congestion on potential benefits is also evaluated.

6.1.6 Components of the Analysis

Two major components of analysis were performed to evaluate the benefits of balancing authority consolidation: (1) production cost analysis and (2) balancing service analysis.

6.1.6.1 Production Cost Analysis

The objective of production cost analysis is to determine the maximum savings that could be achieved in a CBA due to decreasing production cost through more efficient UC and economic dispatch. The expected savings in production costs result from various factors, including (1) the access to a larger pool of generating units; (2) the diversity of DA forecast errors for load and intermittent resources such as wind and solar; and (3) the diversity of variability characteristics of these resources. This type of analysis is conducted using commercial software, Ventyx PROMOD IV [8], to determine the total production costs for individual BAs and a CBA for comparison.

6.1.6.2 Balancing Requirements Analysis

The objective of balancing requirements analysis is to determine the savings of the CBA in load-following and regulation requirements in terms of capacity, ramp rates, and durations, due to the diversity of hour-ahead (HA) and real-time (RT) forecast errors for load, wind and solar, and the diversity of variability characteristics of these resources. Traditionally, each BA works independently to maintain balance among generation and load, wind, solar, and other intermittent resources and interchange within its area. Each BA needs sufficient balancing reserves and proper procedures to comply with NERC control performance standards. Balancing reserves should be sufficient to compensate predicted and unpredicted load variations and changes in wind and solar generation. Significant imbalances caused by insufficient reserves or deficient characteristics of these reserves (i.e., insufficient ramping capability, system inertia, or frequency response) could result in significant interconnection frequency deviations, transmission system violations, stability problems, etc. Ultimately, those problems could result in widespread system blackouts. The task of balancing the system becomes more difficult with increasing penetration of variable resources in the systems. The balancing services, including load-following and regulation, can be very expensive, so each BA tries to minimize these requirements without compromising system reliability and control performance.

With the potential benefits achieved from BA cooperation or full WECC consolidation, it is anticipated that the variance of load and intermittent resources can be leveraged so that the total balancing requirements are reduced. In this study, a detailed modeling procedure developed by PNNL is used to calculate balancing reserves for evaluating the benefits of WECC consolidation. This method has been used for projects conducted by the California Independent System Operator (CAISO) and the Bonneville Power Administration (BPA).

6.2 Production Cost Model

The objectives of production cost analyses are to determine the saving in production cost due to consolidation of WECC BAs as an upper bound for BA cooperation benefits and due to sharing of operating reserve of the CBA in comparison to individual BAs. The analyses use the Transmission Expansion Planning Policy Committee projections for load, generation, and transmission for 2020 [9] as the base and develop scenarios with 11% and 33% of VG penetration (as a percentage of energy demand) in the WECC system for the study as given in Table 6.2. The analyses also examine the effect of transmission congestion on potential benefits and determine the congested transmission paths that limit the achievements of maximum benefits for consolidation.

Table 6.2 shows the amount of renewable penetration for the two cases.

The study model is based on the WECC Transmission Expansion Planning Policy Committee (TEPPC) 2020 PC0 case [9]. It is a WECC-wide nodal model in

Table 6.2 WECC-wide renewable penetrations used in the study

Penetration level (%)	Nameplate wind (MW)	Nameplate solar (MW)
11	26,816	14,917
33	68,361	26,816

the PROMOD IV production cost modeling tool software. The model is based on the WECC 2020 high summer (HS1A) power flow base case.

6.2.1 BA Structure in the Model

WECC currently consists of 38 BAs, with six of these being generation-only BAs. The model used for the TEPPC case has 39 load areas. The BAs in WECC are grouped into seven sub-regions. The six generation-only BAs are not modeled, but instead their generation resources are modeled as belonging to other BAs. In this study, the TEPPC case was modified so that the 32 BAs that are not generation-only BAs are modeled separately. The model has a total of 86 BA-to-BA transmission paths or flowgates on which hurdle rates could be applied.

Operating reserves are divided into contingency and regulation reserves. The contingency spinning reserve and regulation requirement for each BA are assumed to be 3% and 1% of the BA weekly peak load, respectively.

6.2.2 Load, Wind and Solar Data

The 2006 load shapes and the load growth forecast for 2020 are used to derive the 2020 load shapes. The BA total load is split between load buses within the BA based on the WECC load flow base case for 2012 heavy summer.

The wind and solar data for 2020 were collected from the 15.5% renewables case (calculated as a percentage of WECC 2020 total demand) defined by TEPPC [10]. Time series data for wind and solar production were generated based on 2006 weather models. The TEPPC 2020 PC0 case assumes that all Renewable Portfolio Standards (RPSs) in 2020 are met, with the level of the RPSs in 2020 derived based on assuming a linear progression for those RPSs that have target dates later than 2020. Using this approach yielded a WECC-wide (including BAs in Mexico and Canada) RPS of approximately a 15.5% renewables penetration. This 15.5% renewables penetration level was met by wind (8%), solar (3%), geothermal (2.4%), biomass (1.3%) and small hydro (0.77%).

The wind generation data are based on western wind data sets that were developed by 3Tier [11] and used by the National Renewable Energy Laboratory (NREL). In the study presented in this chapter, wind generation models were improved by disaggregating the wind profiles at the bus level to more realistically simulate the diversity of wind generation.

The solar generation data were obtained from NREL. These data are based on hourly, satellite-derived data from the State University of New York–Clean Power Research and a statistical model that synthesized the sub-hourly variations [12].

6.2.3 Hydropower Plant Modeling

Hydro production in the TEPPC 2020 case is based on an average year (which is representative of hydro generation in 2006). For hydro plants that have no 2006 data available, data available from either 2002, 2003, or 2007 are used. Three hydro models are being used. (1) Proportional load following: the generation schedule of this type of hydro unit is related to the shape of load profiles. (2) Hydrothermal coordination: the generation schedule of this type of hydro unit is based on the proportional load following hydro generation shape, and a "P factor" is used to indicate what portion of the plant is responsive to market pricing and other constraints. The designated dispatchable portion is optimized as a thermal unit. (3) Fixed shape: hourly generation profile is an input for this type of hydro unit. PROMOD uses it directly as the schedule of the hydro unit.

6.2.4 Thermal Generation Modeling

Thermal generation models have several characteristics that are considered in production cost modeling. The TEPPC model has values for these characteristics that have been widely used and vetted. Characteristics include minimum and maximum ramp rates, minimum up-times and down-times, minimum and maximum generation capacities, planned and forced outages, heat rate curves, emission rates, operational and maintenance costs and start-up costs. The median Henry Hub gas price is $7.28/MMBtu (2010 dollars). The average coal price is $1.69/MMBtu (2010 dollars).

6.2.5 Transmission Modeling

Only existing transmission, transmission needed for future reliability to integrate generation, and projects that have a high likelihood of being in service in 2020 are included in the model. The transmission topology begins with the TEPPC 2020 case for consistency with the load and generation data. However, if at increasing VG penetration levels the case does not solve the DC power flow with current topology, analysis may include augmentation to the current TEPPC case. Transmission losses are not modeled explicitly but are included in the load forecast.

6.2.6 Simulation Scenarios

Three scenarios are considered in this work.

Scenario 1 is today's WECC BA structure. In this scenario all BAs are separated, and the transmission system limits are enforced. Each BA commits its own units based on load and interchange schedules. Operating reserve requirements are calculated based on the weekly maximum load in each BA. Hurdle rates for power transferred between BAs are imposed. The objective of this scenario is to simulate, as realistically as possible, the existing structure of BAs within the WECC. In other words, it assumes the 2020 BAs' structure will be similar to today's situation.

Scenario 2 is a full consolidation (copper sheet) in which the transmission system is assumed to have infinite capability, and all BAs are aggregated to form a CBA. All units within the CBA are available for commitment and dispatch. Contingency reserve requirements are calculated based on the maximum weekly demand of the CBA. There are no hurdle rates. The objective of this scenario is to evaluate the upper bound of consolidation benefits assuming no transmission constraints within the CBA.

Scenario 3 is also a full consolidation of all BAs to form a CBA, but the transmission system limits are enforced as in Scenario 1. The objective of this scenario is to evaluate consolidation benefits more realistically by considering transmission thermal and security constraints within the CBA.

6.2.6.1 Modeling of Existing BA Operation

It can be quite challenging to model current BA operation in production cost models due to lack of information on long-term power purchase contracts and modeling of joint ownership of power plants and transmission lines. One way to control the level of power transaction between different BAs is imposing hurdle rates on energy flowing on the transmission flowgates between different BAs.

The production cost modeling software PROMOD IV uses two different input hurdle rates in the unit commitment and dispatch process. The commitment hurdle rate is used for the preliminary dispatch, and the dispatch hurdle rate is used for the final dispatch. The commitment hurdle rate input adjusts the commitment of units in a BA against its own net load while the dispatch hurdle rate input adjusts the interchange between BAs. Higher commitment hurdle rates will force BAs to commit more of their own units to meet their own load. This will reduce interchange and significantly increase the use of more expensive units in each BA. Normally, the commitment hurdle rate is set at a higher value than the dispatch hurdle rate. In that way, the units in each BA will be dispatched against the BA's own load first in order to get the unit commitment order right, and then interchange between BAs will be allowed during the final dispatch via the dispatch hurdle rate.

The effect of commitment hurdle rate on the final dispatch was investigated by performing sensitivity analysis using four different commitment hurdle rates: 0, 10,

Table 6.3 Case definitions for different simulation scenarios

Scenario	Case #	Definition
1	1A	Current operation, 39 load areas, 32 BAs, and hurdle rate = $0/MWh; operating reserve of 4% at BA level and transmission constraints enforced
1	1B	Current operation, 39 load areas, 32 BAs, and hurdle rate = $5/MWh; operating reserve of 4% at BA level and transmission constraints enforced
1	1C	Current operation, 39 load areas, 32 BAs, and hurdle rate = $10/MWh; operating reserve of 4% at BA level and transmission constraints enforced
1	1D	Current operation, 39 load areas, 32 BAs, and hurdle rate = $15/MWh; operating reserve of 4% at BA level and transmission constraints enforced
2	2	CBA with operating reserve of 4% at system (WECC) level with no limit on transmission system and interfaces
3	3A	CBA with operating reserve of 4% at BA level, no hurdle rate, and transmission constraints enforced
3	3B	CBA with operating reserve of 4% at system (WECC) level, no hurdle rate, and transmission constraints enforced

15, 20, and 1000 $/MWh while the dispatch hurdle rate is fixed at 0 $/MWh. Based on simulation results looking at generation commitment at the BA level, it was decided to use the fixed commitment hurdle rate of 20 $/MWh to model current WECC BA operation.

Several sensitivity cases within Scenario 1 (current BA operation) are modeled by changing the dispatch hurdle rate; specifically, the hurdle rates considered are $0/MWh, $5/MWh, $10/MWh, and $15/MWh. In each case, the same hurdle rate is applied to all the flow gates between WECC BAs.

6.2.6.2 Modeling of Consolidated BA

Two different cases were modeled for Scenario 3. In the first case (3A), the operating reserves are held at the individual BA level. In the second case (3B), operating reserves are held at the CBA level. Case definitions for all scenarios are shown in Table 6.3.

6.3 Production Cost Modeling Analysis Results

6.3.1 Production Cost Analysis Results for 11% Variable Generation

This section gives the production cost modeling results for the 11% VG peneteration case. A comparison for considered cases along with their production and demand cost is shown in Table 6.4. Annual savings in thermal unit production cost

Table 6.4 Comparison across all cases for the 11% renewable integration

Simulation scenario	Scenario 1: current WECC BA structure				Scenario 2: consolidated WECC (CBA) copper sheet	Scenario 3: consolidated WECC with transmission congestion	
Category	Case 1A	Case1B	Case 1C	Case 1D	Case 2	Case 3A	Case 3B
Assumed transmission hurdle rate	0$/MWh	$5/MWh	$10/MWh	15$/MWh	0$/MWh	0$/MWh	0$/MWh
Nodal model type	Full nodal	Full nodal	Full nodal	Full nodal	Full nodal	Full nodal	Full nodal
Contingency reserve	BA level	BA level	BA level	BA level	Consolidate WECC level	BA level	Consolidate WECC level
Total gen cost (k$)	18,782,162	18,806,197	18,849,962	18,945,887	18,097,148	18,549,078	18,336,693
LMP-based demand cost (k$)	43,910,551	43,110,029	42,608,714	42,480,207	45,147,779	42,813,837	43,145,235

MWh megawatt-hours; *LMP* locational marginal price

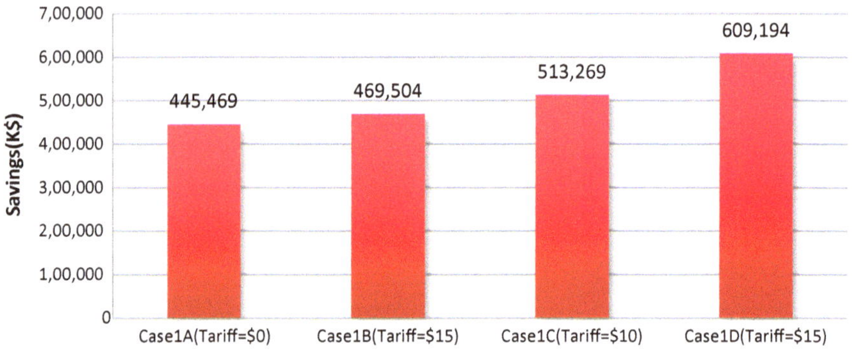

Fig. 6.3 Annual savings in thermal unit production cost between Scenario 1 and Scenario 3 (Case 3B) for the 11% renewable scenario

between Scenario 1 and Scenario 3 (consolidated WECC with transmission congestion and reserve at system level, Case 3B) are listed below:

- Saving ranges from $445 million (2.4%) with no hurdle rates for transmission to $609 million (3.3%) with the $15/MWh flat hurdle rates as shown in Fig. 6.3.
- In addition to that, another $240 million of savings can be achieved in Scenario 2 (copper-sheet consolidation).

Table 6.5 shows the generation dispatch by category for all cases in Scenario 1. Coal is a dominant resource followed by hydro and combined cycle (CC) resources. While hydro output is fixed, coal and CC can vary depending on the hurdle rates. Table 6.6 shows the generation dispatch by category for Scenario 2 and Scenario 3.

Figure 6.4 shows annual generation mix comparison among scenarios. In Scenario 1, when the hurdle rate increases from Case 1A to Case 1D, cheap generation of electricity from coal decreases. The difference is made up by higher-cost combined-cycle (CC) and steam turbine generation resources. The higher hurdle rate limits the transfer of power between BAs, so each BA needs to increase its own generation to meet the load. While many BAs have not used full capacity of coal power, other BAs need to dispatch more expensive resources (CC and steam) because the amount of cheap power transferred from surrounding BAs is reduced by higher hurdle rates.

From Scenario 1 to Scenario 3 (CBA), coal and CC generation are used more while CT and steam are less because the UC is done over the resources of the entire system. In Scenario 3, cheaper coal and CC are used before more expensive CT and steam.

In Scenario 2, because of no transmission limits, cheap coal is maximized, so CC, CT, and steam are reduced; therefore, so this scenario will have the lowest generation cost.

Table 6.5 Generation dispatch by category for Scenario 1 (GWh) for the 11% renewable integration

Category	Case 1A	%	Case 1B	%	Case 1C	%	Case 1D	%
Biomass RPS	16,613	1.6	16,712	1.6	17,097	1.7	17,046	1.7
CC	173,412	16.9	173,545	16.9	173,822	16.9	176,116	17.1
Coal	314,503	30.6	314,008	30.6	313,114	30.5	310,674	30.2
CT	22,970	2.2	23,048	2.2	23,232	2.3	23,327	2.3
DR	127	0.0	130	0.0	126	0.0	127	0.0
Geothermal	36,104	3.5	36,102	3.5	36,101	3.5	36,097	3.5
Hydro	254,726	24.8	254,721	24.8	254,721	24.8	254,722	24.8
IC	394	0.0	404	0.0	411	0.0	445	0.0
Negative bus load	4,640	0.5	4,640	0.5	4,640	0.5	4,640	0.5
Nuclear	81,370	7.9	81,438	7.9	81,277	7.9	81,256	7.9
Other steam	2,209	0.2	2,227	0.2	2,283	0.2	2,333	0.2
Pumped storage	4,117	0.4	4,131	0.4	4,120	0.4	4,123	0.4
Solar	31,580	3.1	31,580	3.1	31,580	3.1	31,580	3.1
Steam	2,182	0.2	2,284	0.2	2,456	0.2	2,648	0.3
Wind	82,243	8.0	82,243	8.0	82,243	8.0	82,243	8.0
Emergency energy	0	0.0	0	0.0	0	0.0	0	0.0
Total generation	**1,027,190**		**1,027,215**		**1,027,223**		**1,027,378**	
Native load	1,011,101	98.4	1,011,101	98.4	1,011,101	98.4	1,011,101	98.4
Dump energy	22	0.0	30	0.0	53	0.0	205	0.0
Pumped storage (load)	5,418	0.5	5,435	0.5	5,420	0.5	5,422	0.5
Pumping load	10,649	1.0	10,649	1.0	10,649	1.0	10,649	1.0
Total load	**1,027,189**		**1,027,214**		**1,027,222**		**1,027,377**	

CT combustion turbine; *DR* demand response; *IC* internal combustion; *RPS* renewable portfolio standard

6.3.2 Production Cost Analysis Results for 33% Variable Generation

This section gives the production cost modeling results for the 33% VG peneteration case. Significant effort went into development of a transmission build for the WECC nodal model to accommodate the 33% VG penetration with the objective of minimizing VG curtailment. A comparison for considered cases is shown in Table 6.7. The total generation cost is much lower than the 11% renewable adoption case because of more free renewable generation in the 33% renewable adoption case.

Table 6.6 Generation dispatch by category for Scenarios 2 and 3 (GWh) for the 11% renewable integration

Category	Case 2	%	Case 3A	%	Case 3B	%
Biomass RPS	18,344	1.8	17,080	1.7	17,386	1.7
CC	169,984	16.6	174,704	17.0	173,878	16.9
Coal	320,544	31.2	315,376	30.7	317,127	30.9
CT	18,838	1.8	20,317	2.0	19,778	1.9
DR		0.0	146	0.0	2	0.0
Geothermal	36,142	3.5	36,107	3.5	36,109	3.5
Hydro	254,818	24.8	254,751	24.8	254,754	24.8
IC	321	0.0	323	0.0	324	0.0
Negative bus load	4,640	0.5	4,640	0.5	4,640	0.5
Nuclear	81,572	8.0	81,426	7.9	81,411	7.9
Other steam	2,386	0.2	2,224	0.2	2,225	0.2
Pumped storage	3,260	0.3	3,325	0.3	3,360	0.3
Solar	31,580	3.1	31,580	3.1	31,580	3.1
Steam	1,282	0.1	1,840	0.2	1,311	0.1
Wind	82,243	8.0	82,243	8.0	82,243	8.0
Emergency energy		0.0		0.0		0.0
Total generation	**1,025,955**		**1,026,082**		**1,026,128**	
Native load	1,011,101	98.6	1,011,101	98.5	1,011,101	98.5
Dump energy	8	0.0	22	0.0	22	0.0
Pumped storage (load)	4,196	0.4	4,310	0.4	4,356	0.4
Pumping load	10,649	1.0	10,649	1.0	10,649	1.0
Total load	**1,025,954**		**1,026,081**		**1,026,127**	

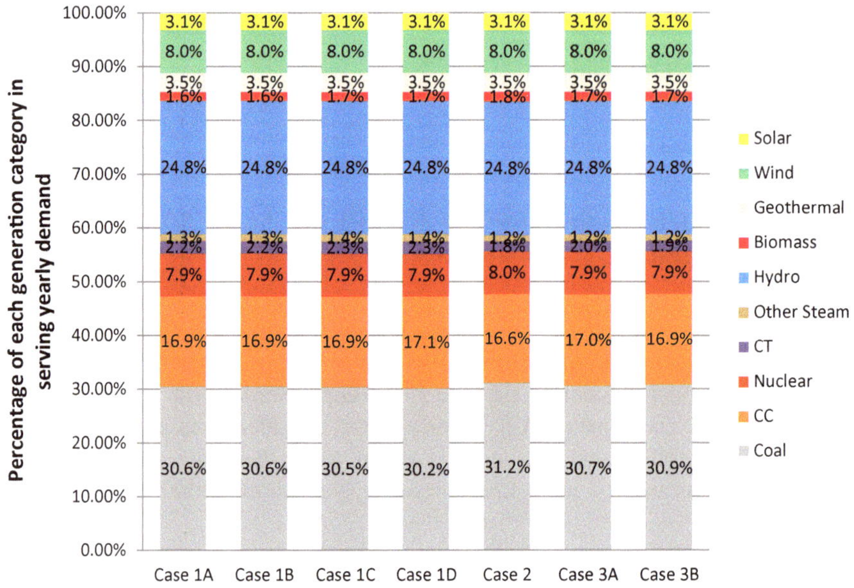

Fig. 6.4 Annual generation mix comparison among Scenarios for 11% renewable integration

Table 6.7 Comparison across all cases for 33% renewable integration

Simulation scenario	Scenario 1: current WECC BA structure (20 $/MWh unit commitment hurdle rate)				Scenario 2: consolidated WECC (CBA) copper sheet	Scenario 3: consolidated WECC with transmission congestion	
Category	Case 1A	Case1 B	Case 1C	Case 1D	Case 2	Case 3A	Case 3B
Assumed dispatch transmission hurdle rate	0$/MWh	$5/MWh	$10/MWh	15$/MWh	0$/MWh	0$/MWh	0$/MWh
Nodal model type	Full nodal	Full nodal	Full nodal	Full nodal	Full nodal	Full nodal	Full nodal
Contingency reserve	BA level	BA level	BA level	BA level	Consolidate WECC level	BA level	Consolidate WECC level
Total gen cost (K$)	12,299,308	12,336,721	12,397,077	12,492,999	10,877,487	12,026,894	11,857,182
LMP based demand cost (K $)	37,652,672	38,569,780	39,881,490	41,281,938	36,375,446	38,474,355	38,888,990

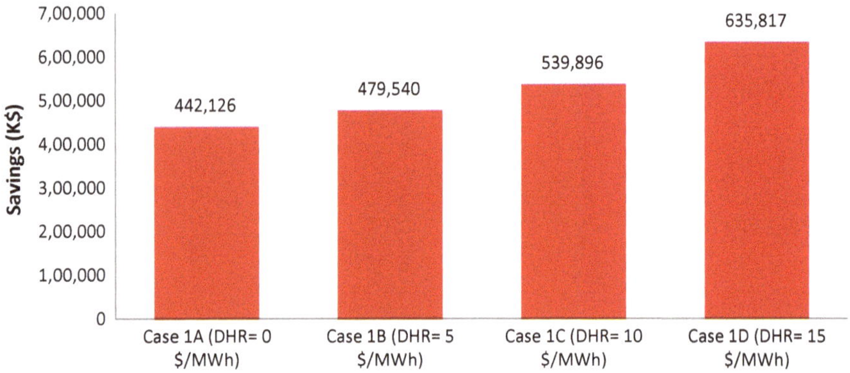

Fig. 6.5 Annual savings in thermal unit production cost between Scenario 1 and Scenario 3 (Case 3B) for 33% VG penetration case

Annual savings in thermal unit production cost between Scenario 1 and Scenario 3 (consolidated WECC with transmission congestion, Case 3B) are described below:

- Savings range from $442 million (3.7%) with no hurdle rates for transmission to $636 million (5.2%) with $15/MWh flat hurdle rates as shown in Fig. 6.5.
- In addition, another $980 million of savings can be achieved in Scenario 2 (copper-sheet consolidation). This saving mainly results from the 4,500 GWh less renewable generation curtailment and higher capacity factor of coal generation that is competing for transmission capacity with renewable generation in the transmission-constrained scenarios because renewable generation resources are geographically close to coal resources.

Table 6.8 shows the generation dispatch by category for all cases in Scenario 1. Hydro is a dominant resource followed by coal and wind. While hydro output is fixed, coal can vary depending on the hurdle rates, and wind varies because wind curtailment varies with different hurdle rates. While baseload resources (coal and nuclear) decrease, the other higher-cost (gas) resources increases when the hurdle rate increases. This is because the higher hurdle rate limits the transfer of the free (i.e., wind) resource and cheaper power (nuclear and coal) resources from one BA to others. As a result, more expensive (gas) resources must be used to meet local load in each individual BA. Table 6.9 shows the generation dispatch by category for Scenarios 2 and 3.

Figure 6.6 shows annual generation mix comparison between scenarios for the 33% renewable integration. In Scenario 1, when hurdle rate increases (from Case 1A to Case 1D), cheap generation from coal and nuclear decreases. The difference is made up by higher-cost CC, CT, biomass, and other resources including steam

Table 6.8 Generation dispatch by category for Scenario 1 (GWh) for the 33% renewable integration

Category	Case 1A	%	Case 1B	%	Case 1C	%	Case 1D	%
Biomass RPS	12,444	1.2	12,806	1.2	13,406	1.3	13,834	1.3
CC	88,947	8.7	89,226	8.7	90,072	8.8	92,259	9.0
Coal	237,807	23.1	237,349	23.1	236,720	23.0	234,681	22.8
CT	17,946	1.7	17,960	1.7	18,045	1.8	18,207	1.8
DR	88	0.0	88	0.0	86	0.0	85	0.0
Geothermal	35,438	3.4	35,454	3.4	35,345	3.4	35,326	3.4
Hydro	254,760	24.8	254,762	24.8	254,760	24.8	254,761	24.8
IC	184	0.0	191	0.0	202	0.0	232	0.0
Negative bus load	4,640	0.5	4,640	0.5	4,640	0.5	4,640	0.5
Nuclear	76,714	7.5	76,568	7.4	75,894	7.4	75,534	7.3
Other steam	1,493	0.1	1,552	0.2	1,648	0.2	1,748	0.2
Pumped storage	4,455	0.4	4,471	0.4	4,456	0.4	4,443	0.4
Solar	61,090	5.9	61,109	5.9	61,100	5.9	61,099	5.9
Steam	1,688	0.2	1,709	0.2	1,738	0.2	1,730	0.2
Wind	230,107	22.4	229,893	22.4	229,625	22.3	229,179	22.3
Emergency energy	–	0.0	–	0.0	–	0.0	–	0.0
Native load	1,011,101	98.4	1,011,101	98.4	1,011,101	98.4	1,011,101	98.4
Dump energy	193	0.0	149	0.0	127	0.0	166	0.0
Pumped storage (load)	5,856	0.6	5,878	0.6	5,858	0.6	5,841	0.6
Pumping load	10,649	1.0	10,649	1.0	10,649	1.0	10,649	1.0
Total load	1,027,798		1,027,777		1,027,734		1,027,756	
Total generation	1,027,800		1,027,778		1,027,736		1,027,757	
Solar curtailment	177	0.3	158	0.3	167	0.3	169	0.3
Wind curtailment	3,890	1.7	4,105	1.8	4,373	1.9	4,818	2.1
Solar total	61,267		61,267		61,267		61,267	
Wind total	233,997		233,997		233,997		233,997	

turbine using fuel other than coal, DR, IC, negative bus load, and pumped storage hydro. The higher hurdle rate limits the transfer of power between pools, so each pool needs to increase its own generation to meet the load. While many pools have not used the full capacity of coal power, other pools need to dispatch energy produced by more expensive resources (CC and steam) because the energy transferred from surrounding pools is reduced by the higher hurdle rate.

Table 6.9 Generation dispatch by category for Scenarios 2 and 3 (GWh) for the 33% renewable integration

Category	Case 3A	%	Case 3B	%	Case 2	%
Biomass RPS	12,714	1.2	12,816	1.2	11,058	1.1
CC	83,898	8.2	83,130	8.1	66,478	6.5
Coal	244,625	23.8	245,815	23.9	259,898	25.3
CT	17,320	1.7	17,021	1.7	15,905	1.5
DR	122	0.0	–	0.0	–	0.0
Geothermal	35,385	3.4	35,426	3.4	36,016	3.5
Hydro	254,753	24.8	254,754	24.8	254,760	24.8
IC	187	0.0	188	0.0	115	0.0
Negative bus load	4,640	0.5	4,640	0.5	4,640	0.5
Nuclear	76,333	7.4	76,434	7.4	76,591	7.5
Other steam	1,511	0.1	1,523	0.1	1,334	0.1
Pumped storage	4,260	0.4	4,281	0.4	4,173	0.4
Solar	61,089	5.9	61,097	5.9	61,261	6.0
Steam	1,600	0.2	1,210	0.1	1,114	0.1
Wind	229,130	22.3	229,255	22.3	233,923	22.8
Emergency energy	0	0.0	0	0.0	–	0.0
Native load	1,011,101	98.4	1,011,101	98.4	1,011,101	98.4
Dump energy	236	0.0	228	0.0	35	0.0
Pumped storage (load)	5,582	0.5	5,611	0.5	5,482	0.5
Pumping load	10,649	1.0	10,649	1.0	10,649	1.0
Total load	1,027,567		1,027,589		1,027,266	
Total generation	1,027,568		1,027,590		1,027,267	
Solar curtailment	178	0.3	171	0.3	6	0.0
Wind curtailment	4,867	2.1	4,743	2.0	74	0.0
Solar total	61,267		61,267		61,267	
Wind total	233,997		233,997		233,997	

From Scenario 1 to Scenario 3 (CBA), coal generation and CC generation are used more while CT and steam are used less because the unit commitment is done over the resources of the entire system. In Scenario 3, cheaper coal is used before more expensive CT and steam. Note that, in Case 3B, curtailment is higher than in Case 1A. Because the reserve is held at the WECC level in Case 3B, there are periods that power transfer between BAs is constrained by the transmission system. Those congestion periods will cause more renewable curtailment in this case.

In Scenario 2, because of no transmission limits, cheap coal and nuclear are maximized and renewable curtailment is minimized, so CC, CT, and steam are reduced. Therefore, this scenario will have the lowest generation cost.

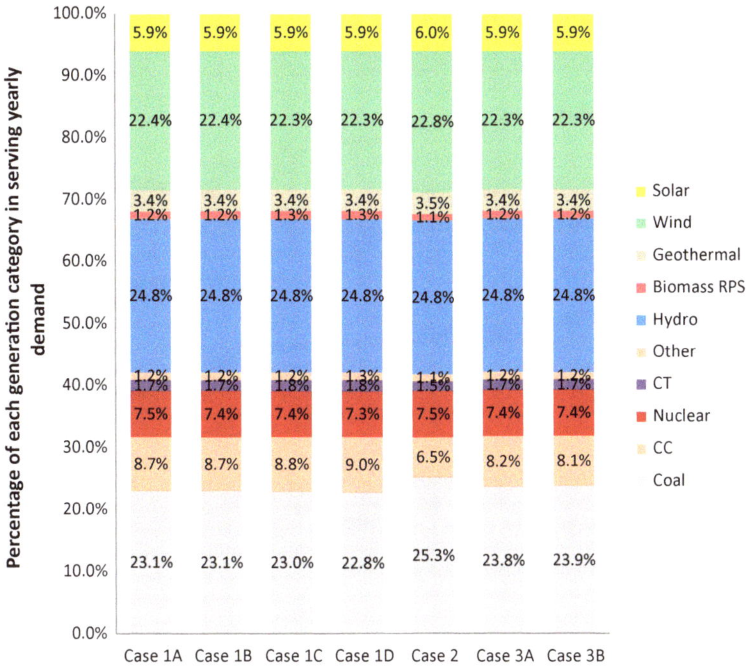

Fig. 6.6 Annual generation mix comparison between Scenarios for the 33% renewable integration

6.4 Balancing Reserve Analysis Approach

This part of the study addresses the reduction in regulation and load-following requirements when operating a single CBA compared to operating individual BAs. Four services necessary to meet system balancing needs were considered: (1) up-regulation, (2) down-regulation, (3) up-load-following, and (4) down-load following. This section introduces the main components of the balancing analysis and provides a summary of the main assumptions used in this study.

6.4.1 Proposed Evaluation Metrics

The main objective of this effort is to identify and evaluate key technical metrics in order to demonstrate the potential gains with respect to savings in balancing requirements that would be accrued by consolidating BAs within a large geographical area and by operating them as a single, consolidated BA. In this study, the savings are calculated in terms of regulation (0–10 min balancing needs) and load following (10 min to 1 h balancing needs). The following subsections present a

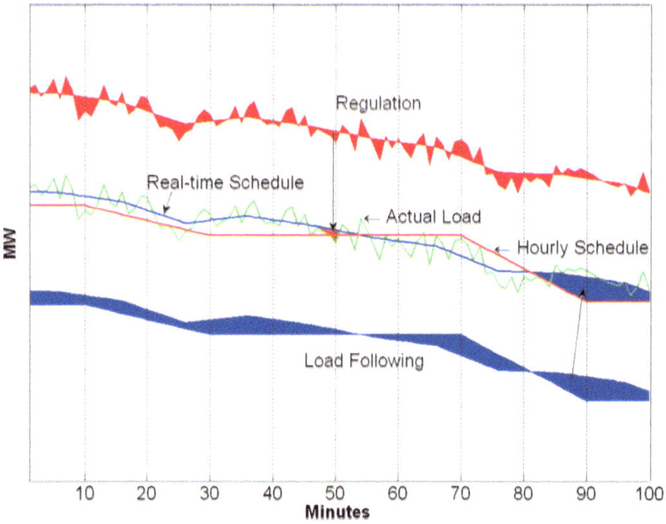

Fig. 6.7 Computation of regulation and load following

detailed procedure and the assumptions used to calculate these two types of balancing requirements.

6.4.1.1 Definition of Regulation and Load Following

The minute-to-minute regulation curve shown in Fig. 6.7 represents the regulation requirements within the 1–10 min time frame appearing as a result of real-time forecast errors and 10 min discretization of the real-time schedule. The regulation signal is calculated by subtracting the real-time schedule from the actual net load. In mathematical terms, regulation is defined as [13–15]:

$$\begin{aligned} Regulation\_curve(m) &= Actual\_net\_load(m) - Real - \text{time\_schedule}(10\text{m}) \\ &= [L_{actual}(m) - W_{actual}(m) + Ramp_{20\text{min}}\{Net\_interchange(h)\}] \\ &\quad - [L_{schedule}(10m) - W_{schedule}(10m) + Ramp_{20min}\{Net\_interchange(h)\}] \end{aligned}$$

thus,

$$\begin{aligned} Regulation\_curve(m) = {} & [L_{actual}(m) - W_{actual}(m)] \\ & - [L_{schedule}(10m) - W_{schedule}(10m)] \end{aligned} \tag{1}$$

where L stands for system load; W stands for wind power output; m represents minutes; and h represents hours. Usually, a 10 min ramp is added between every two continuous 10 min average intervals in the two terms, $L_{schedule}(10m)$ and $W_{schedule}(10m)$.

The minute-to-minute load following curve represents the load following requirements within the hour time frame as a result of hour-ahead forecast errors and 1 h discretization of the hour-ahead schedule. It is calculated by subtracting the hour-ahead schedule from the real-time schedule:

$$Load\_following\_curve(m) = [L_{schedule}(10m) - W_{schedule}(10m)] - [L_{schedule}(h) - W_{schedule}(h)] \quad (2)$$

Usually, a 20 min ramp is added between every two continuous hourly average intervals in the two terms, $L_{schedule}(h)$ and $W_{schedule}(h)$. The definitions of regulation and load following curves are further illustrated in Fig. 6.7.

6.4.1.2 Evaluation Metrics

Potential benefits of BA consolidation will be demonstrated through four basic metrics, which we refer to as the "first performance envelope." They are capacity, ramp duration, ramp rate and energy of ramps encountered in the derived regulation and load following curves. These four metrics are schematically illustrated in Fig. 6.8.

After a regulation/load following curve is obtained using the approach discussed in Sect. II.A, a "swinging door" algorithm is implemented to smooth the curve and reject noises by carefully tuning the width of the "door". The "swinging door" algorithm was initially proposed for data compression, and then was used in [14] to

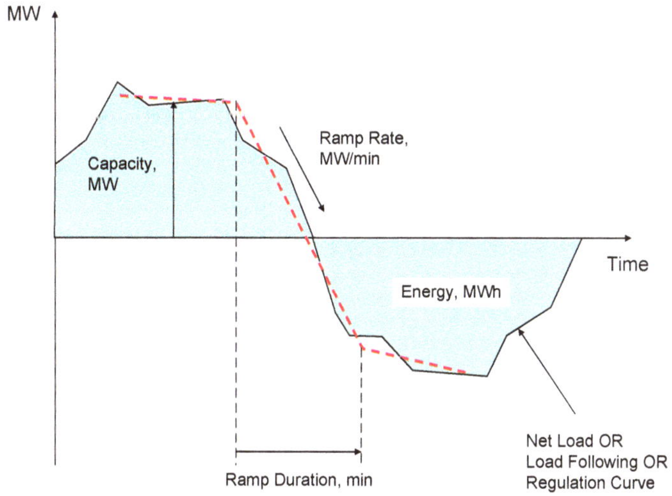

Fig. 6.8 Schematic illustration for the four performance metrics

pre-process regulation and load-following signals. This algorithm can effectively compress time series data by grouping points with a similar ramp rate. A uniform ramp rate value will then be assigned to all the points within the same "door." The idea of a performance envelope is to include all the data points for a time period, e.g., a year, in a four- dimensional space built by the four evaluation metrics. Given a specified percentile threshold (e.g., 95%), a probability box can be determined to include 95% of the points inside the box. If the system can meet the requirements within the performance envelope, it will likely comply with the control performance standard criteria without over-performing the balancing job. In this way, the maximum balancing requirements, such as capacity and ramp rate, can be easily determined. A more detailed explanation of this "performance envelope" algorithm can be found in [14]. In this study, we selected the percentile thresholds to be 95% in the study scenarios.

6.4.1.3 Evaluation of Savings Due to BAs Consolidation

The savings due to BAs consolidation can be identified in terms of saving in load following and regulation requirements of the consolidated BA in comparison with individual BAs in both capacity and ramping.

As an example, the savings in load following capacity requirements differ for the consolidated BA versus the sum of individual BAs as shown in Fig. 6.9. For each

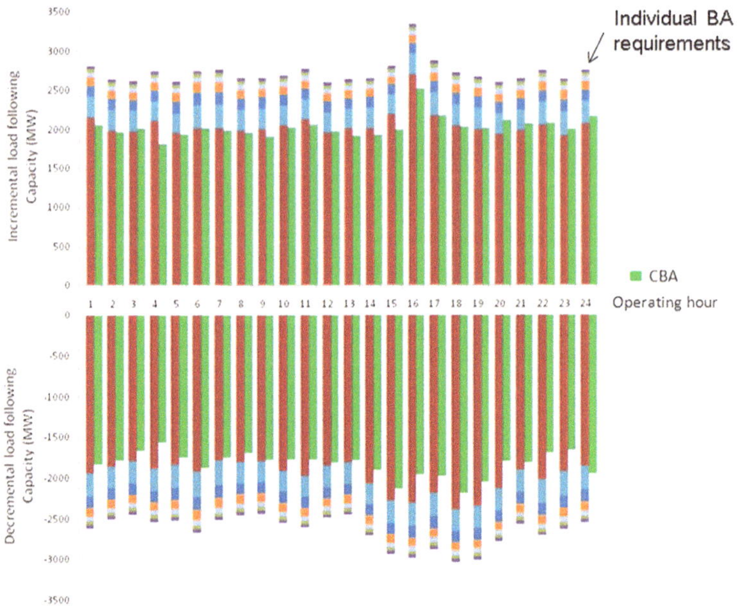

Fig. 6.9 Load following capacity requirements for each operating hour (individual BAs vs. consolidated BA)

operating hour, a pair of upward (incremental) and downward (decremental) bars indicate the differences in load following capacity for that hour between the consolidated BA and separate BA cases. The multi-colored bars show the load following for individual BAs, whereas the solid green bar denotes the load following capacity for the same hour if the BAs were consolidated.

6.4.2 Impact of Forecast Errors and Their Correlation

The results presented in this study are heavily influenced by forecast errors and their correlation among the separate BAs [16–18]. These errors impact the results because BAs rely on these forecasts to schedule and dispatch their generation. When the wind and solar forecast errors are strongly and positively correlated, this would result in simultaneous over-generation or under-generation situations in BAs. This eliminates the possibility of situations where some BAs are over-generating and some are under-generating, and the opportunity for diversity is minimized. Consequently, in such a case, the benefits of consolidation would be minimal. When the forecast errors are non-correlated, the diversity needed for revealing the maximum benefits from BA consolidation will be created. The effect of forecast error correlation is further explained using the following rationale.

For the load forecast errors, ΔL_i, of the WECC balancing authorities, the standard deviation, σ, of the load forecast error for the CBA is calculated as:

$$\sigma^2\left(\sum_i \Delta L_i\right) = \sum_i \sigma_i^2(\Delta L_i) + 2\sum_{\substack{i,j \\ i\neq j}} [\rho_{ij}\,\sigma(\Delta L_i) * \sigma(\Delta L_j)] \tag{3}$$

where i is the BA number and ρ_{ij} is the correlation factor between any two BAs. The percentage standard deviation is defined by

$$\sigma = \sigma_{\%} * \max(L) \tag{4}$$

Thus,

$$\begin{aligned}\sigma_{\%}^2\left(\sum_i \Delta L_i\right) * max^2\left(\sum_i L_i\right) &= \sum_i \sigma_{\%}^2(\Delta L_i) * max^2(L_i) \\ &+ 2\sum_{\substack{i,j \\ i\neq j}} \{\rho_{ij}\,\sigma_{\%}(\Delta L_i) * \max(L_i) * \sigma_{\%}(\Delta L_j) * \max(L_j)\}\end{aligned} \tag{5}$$

Case 1: By assuming 100% correlation between forecast errors, which is $\rho_{ij} = 1$ and $\sigma_{\%}(\Delta_{Li})$ = constant, the new percentage standard deviation of CBA forecast error becomes:

$$\sigma_{\%}\left(\sum_i \Delta L_i\right) * max\left(\sum_i L_i\right) = \sum_i \sigma_{\%}(\Delta L_i) * \max(L_i)$$
$$\sigma_{\%CBA} = \sigma_{\%}(\sum_i \Delta L_i) = \sigma_{\%}(\Delta L_i) * \frac{\sum_i \max(L_i)}{max(\sum_i L_i)} \quad (6)$$

Therefore, the percentage standard deviation of the load forecast error of a CBA equals the percentage standard deviation of the load forecast error of each participating BA multiplied by a factor.

Case 2: Assuming zero correlation factor, which is $\rho_{ij} = 0$ and $\sigma_{\%}(\Delta_{Li})$ being the same for all BAs, results in reduction of the CBA forecast error span due to the error diversity observed in this case. The new CBA standard deviation of load forecast error is:

$$\sigma^2_{\%CBA} = \sigma^2_{\%}(\textstyle\sum_i \Delta L_i) = \sigma^2_{\%}(\Delta L_i) * \frac{\sum_i max^2(L_i)}{max^2(\sum_i L_i)}$$
or
$$\sigma_{\%CBA} = \sigma_{\%}(\Delta L_i) * \sqrt{\frac{\sum_i max^2(L_i)}{max^2(\sum_i L_i)}} \quad (7)$$

The last expression is used to re-calculate the percentage standard deviation of the load forecast error of a CBA to reflect zero correlation among BAs' load forecast errors; it is obvious that the standard deviation of CBA forecast error in this case is lower than that of Eq. (6).

Most forecasting companies use proprietary methods to employ different forecasting algorithms and techniques that are generally well known to industry. The major differences in the forecast results for these companies lie in the details and heuristic techniques involved. Forecast errors might be correlated primarily because of the statistical methods used that imply one or another form of the averaging process. Although different forecasting companies use different types of algorithms, apparently they all include some form of the averaging process. As a result, the forecasts lag the actual generation curve. The lagging effect could create correlation if the actual wind and load changes in participating BAs have similar patterns. This type of correlation is worst among BAs that are geographically closer and less impactful for BAs that are geographically dispersed. Of course, the choice of forecasting services would impact the result.

6.5 Results of Balancing Analysis

Several study cases are developed and tested to investigate the impact of different levels of forecast accuracy and different forecast error generators on the savings in balancing services achieved from BA consolidation. The detailed methodologies of calculating balancing services, the proposed metrics to evaluate savings in

Table 6.10 Definition of four study cases

Cases\forecast error	TND method			Optimization method
	Low	Medium	High	
Low	√			
Medium		√		
High			√	
Historical				√

Table 6.11 Statistics of hour-ahead forecast errors for load and wind, for cases "low," "medium," and "high;" (the percentage values are based on peak load for load and on installed wind capacity for wind)

Statistics of HA forecast errors		High accuracy	Medium accuracy	Low accuracy
Load	Mean	0%	0%	0%
	Standard deviation	1%	2%	2.5%
	Autocorrelation	0.9	0.9	0.9
Wind	Mean	0%	0%	0%
	Standard deviation	4%	7%	12%
	Autocorrelation	0.6	0.6	0.6

balancing services, and several types of forecast error generators are provided in [14–18].

6.5.1 *Study Cases*

In the balancing reserve analysis, four study cases are defined in Table 6.10 to provide a comprehensive evaluation of the savings in balancing services, to compare the performance of different forecast error generators, and to estimate the sensitivity of savings to forecast accuracy. Cases "Low," "Medium," and "High" are the cases with HA forecast errors generated using the Truncated Normal Distribution (TND) method; the forecast errors are at low, medium, and high accuracy levels, respectively. Case "Historical" has HA forecast errors generated by the optimization method. Table 6.11 lists the statistics of HA forecast errors for load and wind used to generate Cases "Low," "Medium," and "High." In Table 6.12, the statistics of RT forecast errors for load are shown; a persistence model is used to generate RT wind forecast data. Because solar forecast errors are highly affected by solar clear clearness index, the statistics of HA solar forecast errors are shown separately in Table 6.13. The same persistence model described in Table 6.12 is used to generate the RT solar forecast.

Table 6.14 illustrates the savings for two study scenarios: (1) the 11% VG case and (2) the 33% VG case. The term capacity is defined as the amount of unit commitment required to meet regulation or load-following requirements.

Table 6.12 Statistics of RT forecast errors for load and wind, for cases "low," "medium," and "high"

Statistics of RT load forecast	Mean error	0%
	Standard deviation	0.15%
	Auto correlation	0.6
Statistics of RT wind forecast	Mean error	Persistence model assumes that the RT wind forecast equals the actual wind production observed 7.5 min before the beginning of the dispatch interval
	Standard deviation	
	Autocorrelation	

Note The percentage values are based on peak load

Table 6.13 Statistics of HA solar forecast errors for cases "low," "medium," and "high"

Clearness index (CI)	Mean	Standard deviation (%)			Autocorrelation
		Low	Medium	High	
$0 < CI \leq 0.2$	0	10	5	4	0.6
$0.2 < CI \leq 0.5$	0	30	20	9	0.6
$0.5 < CI \leq 0.8$	0	25	15	6.5	0.6
$0.8 < CI \leq 1.0$	0	10	5	4	0.6

Note The percentage values are based on solar installed capacity

Table 6.14 Benefits, or savings, of CBAs based on the 11 and 33% VG cases

		11% VG case		33% VG case	
		Up	Down	Up	Down
Regulation capacity, GW	Individual BAs	1.76	–1.82	3.65	–3.78
	CBA	0.74	–0.75	1.05	–1.09
	Savings in %	58	59	71	71
Regulation ramp rate, MW/min	Individual BAs	566	–597	962	–993
	CBA	138	–143	222	–220
	Savings in %	75	76	76	77
Load-following capacity, GW	Individual BAs	12	–11	18	–18
	CBA	4.1	–3.9	5.2	–5.0
	Savings in %	64	66	70	72
Load-following ramp rate, MW/min	Individual BAs	356	–357	708	–715
	CBA	186	–190	245	–254
	Savings in %	48	47	65	65

The savings are calculated as the difference between the sum of individual BA balancing requirements and the combined BA balancing requirements.

The study identifies the savings in regulation and load-following for each case as the differences between separate BA balancing needs versus CBA balancing needs. These balancing needs are divided into regulation and load-following. Regulation is defined as the capacity and energy that are available within the 1–10 min time frame needed to balance load and generation. Load-following is the capacity and energy that are available within the 10 min to 1 h time frame needed for the balance. The capacity savings is the capacity that what would otherwise have been set aside for regulation and load-following.

6.5.1.1 Individual Versus Combined Savings Due to Balancing Authority Consolidation

Benefits of consolidations are shown in several different ways in this report. Figure 6.10 illustrates how the load-following capacity requirements differ for the CBA versus the sum of individual BAs. For each operating hour, a pair of upward (i.e., Incremental) and downward (i.e., Decremental) bars indicate the differences in load-following capacity for that hour between the CBA and separate BA cases. The blue bars show load-following for individual BAs, whereas the red bars denote the load-following capacity for the same hour if the BAs were consolidated.

Figure 6.10 reflects study results for a single month, illustrating savings in load-following due to BA consolidation. Figures 6.11 and 6.12 illustrate the monthly savings in the capacity component of load-following for the 11% VG Case in 2020, while Figs. 6.13 and 6.14 show savings in the capacity component of load-following for the 33% VG Case in 2020.

This study assumes that the load-following reserves during each month were permanently committed for that month, regardless of whether they are fully used. Stated another way, the monthly capacity savings values shown are not the expected changes in reserves' utilization but rather their commitments. The same assumptions were used for regulation reserves as well.

6.5.1.2 Additional Benefits Due to Reduced Ramp Rates When Balancing Authorities Are Consolidated

To comprehensively quantify the savings in regulation and load-following capacity due to BA consolidation, the metric of *ramp rate* was introduced. This metric reveals a few different types of savings that will be realized by the generation dispatch although it is very difficult to monetize these savings. These savings imply that the dispatched generation fleet will not need to ramp as steeply; therefore, saving on equipment wear-and-tear is realized. Savings could also be realized through a dispatch that does not require as much ramping capacity; in some instances, this could result in more economic dispatch.

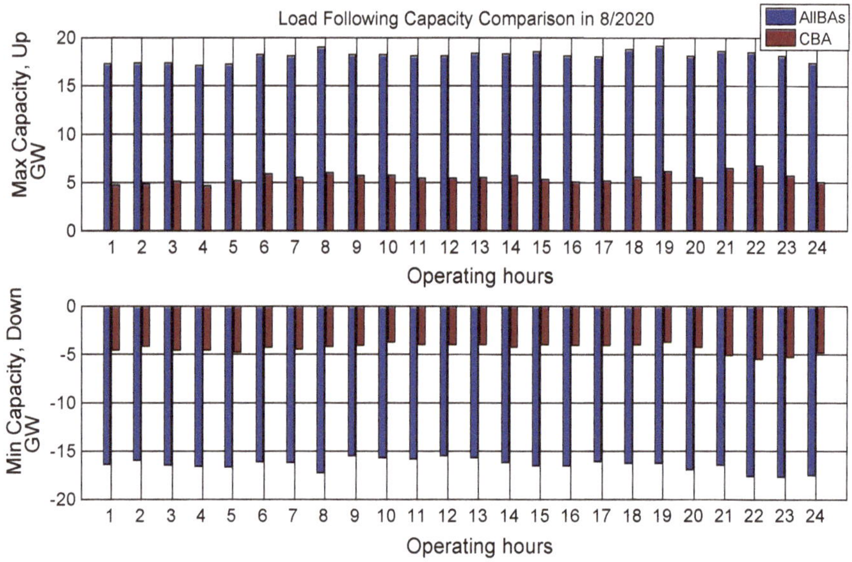

Fig. 6.10 Differences between consolidated and individual BAs' up and down-load-following capacity during August of 2020

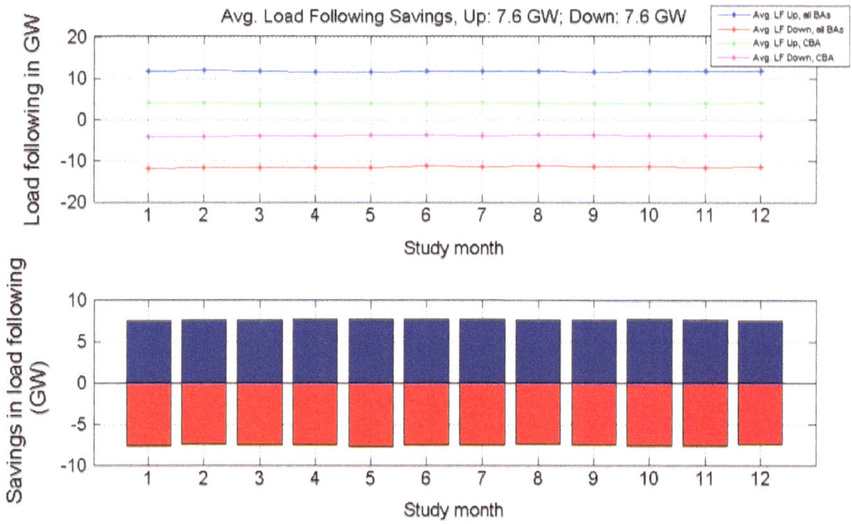

Fig. 6.11 Differences in load-following capacity between CBAs and the sum of individual BAs for the 11% VG case

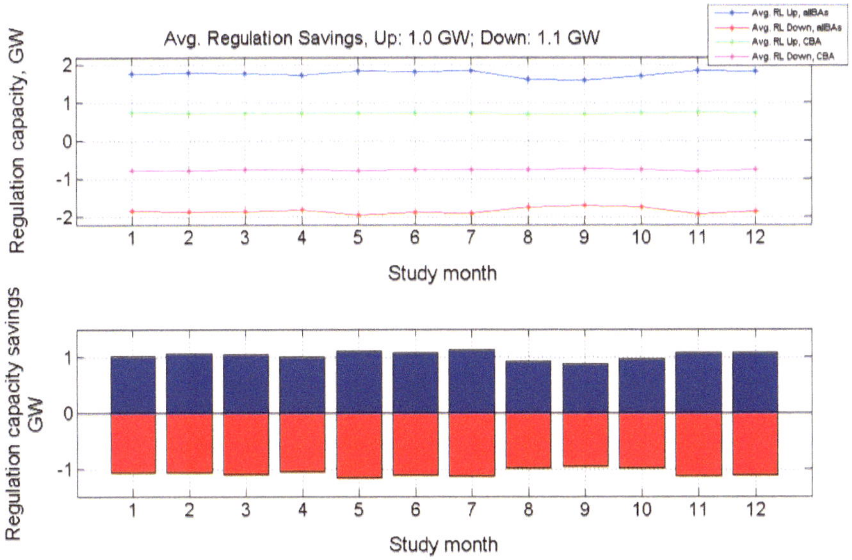

Fig. 6.12 Differences in regulation capacity between CBAs and the sum of individual BAs for the 11% VG case

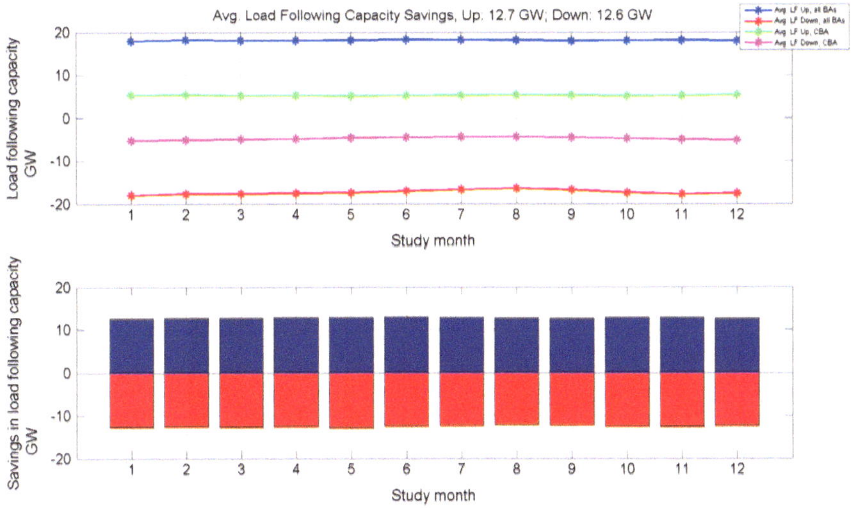

Fig. 6.13 Differences in load-following capacity between CBAs and the sum of individual BAs for the 33% VG case

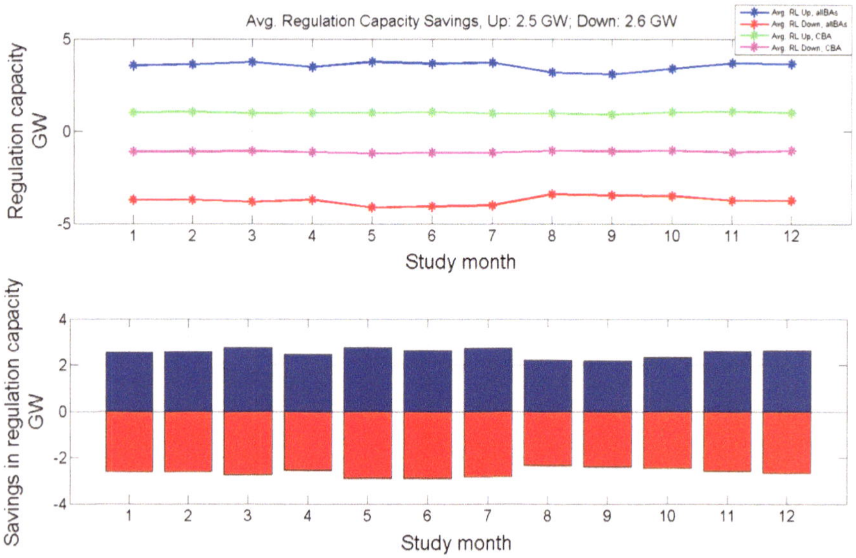

Fig. 6.14 Differences in regulation capacity between CBAs and the sum of individual BAs for the 33% VG case

6.5.1.3 Sensitivity of Balancing Reserve Reduction to Forecast Errors and Variable Generation Penetration Levels

The increase of renewable penetration increases the need for regulation and load-following capacity and ramp rate; meanwhile, the reduction (savings) resulting from BA consolidation increase as summarized in Table 6.15.

Figures 6.15 and 6.16 compare the load-following capacity and ramp-rate savings for cases with different hour-ahead (HA) forecast accuracy levels and with two different VG penetration cases (11 and 33%). Figures 6.17 and 6.18 compare regulation capacity and ramp-rate savings for cases with different HA forecast accuracy levels and with two different VG penetration cases (11 and 33%).

It can be seen that more savings in load-following capacity and ramp-rate are achieved as HA forecast error accuracy for individual BAs decreases. In other words, as forecast error accuracy of individual BAs increases, benefits of consolidation decrease. If forecast error was zero (a hypothetical example), there consolidation would still be beneficial because the diversity of variability in the CBA would still exist.

6.5.1.4 Monetizing the Reduction in Balancing Reserve

There are several ways to monetize the reduction in regulation and load-following. One approach would be to use information obtained from past regional

Table 6.15 Comparison of balancing reserve reduction of CBA versus VG penetration level

		11% case		33% case	
		Up	Down	Up	Down
Regulation capacity, GW	Individual BAs	1.76	−1.82	3.65	−3.78
	CBA	0.74	−0.75	1.05	−1.09
	Savings in %	58	59	71	71
Regulation ramp rate, MW/min	Individual BAs	566	−597	962	−993
	CBA	138	−143	222	−220
	Savings in %	75	76	76	77
Load-following capacity, GW	Individual BAs	12	−11	18	−18
	CBA	4.1	−3.9	5.2	−5.0
	Savings in %	64	66	70	72
Load-following ramp rate, MW/min	Individual BAs	356	−357	708	−715
	CBA	186	−190	245	−254
	Savings in %	48	47	65	65

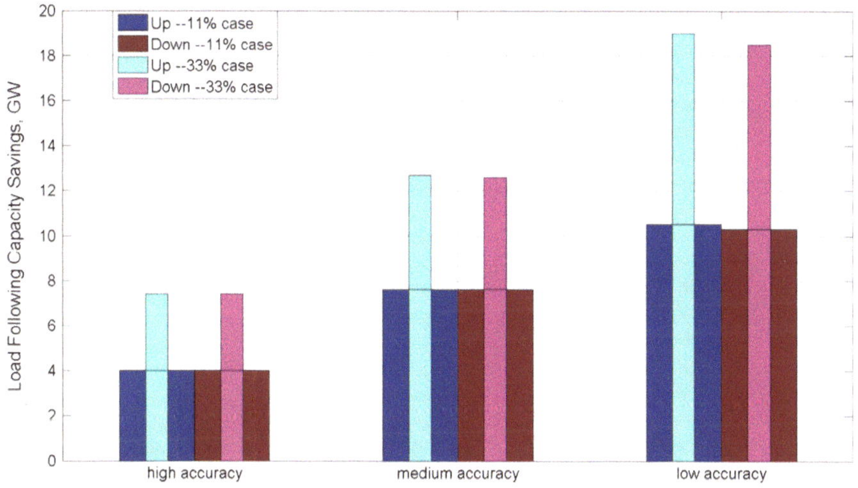

Fig. 6.15 Comparison of the savings in load-following capacity between the 11 and 33% cases

consolidation studies. Another method would be to apply the balancing reserve capacity values used in Ancillary Service Markets across North America. A third alternative would be to develop a proxy price using a gas-fired CT. This type of analysis was out of the scope of this study.

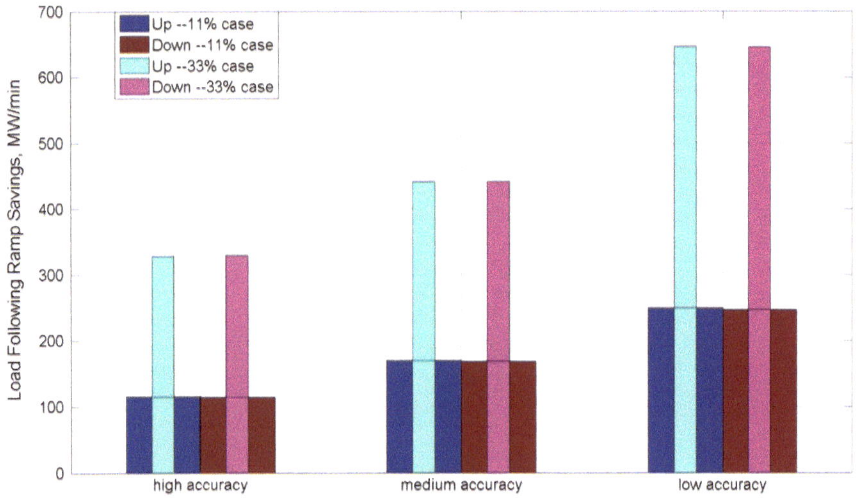

Fig. 6.16 Savings in load-following ramp rate versus hour-ahead forecast accuracy for the 11 and 33% cases

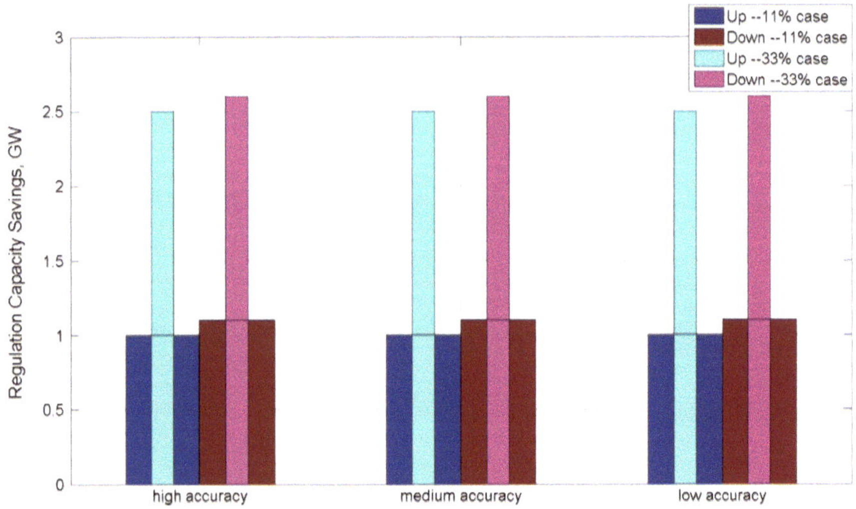

Fig. 6.17 Savings in regulation capacity versus hour-ahead forecast accuracy for the 11 and 33% cases

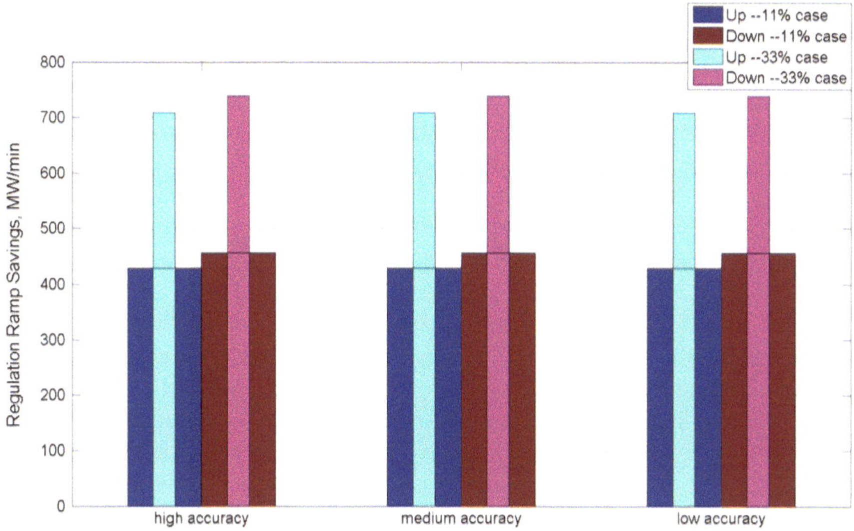

Fig. 6.18 Savings in regulation ramp rate versus hour-ahead forecast accuracy for the 11 and 33% cases

6.6 Conclusions

The research team found that effective use of the diversity in load and renewable generation over a wide area can indeed achieve significant savings. The team tested its detailed procedure for computing the savings derived from CBAs, and its evaluation metrics for demonstrating the benefits of CBAs using several study scenarios designed for the set of BAs in the western United States. Study results showed significant reduction in the required capacity and ramp rates of balancing services. The CBA, with a much wider geographical boundary, results in significantly less load-following, and regulation reserve requirements compared to the sum carried by the individual BAs. It allows for the sharing of variability at a less than hourly timescale among BAs and can provide access to additional existing physical response capability. The savings could be realized immediately by reallocating regulation and load-following reserve requirements among the BAs to reduce the operational burden.

Balancing authority consolidation is one approach that can be taken to mitigate challenging problems for operators. A CBA will provide operators with many advantages for dispatching generation more economically than in the system with many individual BAs. These advantages are the smaller reserve requirements, more use of cheaper units, and smaller load and renewable resource forecast errors. This work shows that the effective use of the diversity in load and renewable generation over a wide area can indeed achieve significant savings.

For the 11% VG penetration case, the study shows that the yearly production cost savings of thermal units for a consolidated WECC ranges between $440 million (2.4% of total annual production cost) and $610 million (3.2%), depending on how the current BA structure is modeled. The implementation cost for the CBA is not considered in these estimates. In addition to these savings, the copper-sheet consolidation scenario shows an extra savings of $240 million (1.4%) per year.

For the 33% VG penetration case, yearly production cost savings range from $442 million (3.7% of total annual production cost) with no hurdle rates for transmission to $636 million (5.4%) with $15/MWh flat hurdle rates. In addition, another $980 million (8.3%) of savings can be achieved in the copper-sheet consolidation scenario. This saving mainly results from the 4,500 GWh less renewable generation curtailment and higher capacity factor of coal generation.

Increasing ramping and ramp duration requirements, as well as increasing uncertainty surrounding the size and timing of these ramps, significantly impacts the generation capacity needed for providing essential grid balancing functions, overall efficiency and costs of the generation commitment and dispatches, operating reserve requirement, emissions from thermal plants, scheduling of hydro power cascades, and on fish preservation issues, as well as on wear-and-tear on conventional generation units participating in the balancing services. In this chapter, the proposed method to quantify load-following and regulation is used to evaluate the benefits of BA consolidation. The results for different study cases clearly show that significant savings in the regulation and load-following requirements can be achieved by BA consolidation. In addition, the savings are highly sensitive to HA forecast accuracy. The savings in load-following and regulation, both capacity and ramp, play an important role in operating the power system more economically. Although a quantification of these impacts in many instances becomes a challenge, especially their dollar quantification, existing expert opinions and studies demonstrate the existence of these impacts quite clearly. The dollar value of varying ramping requirements could be evaluated in the future by applying UC and dispatch procedures incorporating environmental, technical, and generator flexibility constraints. Several ways to monetize the savings in regulation and load-following are described below:

1. Use information obtained from past regional consolidation studies.
2. Monetize the savings based on one of many ancillary service markets across North America.
3. Develop a proxy price using a gas-fired CT.
4. Develop a hybrid approach, using a current price for capacity and a market index price for energy used in providing regulation and load-following services.

References

1. GE Energy, Western wind and solar integration study. NREL/SR-550-47434 (National Renewable Energy Laboratory, Golden, Colorado, 2010), http://www.nrel.gov/docs/fy10osti/47434.pdf. Accessed 15 Sept 2013
2. Y.M.R. Diao, P.V. Etingov, S. Malhara, N. Zhou, R.T. Guttomson, J. Ma, P. Du, N.A. Samaan, C. Sastry, Analysis methodology for balancing authority cooperation in high penetration of variable generation. PNNL-19229 (Pacific Northwest National Laboratory, Richland, Washington, 2010), http://www.pnl.gov/main/publications/external/technical_reports/PNNL-19229.pdf. Accessed 15 Sept 2014
3. M. Milligan et al., Examination of potential benefits of an energy imbalance market in the western interconnection. NREL/TP-5500-57115 (National Renewable Energy Laboratory, Golden, CO, 2013, http://www.nrel.gov/docs/fy13osti/57115.pdf. Accessed 15 Sept 2014
4. E3, *PacifiCorp-ISO Energy Imbalance Market Benefits* (Energy and Environmental Economics, San Francisco, California, 2013), http://www.caiso.com/Documents/PacifiCorp-ISOEnergyImbalanceMarketBenefits.pdf. Accessed 15 Sept 2014
5. R. Orans, A. Olson, J. Moore, *WECC EDT Phase 2 EIM Benefits Analysis & Results (October 2011 Revision)* (Energy and Environmental Economics, San Francisco, California, 2011), http://www.wecc.biz/committees/EDT/Documents/E3_EIM_Benefits_Study-Phase_2_Report_RevisedOct2011_CLEAN2[1].pdf. Accessed 15 Sept 2014
6. N.A. Samaan, R. Schellberg, D. Warady, S. Williams, R. Bayless, S. Conger, R. Brush, T. Gossa, M. Symonds, K. Harris, J. Newkirk, C. Kalich, T.B. Nguyen, M. Rarity, P. Williams, P. Damiano, C. Jin, S. Wallace, M. Landauer, C. Macarthur, D. Wu, J. Austin, H. Owen, T. Martin, R. Diao, R. Noteboom, W. Morter, J. Hoerner, Y.V. Makarov, T. Van Blaricom, K. Haraguchi, S. Knudsen, L. Kannberg, K. McRunnel, J. Portouw, A. Johnson, T. Guo, J. Apperson, K. Downey, R. Link, S. Dennison-Leonard, M. Empey, S. Sorey, D. Holcomb, M. Goodenough, P. Etingov, Analysis of benefits of an energy imbalance market in the NWPP. PNNL-22877 (Pacific Northwest National Laboratory, Richland, Washington, 2013), http://www.pnnl.gov/main/publications/external/technical_reports/PNNL-22877.pdf. Accessed 15 Nov 2013
7. M. Hunsaker, N. Samaan, M. Milligan, T. Guo, G. Liu, J. Toolson, Balancing authority cooperation concepts to reduce variable generation integration costs in the western interconnection: intra-hour scheduling. Final WECC Project Report (2013), http://www.wecc.biz/committees/StandingCommittees/JGC/VGS/Shared%20Documents/BA%20Cooperation%20Study/Final%20Report/Balancing%20Authority%20Cooperation%20Concepts%20-%20Intra-Hour%20Scheduling.pdf. Accessed 15 Sept 2014
8. Ventyx PROMOD IV software package, http://www.ventyx.com/en/enterprise/business-operations/business-products/promod-iv. Accessed 15 Sept 2014
9. Western Electricity Coordinating Council, *Assumptions Matrix for the 2020 TEPPC Dataset* (Salt Lake City, Utah, 2010), http://www.wecc.biz/library/StudyReport/Documents/Assumptions%20Matrix%20for%20the%202020%20TEPPC%20Dataset.pdf. Accessed 15 Sept 2014
10. B.M. Nickell, *TEPPC Renewable Energy Cases* (Salt Lake City, Utah, 2008), http://www.wecc.biz/committees/StandingCommittees/PCC/LRS/Shared%20Documents/Wind%20Profiles/Renewable%20Energy%20Generation%20Paper.pdf. Accessed 15 Sept 2014
11. TIER, Development of regional wind resource and wind plant output datasets. NREL/SR-550-47676 (3 TIER for the National Renewable Energy Laboratory, Golden, Colorado, 2010), http://www.nrel.gov/docs/fy10osti/47676.pdf. Accessed 15 Sept 2014
12. K. Orwig, M. Hummon, B.M. Hodge, D. Lew, Solar data inputs for integration and transmission planning studies. NREL/PO-5500-52985 (National Renewable Energy Laboratory, Golden, Colorado, 2011)

13. J. Ma, Y.V. Makarov, C. Loutan, Z. Xie, Impact of wind and solar generation on the California ISO's intra-hour balancing needs, in *IEEE 2011 Power and Energy Society General Meeting, 24–29, July 2011*, pp. 1–6. doi:10.1109/PES.2011.6039410
14. Y.V. Makarov, C. Loutan, J. Ma, P. de Mello, Operational impacts of wind generation on California power systems. IEEE Trans. Power Syst. **24**(2), 1039–1050 (2009). doi:10.1109/TPWRS.2009.2016364
15. R. Diao, N. Samaan, Y.V. Makarov, R. Hafen, J. Ma, Planning for variable generation integration through balancing authorities consolidation, in *Proceedings of IEEE/PES General Meeting* (San Diego, California, 2012)
16. R. Hafen, N.A. Samaan, R. Duo, Y.V. Makarov, N. Lu, Joint seasonal ARMA approach for modeling of load forecast errors in planning studies, in *Proceedings of the IEEE/PES Transmission and Distribution Conference and Exposition* (Chicago, Illinois, 2014)
17. N. Lu, R. Diao, R. Hafen, N.A. Samaan, Y.V. Makarov, A comparison of forecast error generators for modeling wind and load uncertainty, in *Proceedings of the IEEE/PES General Meeting* (Vancouver, Canada, 2013)
18. Y.V. Makarov, P.V. Etingov, J. Ma, Incorporating forecast uncertainty in utility control center, in *Renewable Energy Integration: Practical Management of Variability, Uncertainty and Flexibility in Power Grids*, ed. by L.E. Jones (Elsevier Academic Press, Waltham, Massachusetts, 2014)

Chapter 7
Robust Optimization in Electric Power Systems Operations

X. Andy Sun and Álvaro Lorca

7.1 Overview

Electric power systems operation has heavily relied on advanced optimization models and algorithms. For example, one of the most important daily operations, the unit commitment (UC) problem, has been formulated as a large-scale mixed-integer optimization (MIO) problem with complicated constraints on generation cost, generators' production levels, power flow equations in the network, and various reliability and security constraints. Such MIO-based UC models are routinely solved everyday in almost all major independent system operators (ISO) in the US, such as the ISO-New England, PJM, MISO, CAISO, etc. Solving these models to optimality or near-optimality by the state-of-the-art mixed-integer optimization solvers has been crucial to reduce economic costs of operation, ensure a fair market outcome, and maintain a high security and reliability level in power systems.

In the past decade or so, power systems have experienced fundamental changes. The establishment of deregulated electricity markets and above-mentioned ISO's is one of the driving forces of using more accurate and sophisticated optimization models and methods. Another significant driving force is the integration of renewable energy resources into the power systems, especially wind power and solar power. With federal mandates and state-level initiatives, the penetration of renewable energy

X.A. Sun (✉)
H. Stewart School of Industrial and Systems Engineering,
Georgia Institute of Technology, Atlanta 30332, USA
e-mail: andy.sun@isye.gatech.edu

Á. Lorca
Department of Electrical Engineering,
Department of Industrial and Systems Engineering,
Pontificia Universidad Católica de Chile, Vicuña Mackenna 4860,
7820436 Macul, Santiago, Chile
e-mail: alvarolorca@uc.cl

P. Du et al. (eds.), *Integration of Large-Scale Renewable Energy into Bulk Power Systems*, Power Electronics and Power Systems,
DOI 10.1007/978-3-319-55581-2_7

resources in the US power systems has been increasing rapidly [24]. Such an increase of wind and solar power, with their stochastic and intermittent characteristics, poses an increasing challenge to the standard operation methodology in today's electricity systems. In particular, the uncertainty in wind and solar power output is much more difficult to predict, even for real-time operations, than the traditional load; and the UC and dispatch models used in today's systems are deterministic optimization models, which do not handle supply and load uncertainty in an effective and efficient way. In order to fully exploit the benefit of installed renewable energy capacity and adapt to a system with increasing penetration of such variable energy resources, it is critical to rethink the current operational methodology and propose a new generation of operation models and solution methods.

Operating a large-scale power system with significant stochastic supply and demand can be modeled as an instance of optimization under uncertainty. In particular, the optimization problem under study contains uncertain parameters, such as electricity load, wind and solar power output, and the availability of generators and transmission lines. Optimization under uncertainty has been extensively studied in operations research since the 1960s. A fundamental modeling framework is stochastic optimization, where the uncertainty in parameters is treated as random variables and characterized by probability distributions and scenario samples. Stochastic optimization has also been widely applied to electric power system operations and planning, such as in unit commitment and long-term generation and transmission capacity expansions.

In the past 5 years, an alternative modeling paradigm, the so-called robust optimization, has attracted increasing attention in both academia and the power system industry. A robust optimization model requires less accurate information on probability distributions of uncertain parameters, guarantees a higher level of feasibility (i.e. robustness) of the resulting solutions in the face of uncertainty, and leads to computationally more tractable and scalable models. In this paper, we will study this new optimization paradigm for power system operational problems.

The paper is organized as follows. In Sect. 7.2, we provide a brief introduction to the methodology of robust optimization, with an emphasis on introducing fundamental concepts and models. In Sect. 7.3, we review the most recent development in applying two-stage robust optimization to daily power system operations, including two-stage robust UC and economic dispatch models, and techniques to model spatial and temporal correlations in wind and solar power in them. In Sect. 7.4 we review the area of multistage robust optimization to power system problems. In Sect. 7.5 we provide some literature review for bidding models that make use of robust optimization. In Sect. 7.6, we present some interesting developments on distributionally robust optimization. In Sect. 7.7 we summarize recent results on chance-constrained optimization for real-time power system operations. In Sect. 7.8, we close with some concluding remarks.

7.2 Brief Introduction to Robust Optimization

A generic deterministic optimization problem can be formulated as follows:

$$\begin{aligned} \min_{x \in X} \quad & f(\boldsymbol{x}, \boldsymbol{u}_0) \\ \text{s.t.} \quad & f_i(\boldsymbol{x}, \boldsymbol{u}_0) \le 0 \quad \forall i = 1, ..., m, \end{aligned} \tag{7.1}$$

where $X \subseteq \mathbb{R}^n$, $\boldsymbol{x}$ is the decision variable, and $\boldsymbol{u}_0$ represents the nominal value of some parameters $\boldsymbol{u}$ in the optimization problem. For example, in a deterministic unit commitment problem, a UC schedule $\boldsymbol{x}$ is determined so as to minimize the total operational cost and to satisfy the forecast load with the forecast wind and solar output, represented by $\boldsymbol{u}_0$. However, the load and renewable output forecast $\boldsymbol{u}_0$ is never exact. The solution $\boldsymbol{x}^*$ obtained from (7.1) has to stand the test when the realized value of $\boldsymbol{u}$ is different from the forecast value $\boldsymbol{u}_0$. For example, $\boldsymbol{x}^*$ may not be feasible for $f_i(\boldsymbol{x}, \boldsymbol{u}) \le 0$ when $\boldsymbol{u} \neq \boldsymbol{u}_0$; or even if $\boldsymbol{x}^*$ is feasible for some realizations of $\boldsymbol{u}$, there is no control on the cost $f(\boldsymbol{x}, \boldsymbol{u})$ for different $\boldsymbol{u}$.

Robust optimization (RO) is motivated by the above limitations of the deterministic formulation. In particular, a RO model guarantees to obtain a solution that is feasible for *any* realization of the uncertain parameters within a pre-determined uncertainty model; and the *worst-case* cost is minimized. The following is a RO formulation of (7.1).

$$\min_{x \in X} \max_{u \in \mathcal{U}} \quad f(\boldsymbol{x}, \boldsymbol{u}) \tag{7.2a}$$

$$\text{s.t.} \quad f_i(\boldsymbol{x}, \boldsymbol{u}) \le 0 \quad \forall \boldsymbol{u} \in \mathcal{U}, \forall i = 1, ..., m. \tag{7.2b}$$

In (7.2), $\mathcal{U}$ is called the *uncertainty set* of the parameter $\boldsymbol{u}$. It is a deterministic representation of the uncertainty in the original problem (7.1). The objective is to minimize the worst-case cost in (7.2a), and the constraint (7.2b) ensures that a solution $\boldsymbol{x}$ is feasible for any realization of the uncertainty within the uncertainty set. Notice that the robust constraint (7.2b) enforces a potentially infinite number of constraints, one for each realization of $\boldsymbol{u}$.

An important type of RO problems is the robust linear optimization problem, where $f(\boldsymbol{x}, \boldsymbol{u})$ and $f_i(\boldsymbol{x}, \boldsymbol{u})$ are bi-affine functions in $\boldsymbol{x}$ and $\boldsymbol{u}$, i.e., fixing one set of variables of $\boldsymbol{x}$ and $\boldsymbol{u}$, they are affine functions in the other set. For example, consider the following linear RO constraint:

$$\boldsymbol{a}^\top \boldsymbol{x} \le b \quad \forall \boldsymbol{a} \in \mathcal{U}. \tag{7.3}$$

Here $\boldsymbol{a}$ is the uncertain parameter. The above linear RO constraint is equivalent to

$$\max_{a \in \mathcal{U}} \boldsymbol{a}^\top \boldsymbol{x} \le b. \tag{7.4}$$

Depending on the structure of the uncertainty set $\mathscr{U}$, (7.4) can be further transformed using convex optimization duality. For instance, if $\mathscr{U}$ can be represented as a polyhedron $\{\boldsymbol{a} \in \mathbb{R}^n \mid \boldsymbol{M}\boldsymbol{a} \leq \boldsymbol{g}\}$, (7.4) is equivalent to the following set of linear inequalities by using strong duality of linear programming.

$$\min_{\boldsymbol{\pi}\geq \boldsymbol{0}}\{\boldsymbol{\pi}^\top \boldsymbol{g} : \boldsymbol{\pi}^\top \boldsymbol{M} = \boldsymbol{x}\} \leq b,$$

which is further equivalent to

$$\boldsymbol{\pi}^\top \boldsymbol{g} \leq b, \quad \boldsymbol{\pi}^\top \boldsymbol{M} = \boldsymbol{x}, \quad \boldsymbol{\pi} \geq \boldsymbol{0}. \tag{7.5}$$

We call (7.5) the *robust counterpart* of constraint (7.3).

Other structure of the uncertainty set is also possible. For example, an ellipsoidal uncertainty $\mathscr{U} = \{\boldsymbol{a} \in \mathbb{R}^n : \|\boldsymbol{M}\boldsymbol{a} - \boldsymbol{q}\|_2 \leq r\}$ is convenient to model correlations in the uncertainty vector $\boldsymbol{a}$. The robust constraint (7.2) can be similarly transformed by taking the dual of the convex optimization problem (7.4). The resultant robust counterpart involves a convex quadratic constraint. Constructing a proper uncertainty set is a very important aspect of a robust optimization model. It determines how well the model captures the essential features of the uncertainty, and how tractable the robust counterpart is. We will have more discussion on this issue in later sections.

A key structural property of the RO model (7.2) is that the decision $\boldsymbol{x}$ does not adapt to specific values of $\boldsymbol{u}$. Such RO models are called *static*. An important extension of the RO models is to make the decision $\boldsymbol{x}$ *adaptive* to the realization of uncertainty, i.e. $\boldsymbol{x}(\boldsymbol{u})$. In this way, different realization of uncertainty can be dealt with differently. This is called the *adaptive robust optimization* or adjustable robust optimization model.

$$\min_{\boldsymbol{x}(\cdot)\in X}\max_{\boldsymbol{u}\in\mathscr{U}} \quad f(\boldsymbol{x}(\boldsymbol{u}), \boldsymbol{u}) \tag{7.6a}$$

$$\text{s.t.} \quad f_i(\boldsymbol{x}(\boldsymbol{u}), \boldsymbol{u}) \leq 0 \quad \forall \boldsymbol{u} \in \mathscr{U}, \forall i = 1, ..., m. \tag{7.6b}$$

Note that in (7.6), the decision variable is a function $\boldsymbol{x}(\cdot)$, therefore the decision space is of infinite dimension. The adaptive robust optimization model is especially suited to the application in unit commitment problem. The unit commitment decision has to be determined before the uncertainty in load and renewable output is realized, then the dispatch decision can be adaptive to the specific realization of uncertainty. This is exactly a *two-stage* decision making process. In the next Section, we will introduce a two-stage robust UC model, where the dispatch decision at any time t is a function of the realization of uncertainty over the *entire* horizon. If we further consider the fact that a dispatch decision at time t can only depend on the information before and up to time t (not after time t), then we come to the so-called *multi-stage* RO model, where time causality is modeled by the so-called *non-anticipativity* constraints, which we will discuss in Sect. 7.4.

7.3 Two-Stage Robust Optimization in Power System Operations

A deterministic UC model is given below.

$$\min_{\boldsymbol{x},\boldsymbol{u},\boldsymbol{v},\boldsymbol{p}} \quad \sum_{t\in\mathcal{T}}\sum_{i\in\mathcal{G}} \left(f_{it}(x_{it},u_{it},v_{it}) + c_{it}(p_{it})\right) \tag{7.7a}$$

$$\text{s.t.} \quad x_{i,t-1} - x_{it} + u_{it} \geq 0, \quad \forall i \in \mathcal{G}, t \in \mathcal{T}, \tag{7.7b}$$

$$x_{it} - x_{i,t-1} + v_{it} \geq 0, \quad \forall i \in \mathcal{G}, t \in \mathcal{T}, \tag{7.7c}$$

$$x_{it} - x_{i,t-1} \leq x_{i\tau}, \quad \forall \tau \in [t+1, \min\{t + \text{MinUp}_i - 1, T\}], \quad t \in [2,T], i \in \mathcal{G}, \tag{7.7d}$$

$$x_{i,t-1} - x_{i,t} \leq 1 - x_{i\tau}, \quad \forall \tau \in [t+1, \min\{t + \text{MinDw}_i - 1, T\}], \quad t \in [2,T], i \in \mathcal{G} \tag{7.7e}$$

$$\sum_{i\in\mathcal{G}} p_{it} = \sum_{j\in\mathcal{D}} \bar{d}_{jt}, \qquad \forall t \in \mathcal{T}, \tag{7.7f}$$

$$p_{it} - p_{i,t-1} \leq RU_i x_{i,t-1} + SU_i u_{it}, \quad \forall i \in \mathcal{G}, t \in \mathcal{T}, \tag{7.7g}$$

$$p_{i,t-1} - p_{it} \leq RD_i x_{it} + SD_i v_{it}, \quad \forall i \in \mathcal{G}, t \in \mathcal{T}, \tag{7.7h}$$

$$-f_{l,k}^{\max} \leq \boldsymbol{a}_{l,k}^{\top}(\boldsymbol{p}_t - \boldsymbol{d}_t) \leq f_{l,k}^{\max}, \quad \forall t \in \mathcal{T}, l \in \mathcal{C}_k, k \in \mathcal{L}, \tag{7.7i}$$

$$p_i^{\min} x_{it} \leq p_{it} \leq p_i^{\max} x_{it}, \quad \forall i \in \mathcal{G}, t \in \mathcal{T}, \tag{7.7j}$$

$$x_{it}, u_{it}, v_{it} \in \{0,1\}, \quad \forall i \in \mathcal{G}, t \in \mathcal{T}. \tag{7.7k}$$

The unit commitment decisions include binary variables x_{it}, u_{it}, v_{it}, where $x_{it} = 1$ if the thermal generator i is on at time t, and $x_{it} = 0$ otherwise; $u_{it} = 1$ if generator i is *turned* on from the off state at time t; $v_{it} = 1$ if generator i is turned off at time t. The dispatch decision is p_{it} of a thermal generator $i \in \mathcal{G}$ at time $t \in \mathcal{T}$, where $\mathcal{G}$ is the set of thermal generators, and $\mathcal{T}$ is the set of time periods in the decision horizon. Note in this UC model, we assume the wind output can not be curtailed. In the robust economic dispatch model introduced later, we will also allow the wind output to be curtailed.

The fixed cost $f_{it}(x_{it}, u_{it}, v_{it})$ of each generator includes start-up and shut-down costs and other fixed costs. The variable cost $c_{it}(p_{it})$ is usually approximated by a convex piecewise linear function of the active power output p_{it}. The forecast demand $\bar{d}_{jt}$ is the net load at bus j, time t, which is the traditional load minus wind output at this bus, if there is any. Constraints (7.7b) and (7.7c) represent logic relations between on and off status and the turn-on and turn-off actions. Constraints (7.7d) and (7.7e) restrict the minimum up and down times for each generator. Constraint (7.7f) enforces system wide energy balance in each time period. Constraints (7.7g)–(7.7h) limit the rate of production changes over a single period, where RU_i and RD_i are limits for ramp-up and ramp-down rates when the generator is already running, and SU_i and SD_i are ramping limits when generator i is just starting up and shutting down. Constraint (7.7i) expresses the power flow in the transmission lines as a linear function of power production and load in the entire system, where the coefficients of the linear function, $\boldsymbol{a}_{l,k}$, are called the shift factors of line l, and the index k represents the k-th contingency, i.e., when the line k is tripped offline, $\mathscr{C}_k$ is the set of remaining lines. Constraint (7.7j) represents the physical limits on the production levels of each generator.

Compactly, the deterministic UC model (7.7) can be written as

$$\min_{\boldsymbol{x},\boldsymbol{p}} \quad f(\boldsymbol{x}) + c(\boldsymbol{p}) \tag{7.8a}$$

$$\text{s.t.} \quad \boldsymbol{x} \in X \tag{7.8b}$$

$$\boldsymbol{p} \in \Omega(\boldsymbol{x}, \boldsymbol{d}), \tag{7.8c}$$

where $\boldsymbol{x}$ contains all the commitment related binary variables, and $\boldsymbol{p}$ contains the dispatch level of generators, and $\boldsymbol{d}$ is the vector of net load at each bus, i.e. traditional load minus the renewable power output. Note that (7.8c), defined by (7.7f)–(7.7j), includes all the coupling constraints between $\boldsymbol{x}$ and $\boldsymbol{p}$ as well as all the dispatch constraints. To handle the uncertainty in nodal net load, we present a two-stage robust UC model in the next subsection.

7.3.1 Two-Stage Robust Unit Commitment Model

A two-stage robust optimization model is given blow:

$$\begin{aligned} \min_{\boldsymbol{x}} \quad & \left\{ f(\boldsymbol{x}) + \max_{\boldsymbol{d} \in \mathscr{D}} \min_{\boldsymbol{p} \in \Omega(\boldsymbol{x},\boldsymbol{d})} c(\boldsymbol{p}) \right\} \\ \text{s.t.} \quad & \boldsymbol{x} \in \mathscr{F}, \end{aligned} \tag{7.9}$$

where $\mathscr{D}$ is the uncertainty set for the net load. In Sect. 7.3.3, we will study in details how to construct uncertainty sets. Now let us understand the meaning of the two-stage robust UC formulation (7.9).

It is called a *two-stage* robust optimization model because there are two types of decisions in (7.9): the unit commitment decision $\boldsymbol{x}$, which needs to be determined *before* the realization of uncertainty, is called the first-stage decision; the dispatch decision $\boldsymbol{p}$ is called the second-stage decision, which can be adjusted *after* observing the uncertainty realization. Note that $\boldsymbol{p}$ needs to satisfy all the dispatch related constraints in $\Omega(\boldsymbol{x}, \boldsymbol{d})$ given a UC solution $\boldsymbol{x}$ and a net load realization $\boldsymbol{d}$.

The objective of the two-stage robust UC model is also composed of two parts: the cost associated with the commitment decision $f(\boldsymbol{x})$, and the cost associated with the dispatch cost $c(\boldsymbol{p})$. Note that for every net load $\boldsymbol{d} \in \mathscr{D}$, the robust model finds the most economic dispatch decision $\boldsymbol{p}$ by solving the inner minimization problem, which by itself is a standard economic dispatch problem. However, the robust model wants to find the worst possible dispatch cost by maximizing the dispatch cost over all possible net load realization $\boldsymbol{d}$ in the uncertainty set $\mathscr{D}$. Finally, the overall objective of the two-stage robust UC model is to minimize the commitment cost plus the worst-case economic dispatch cost.

In summary, the two-stage robust UC model obtains a UC solution that will guarantee the dispatch process to be feasible for any realization of uncertain net load $\boldsymbol{d}$ in the uncertainty set $\mathscr{D}$, therefore achieving *robustness*, where the feasibility of the dispatch problem to any $\boldsymbol{d} \in \mathscr{D}$ is achieved by making the dispatch decision adjustable to $\boldsymbol{d}$, hence *adaptive*. Both these two features, robustness and adaptiveness, is in contrast to the traditional deterministic UC model, where the commitment decision is only solved for a fixed, forecast load level, and the dispatch solution is only feasible for this level of load also.

Such a two-stage robust UC model has been proposed in [4, 11, 27].

7.3.2 Two-Stage Robust Economic Dispatch model

Note that in the two-stage robust UC model (7.9), the first-stage decision is the commitment decision for all 24 h, and the second-stage decision is the dispatch decision also over the 24 h. Such a two-stage decision structure is quite flexible. We can also use it in the real-time economic dispatch process. For real-time operation, the first-stage decision $\boldsymbol{x}$ in (7.9) is the dispatch level $\boldsymbol{p}_1$ for the current time period $t = 1$, which needs to be made now and before the observation of any future uncertainty. The dispatch process looks forward a few periods $t = 1, ..., T$, and consider these future dispatch decisions $\boldsymbol{p} = (\boldsymbol{p}_2, ..., \boldsymbol{p}_T)$ as the second-stage recourse. The uncertainty is again the load and renewable power in periods $t = 2, ..., T$. With these setup, the two-stage robust economic dispatchmodel is given below.

$$\min_{p_1 \in \Omega_1(p_0, d_1, \bar{p}_1^w)} \left\{ c_1(\boldsymbol{p}_1) + \max_{d \in \mathscr{D}, \bar{p}^w \in \mathscr{P}^w} \min_{p \in \Omega(p_1, d, \bar{p}^w)} c(\boldsymbol{p}) \right\}, \tag{7.10}$$

The feasible region $\Omega_1(p_0, d_1, \bar{p}^w)$ of the first-stage decision variables corresponds to the constraints of a single-period dispatch problem, that is

$$\Omega_1(p_0, d_1, \bar{p}_1^w) = \Big\{ \boldsymbol{p}_1 = (\boldsymbol{p}_1^g, \boldsymbol{p}_1^w) : \underline{p}_{i1}^g \le p_{i1}^g \le \bar{p}_{i1}^g \quad \forall i \in \mathscr{G} \tag{7.11a}$$

$$- RD_i^g \le p_{i1}^g - p_{i0}^g \le RU_i^g \quad \forall i \in \mathscr{G} \tag{7.11b}$$

$$0 \le p_{i1}^w \le p_i^{w,max} \quad \forall i \in \mathscr{G}_w \tag{7.11c}$$

$$p_{i1}^w \le \bar{p}_{i1}^w \quad \forall i \in \mathscr{G}_w \tag{7.11d}$$

$$- RD_i^w \le p_{i1}^w - p_{i0}^w \le RU_i^w \quad \forall i \in \mathscr{G}_w \tag{7.11e}$$

$$\left|\boldsymbol{\alpha}_l^\top(\boldsymbol{p}_1-\boldsymbol{d}_1)\right| \le f_l^{max} \quad \forall l \in \mathcal{N}^l \tag{7.11f}$$

$$\left.\sum_{i\in\mathcal{G}} p_{i1}^g + \sum_{i\in\mathcal{G}_w} p_{i1}^w = \sum_{j\in\mathcal{D}} d_{j1}\right\}, \tag{7.11g}$$

where p_1^g is the dispatch level of thermal units, and p_1^w is the dispatch level of wind farms. Here we assume wind farm output can be curtailed. Constraints (7.11a) and (7.11b) are dispatch level and ramping constraints for thermal generators, respectively, where p_0 is the initial output level of generators before time $t = 1$. Constraints (7.11c) is the production range for wind farm i, where $p_i^{w,max}$ is the capacity rating of wind farm i. Constraint (7.11d) ensures that the generation of wind farms does not exceed the available wind power $\overline{p}_{i1}^w$ at time $t = 1$. Constraint (7.11e) imposes ramping limits on wind farms. Constraints (7.11f) and (7.11g) are flow limits and system-wide power balance.

Constraints in the second-stage problem are parameterized by the first-stage dispatch $\boldsymbol{p}_1$ and uncertain demand $\boldsymbol{d} \in \mathcal{D}$ and available wind power $\overline{\boldsymbol{p}}^w \in \overline{\mathcal{P}}^w$ realized in the uncertainty sets $\mathcal{D}$ and $\overline{\mathcal{P}}^w$. The feasible region $\Omega(\boldsymbol{p}_1, \boldsymbol{d}, \overline{\boldsymbol{p}}^w)$ is similarly defined as in (7.11) for $\boldsymbol{p} = \left(\boldsymbol{p}_t^g, \boldsymbol{p}_t^w, \forall t = 2, \ldots, T\right)$. See [14] for more details.

7.3.3 Uncertainty Set Modeling of Renewable Energy and Demand

A crucial component of the robust UC and robust ED model is the uncertainty set, which determines how much uncertainty is considered in the robust model. We introduce two types of uncertainty sets: a static uncertainty set, where the uncertainty in each period is not explicitly coupled to uncertainty in other periods; and then a dynamic uncertainty set, which explicitly describes temporal and spatial correlations of the uncertainty. Both of these models are proposed for robust unit commitment and dispatch models (see e.g. [4, 11, 27] for the static uncertainty sets, and [14] for the dynamic uncertainty sets.).

7.3.3.1 Static Budget Uncertainty Set

The most commonly used uncertainty sets are the following so-called budget uncertainty set. It assigns an interval for each component of the uncertain vector to model the range of the variation of each uncertain parameter; then it has a coupling constraint that restricts the total variation of the uncertainty realization from the nominal value. A commonly used uncertainty set is the following budget uncertainty set [4, 5].

$$\mathscr{D}^t(\bar{\boldsymbol{d}}^t, \hat{\boldsymbol{d}}^t, \Delta^t) := \left\{ \boldsymbol{d}^t : \sum_{i \in \mathscr{N}_d} \frac{|d_i^t - \bar{d}_i^t|}{\hat{d}_i^t} \le \Delta^t, d_i^t \in \left[\bar{d}_i^t - \hat{d}_i^t, \bar{d}_i^t + \hat{d}_i^t\right], \forall i \right\}, \tag{7.12}$$

where $\mathscr{N}_d$ is the set of nodes that have uncertain net load, $\boldsymbol{d}^t = (d_i^t, i \in \mathscr{N}_d)$ is the vector of uncertain net load at time t, $\bar{d}_i^t$ is the nominal value of the net load of node i at time t, $\hat{d}_i^t$ is the deviation from the nominal net load, and the interval $\left[\bar{d}_i^t - \hat{d}_i^t, \bar{d}_i^t + \hat{d}_i^t\right]$ is the range of the uncertain d_i^t; the inequality in (7.12) controls the deviation of all net loads from the nominal value. The parameter Δ^t is the "budget of uncertainty". With $\Delta^t = 0$, the uncertainty set $\mathscr{D}^t = \{\bar{\boldsymbol{d}}^t\}$ is a singleton, corresponding to the nominal deterministic case. As Δ^t increases, the size of the uncertainty set $\mathscr{D}^t$ enlarges. This means that larger total deviation from the expected net load is considered, so that the resulting robust UC solutions are more conservative and the system is protected against a higher degree of uncertainty. With $\Delta^t = N_d$, $\mathscr{D}^t$ equals to the entire hypercube defined by the intervals for each d_i^t.

In addition to the uncertainty budget over all net loads at each time t, we can also construct a budget constraint over all time periods, such as the following

$$\sum_{t \in \mathscr{T}} \sum_{i \in \mathscr{N}_d} \frac{|d_i^t - \bar{d}_i^t|}{\hat{d}_i^t} \le \Delta. \tag{7.13}$$

For $\Delta < \sum_{t \in \mathscr{T}} \Delta^t$, the budget constraint over time (7.13) further cuts off points from the Cartesian product uncertainty set $\Pi_{t \in \mathscr{T}} \mathscr{D}^t$. Such uncertainty sets are considered in [11, 27].

7.3.3.2 Dynamic Uncertainty Set

The budget uncertainty set (7.12) represents a rather simplified model of net load uncertainty. In reality, traditional load uncertainty and wind or solar power uncertainty has quite substantially different characteristics. The load uncertainty can be quite accurately modeled for the day-ahead operation, whereas wind or solar uncertainty has quite substantial temporal and spatial correlations, which are very important for real-time operations. In the following, we present a way to construct a dynamic uncertainty set that captures the dynamic evolution of wind speed at different wind farms, akin to a linear dynamical model [14].

We denote the wind speed vector of multiple wind farms at time t as $\boldsymbol{r}_t = (r_{1t}, \ldots, r_{N^w t})$, where r_{it} is the wind speed at wind farm i and time t. Define the dynamic uncertainty set for $\boldsymbol{r}_t$ as:

$$\mathscr{R}_t(\boldsymbol{r}_{[t-L:t-1]}) = \Big\{ \boldsymbol{r}_t : \exists \tilde{\boldsymbol{r}}_{[t-L:t]}, \boldsymbol{u}_t \quad \text{s.t.}$$

$$\boldsymbol{r}_\tau = \boldsymbol{g}_\tau + \tilde{\boldsymbol{r}}_\tau \quad \forall \tau = t-L, \ldots, t \tag{7.14a}$$

$$\widetilde{\boldsymbol{r}}_t = \sum_{s=1}^{L} \boldsymbol{A}_s \widetilde{\boldsymbol{r}}_{t-s} + \boldsymbol{B}\boldsymbol{u}_t \tag{7.14b}$$

$$\sum_{i\in\mathcal{N}^w} |u_{it}| \le \Gamma^w \sqrt{N^w} \tag{7.14c}$$

$$|u_{it}| \le \Gamma^w \qquad \forall i \in \mathcal{N}^w \tag{7.14d}$$

$$\boldsymbol{r}_t \ge \boldsymbol{0}\Big\}, \tag{7.14e}$$

where vectors $\boldsymbol{r}_{t-L}, \ldots, \boldsymbol{r}_{t-1}$ are the realizations of wind speeds in periods $t-L, \ldots, t-1$. Equation (7.14a) decomposes wind speed vector $\boldsymbol{r}_\tau$ as the sum of a seasonal pattern $\boldsymbol{g}_\tau$, which is estimated from wind data, and a residual component $\widetilde{\boldsymbol{r}}_\tau$ which is the deviation from $\boldsymbol{g}_\tau$. Equation (7.14b) represents a linear dynamic relationship involving the residual $\widetilde{\boldsymbol{r}}_t$ at time t, residuals realized in earlier periods $t-L$ to $t-1$, and an error term $\boldsymbol{u}_t$. The parameter L sets the time lags considered in the model. In Eq. (7.14b), matrices $\boldsymbol{A}_s$'s capture the temporal correlation between $\boldsymbol{r}_t$ and $\boldsymbol{r}_{t-s}$, and $\boldsymbol{B}$ specifically captures the spatial relationship of wind speeds at adjacent wind farms at time t. Equation (7.14c)–(7.14d) describe a budget uncertainty set for the error term $\boldsymbol{u}_t$, where Γ^w controls its size, and (7.14e) avoids negative wind speeds. $\mathcal{N}^w$ and N^w denote the set and number of wind farms, respectively.

Using the above uncertainty sets (7.14) for wind speeds, we can further construct dynamic uncertainty sets for wind power through power curves. In particular, we denote the *available wind power* of wind farm i at time t as $\overline{p}^w_{it}$. Given the wind speed r_{it}, $\overline{p}^w_{it}$ is described by the following constraints

$$\overline{p}^w_{it} \ge \max_{k=1,\ldots,K} h^0_{ik} + h_{ik}\, r_{it} \quad \forall i \in \mathcal{N}^g, \tag{7.15}$$

where parameters h^0_{ik}, h_{ik} are determined based on a convex piecewise linear approximation with K pieces of the increasing part of the power curve at wind farm i. Although (7.15) allows available wind power to exceed the maximum cut-off power, the robust optimization model described in Sect. 7.3.2 always ensures that the available wind power lies on the power curve including the plateau part for wind speed exceeding a cut-off value.

The dynamic uncertainty set of the trajectory of available wind power in time periods 2 through T, namely $\overline{\boldsymbol{p}}^w = (\overline{\boldsymbol{p}}^w_2, \ldots, \overline{\boldsymbol{p}}^w_T)$, is given as

$$\overline{\mathscr{P}}^w = \Big\{ (\overline{\boldsymbol{p}}^w_2, \ldots, \overline{\boldsymbol{p}}^w_T) : \exists (\boldsymbol{r}_2, \ldots, \boldsymbol{r}_T) \text{ s.t. } \boldsymbol{r}_t \in \mathscr{R}_t(\boldsymbol{r}_{[t-L:t-1]}) \\ \text{and (7.15) is satisfied for } t = 1, \ldots, T \Big\}, \tag{7.16}$$

which is used in the robust ED model described in Sect. 7.3.2.

7.3.4 Solution Algorithms

The two-stage robust UC model (7.9) and the two-stage robust ED model (7.10) can be written in the following form:

$$\min_{x \in \mathscr{F}} \left\{ f^\top \boldsymbol{x} + \max_{d \in \mathscr{D}} \min_{y \in \Omega(x,d)} c^\top \boldsymbol{y} \right\}, \tag{7.17}$$

where the first-stage decision $\boldsymbol{x}$ may contain integer variables, but the second-stage problem only has continuous variables, and

$$\Omega(\boldsymbol{x}, \boldsymbol{d}) = \{\boldsymbol{y} : \boldsymbol{By} \geq \boldsymbol{Md} - \boldsymbol{Ax} + \boldsymbol{h}\}, \tag{7.18}$$

that is, the uncertainty $\boldsymbol{d}$ is on the right-hand side of $\Omega(\boldsymbol{x}, \boldsymbol{d})$. Using linear programming duality, the second-stage problem in (7.17) can be reformulated as

$$\begin{aligned} Q(\boldsymbol{x}) := \max_{\boldsymbol{d},\lambda} \ & \lambda^\top (\boldsymbol{Md} - \boldsymbol{Ax} + \boldsymbol{h}) \\ \text{s.t.} \ & \lambda \in \Lambda \\ & \boldsymbol{d} \in \mathscr{D}, \end{aligned} \tag{7.19}$$

where $\Lambda := \{\lambda : \lambda^\top \boldsymbol{B} = \boldsymbol{c}^\top, \ \lambda \geq \boldsymbol{0}\}$. By adding penalty variables to the inner minimization problem (i.e., the dispatch problem in the two-stage robust UC or the look-ahead dispatch in the robust ED model), we can always assume the second-stage problem of (7.17) is feasible and achieves a bounded optimal solution for any $\boldsymbol{x} \in \mathscr{F}$. With these assumptions, we can characterize the structure of $Q(\boldsymbol{x})$ as follows.

Proposition 1 *Assume the uncertainty set $\mathscr{D}$ and Λ are polytopes with extreme points $\{\boldsymbol{d}^1, \ldots, \boldsymbol{d}^r\}$ and $\{\lambda^1, \ldots, \lambda^s\}$, respectively. The second-stage value function $Q(\boldsymbol{x})$ is a convex piecewise-linear function represented as*

$$Q(\boldsymbol{x}) = \max_{i=1,\ldots,r, j=1,\ldots,s} \lambda_j^\top (\boldsymbol{Md}_i - \boldsymbol{Ax} + \boldsymbol{h}). \tag{7.20}$$

7.3.4.1 Benders Decomposition

From the above proposition we can see that a Benders decomposition type algorithm can be readily applied to solve (7.17). In particular, the restricted master problem is given as below

$$\begin{aligned} \max \ & f^\top \boldsymbol{x} + \eta \\ \text{s.t.} \ & \boldsymbol{x} \in \mathscr{F} \\ & \eta \geq \lambda_j^\top (\boldsymbol{Md}_j - \boldsymbol{Ax} + \boldsymbol{h}) \quad \forall j = 1, \ldots, k. \end{aligned}$$

The subproblem solves the second-stage problem (7.19) for the $\boldsymbol{x}$ obtained in the restricted master problem. Given the assumption of Proposition (1), it can be seen that the above Benders decomposition algorithm converges in at most rs number of iterations, i.e., enumerating all the combinations of extreme points of $\mathscr{D}$ and Λ.

Such a Benders decomposition algorithm is routinely used for solving two-stage *stochastic* optimization problem. However, it turns out that, at least for the two-stage robust UC and ED models, this Benders decomposition is quite slow to converge as observed in e.g. [4, 11, 27]. A more efficient decomposition algorithm can be devised for the two-stage robust model.

7.3.4.2 Constraint Generation

The two-stage robust model (7.17) is equivalent to

$$\begin{aligned} \min_{\boldsymbol{x},\eta} \quad & \boldsymbol{f}^\top \boldsymbol{x} + \eta \\ \text{s.t.} \quad & \boldsymbol{x} \in \mathscr{F} \\ & \eta \geq \min_{\boldsymbol{y} \in \Omega(\boldsymbol{x},\boldsymbol{d})} \boldsymbol{c}^\top \boldsymbol{y} \quad \forall \boldsymbol{d} \in \mathscr{D}, \end{aligned} \tag{7.21}$$

where constraint (7.21) can be reformulated using the extreme points of $\mathscr{D}$ as

$$\begin{aligned} \min_{\boldsymbol{x},\eta,\boldsymbol{y}} \quad & \boldsymbol{f}^\top \boldsymbol{x} + \eta \\ \text{s.t.} \quad & \boldsymbol{x} \in \mathscr{F} \\ & \eta \geq \boldsymbol{c}^\top \boldsymbol{y}^i \quad \forall i = 1, \dots, r && (7.22) \\ & \boldsymbol{y}^i \in \Omega(\boldsymbol{x}, \boldsymbol{d}^i) \quad \forall i = 1, \dots, r. && (7.23) \end{aligned}$$

Now it is clear to see that the above formulation has a large number of constraints (7.22)–(7.23). A constraint generation framework is a natural thing to try. The restricted master problem has a subset I of constraints (7.22)–(7.23) for $I \subset \{1, \dots, r\}$. The subproblem is to solve the second-stage problem (7.19) for a given $\boldsymbol{x}$, just as in the Benders decomposition algorithm (see exact and heuristic methods for solving these type of bilinear problems in [12], [19]). The worst-case uncertainty scenario and its associated cost and dispatch constraints (7.22)–(7.23) found in the subproblem are then added to the restricted master problem. In each iteration of the constraint generation, an extreme point of $\mathscr{D}$ is added to the master problem. Therefore, the algorithm terminates with at most r steps, i.e. the number of extreme points of the uncertainty set $\mathscr{D}$.

Since in each iteration both the constraints of the form (7.22)–(7.23) and the associated dispatch variable $\boldsymbol{y}^i$ are generated and added to the master problem, such a procedure is also termed column-and-constraint generation, as proposed in [25]. A similar procedure is also proposed in [4] to accelerate the Benders decomposition algorithm.

7.3.5 Computational Experiments

In this Section, we present some computational results for the two-stage robust UC and ED models using the constraint-and-column generation algorithm.

7.3.5.1 Two-Stage Robust UC with Budget Uncertainty Sets

In [4], the two-stage robust security-constrained UC problem is solved for the ISO-NE's power system, which has 312 generating units, 174 loads, 2816 buses, and 90 representative lines.

To understand the potential benefits of the two-stage robust UC, the performance of its solution is compared to that obtained by a traditional deterministic UC with reserve-based rules. For this purpose, the solution of the two-stage robust UC is obtained using different levels of the uncertainty budget Δ^t in the uncertainty set, and the solution of the deterministic UC is obtained using different levels of reserve, also parameterized by Δ^t (the higher Δ^t, the more reserve is considered). Once the UC solutions are obtained, these are tested under simulated nodal net loads, using a normal distribution for sampling them. Tables 7.1 and 7.2 present the results. Here, "AdptRob" is the two-stage robust UC model, "ResAdj" is the deterministic UC model with reserves, and "std" refers to standard deviation. Several observations are in order. First, we can see that when $\Delta^t = 0$ both models obtain the same results, since both models are the same under this parameter (no uncertainty in the uncertainty set, nor in reserves). Second, as Δ^t increases, for both approaches the total cost decreases until some best point after which it starts to increase again due to an excess of conservativeness. Third, the two-stage robust UC model achieves a better cost under every tested Δ^t, and a significantly reduced cost standard deviation. From these results we can appreciate a promising potential of the two-stage robust UC model for effectively managing uncertainty in power system operations.

Table 7.1 The average dispatch costs and total costs of the AdptRob and ResAdj for normally distributed load $\Delta^t/\sqrt{N_d} = 0.5, 1, \ldots, 3$ and $\hat{d}_j^t = 0.1\bar{d}_j^t$ [4]

	AdptRob		ResAdj	
Budget $\Delta^t/\sqrt{N_d}$	Dispatch cost (M$)	Total cost (M$)	Dispatch cost (M$)	Total cost (M$)
0.0	19.3530	20.8503	19.3530	20.8503
0.5	16.9195	18.6050	18.1855	19.6837
1.0	16.9650	18.6688	17.4907	18.9942
1.5	16.9815	18.7365	17.3027	18.8006
2.0	17.0297	18.7937	17.7403	19.2415
2.5	17.0586	18.8366	17.6567	19.1618
3.0	17.0745	18.8526	18.0804	19.5889

Table 7.2 Standard deviation of the dispatch costs of the two approaches and their ratio for normally distributed load [4]

	AdptRob	ResAdj	ResAdj
Budget Δ^t/N_d	Std dispatch cost (k\$)	Std dispatch cost (k\$)	/ AdptRob
0.0	1,769.5107	1,769.5107	1
0.1	47.4900	687.5098	14.48
0.2	46.3647	687.5098	8.62
0.3	45.4248	377.7901	8.32
0.4	44.2397	366.7359	8.29
0.5	44.1075	377.1875	8.55
0.6	43.9936	370.8673	8.43
0.7	43.9263	377.0631	8.58
0.8	43.9338	370.7203	8.44
0.9	43.9023	357.9338	8.15
1.0	43.9431	361.0376	8.22

7.3.5.2 Two-Stage Robust ED with Dynamic Uncertainty Sets

Here, we show some results for the two-stage robust ED problem presented in Sect. 7.3.2 with the dynamic uncertainty sets presented in Sect. 7.3.3. In [14] this model is tested within a simulation platform, by successively solving ED over 35 days with a temporal granularity of 10 min. A 14-bus system with 3 conventional generators and 4 wind farms presenting a high wind power penetration is employed. In order to understand the potential benefits of the dynamic uncertainty sets, the two-stage robust ED is solved using both dynamic uncertainty sets and static uncertainty sets, testing different sizes of these sets parameterized by Γ^w.

Figure 7.1 presents the average and standard deviation of the cost obtained over all 10-min time periods over the 35 days, for different values of Γ^w, under dynamic uncertainty sets (DUS), static uncertainty sets ignoring temporal correlations (SUS1), and static uncertainty sets ignoring temporal and spatial correlations (SUS2). Let's first observe the blue curve, obtained by DUS. On the top right, we have the performance under the deterministic case with $\Gamma^w = 0$. Then, as Γ^w increases, we move to the left of this blue curve, reducing both cost average and cost standard deviation, until we reach $\Gamma^w = 0.5$ after which cost average starts to increase, and cost standard deviation keeps decreasing. This curve shows that Γ^w acts as an effective conservativeness parameter. If Γ^w is too small, the ED process is not well prepared for wind power deviations, and if Γ^w is too large, then the ED process is unnecessarily prepared for very strong deviations. The left side of this "U" shaped curve corresponds to the Pareto-frontier, since any point here dominates all points on the right side of this curve. Similarly, SUS1 and SUS2 present similar "U" shaped curves, and we can observe that the Pareto-frontier of DUS dominates both Pareto-frontiers of SUS1 and SUS2. With this we can see that capturing temporal

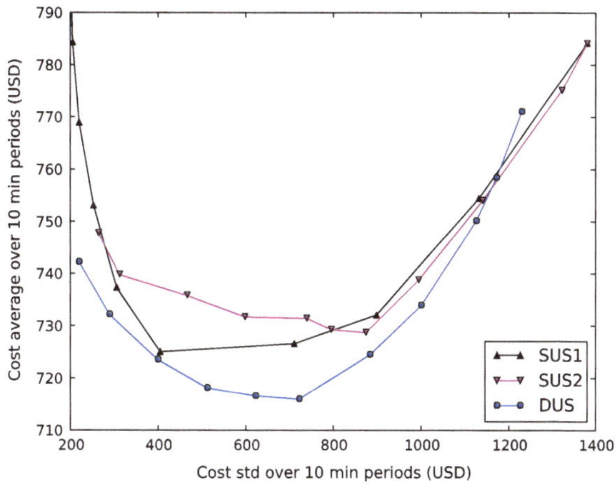

Fig. 7.1 Cost std and cost average obtained for the policies determined by the different models with $\Gamma^w = 0.0, 0.1, ..., 1.0$ [14]

and spatial correlations in a dynamic uncertainty set is effective at improving both economic effectiveness and system reliability in the power system ED process.

7.4 Multistage Robust Optimization in Power System Operations

7.4.1 Multistage Robust Optimization Models

In Sect. 7.3, we have considered *two-stage* robust optimization models where first-stage decisions are decided "right now", and second-stage decisions are to be decided after uncertain parameter realizations are revealed. This is a very versatile framework that we have seen can be applied to day-ahead UC and real-time ED, and has also been applied for long-term power system planning problems. However, a closer look at sequential decision making processes reveals that there is an *information* aspect where this framework is rather limited. Let's consider for example the UC and ED processes in most modern power systems. Consider a time horizon of one whole day, from midnight to midnight. Typically, the UC problem is solved before this day begins to fix on/off decisions for generating units. Then, given these fixed decisions, the ED problem will be repeatedly solved throughout the day, say every 5 or 10 min, to determine dispatch decisions. Every time the ED problem is solved, the relevant information at the time will be the dispatch decisions implemented before, and the current conditions of the power system, that is, the current realization of uncertain parameters. Critically, when ED is solved, there is no available information on the realization of future uncertain parameters. This suggests a fundamental limitation of the two-stage robust optimization framework, since all second-stage decisions are decided based on the full knowledge of the realization of all uncertain parameters, without distinction between early decisions and later decisions. In other words, the information process is misrepresented by the two-stage robust optimization framework when applying it to model real-world sequential decision-making problems, in which decisions can only adapt to the current portion of information revealed up to their decision time.

The above discussion suggests the consideration of *multiple* decision stages, rather than just two, giving rise to *multistage* robust optimization problems. To be precise, let's start from the two-stage modeling framework from Sect. 7.3. We can write the two-stage robust UC problem as follows:

$$\min_{x\in\mathscr{F}} \left\{ f(\boldsymbol{x}) + \max_{d\in\mathscr{D}} \min_{\{p:\, p_t\in\Omega_t(x,d_t,p_{t-1})\,\forall t\in\mathscr{T}\}} \sum_{t\in\mathscr{T}} c_t(\boldsymbol{p}_t) \right\}, \tag{7.24}$$

where x represents first-stage on/off decisions, and p_t represents second-stage dispatch decisions at time t. This problem can also be equivalently written in the following *functional-space* representation [4, 7]:

$$\min_{x,p(\cdot)} \left\{ f(\boldsymbol{x}) + \max_{d\in\mathscr{D}} \sum_{t\in\mathscr{T}} c_t(\boldsymbol{p}_t(\boldsymbol{d})) \right\} \tag{7.25a}$$

$$\text{s.t. } \boldsymbol{x} \in \mathscr{F} \tag{7.25b}$$

$$\boldsymbol{p}_t(\boldsymbol{d}) \in \Omega_t(\boldsymbol{x}, \boldsymbol{d}_t, \boldsymbol{p}_{t-1}(\boldsymbol{d})) \quad \forall t \in \mathscr{T}, \boldsymbol{d} \in \mathscr{D}, \tag{7.25c}$$

where we can see that dispatch decision $\boldsymbol{p}_t(\boldsymbol{d})$ at time t is now an arbitrary function of any realization of net load vector $\boldsymbol{d}$ in the uncertainty set $\mathscr{D}$, thus making the dispatch decision explicitly adaptive to the whole vector $\boldsymbol{d}$. This immediately suggests how to formulate a multistage robust UC problem, by simply limiting the portion of information that $\boldsymbol{p}_t(\cdot)$ can adapt to. In the multistage model, instead of dependence on the whole vector $\boldsymbol{d}$, we need to restrict the dependence of $\boldsymbol{p}_t(\cdot)$ to $\boldsymbol{d}_{[t]} = (\boldsymbol{d}_1, \dots, \boldsymbol{d}_t)$, that is, to the information revealed up to time t, which we can mathematically enforce by replacing $\boldsymbol{p}_t(\boldsymbol{d})$ with $\boldsymbol{p}_t(\boldsymbol{d}_{[t]})$. The multistage robust UC problem is formulated below:

$$\min_{x,p(\cdot)} \left\{ f(\boldsymbol{x}) + \max_{d\in\mathscr{D}} \sum_{t\in\mathscr{T}} c_t(\boldsymbol{p}_t(\boldsymbol{d}_{[t]})) \right\} \tag{7.26a}$$

$$\text{s.t. } \boldsymbol{x} \in \mathscr{F} \tag{7.26b}$$

$$\boldsymbol{p}_t(\boldsymbol{d}_{[t]}) \in \Omega_t(\boldsymbol{x}, \boldsymbol{d}_t, \boldsymbol{p}_{t-1}(\boldsymbol{d}_{[t-1]})) \quad \forall t \in \mathscr{T}, \boldsymbol{d} \in \mathscr{D}. \tag{7.26c}$$

The restriction that decisions at time t depend only on information revealed up to time t is called the *non-anticipativity* constraint, since decisions are not allowed to anticipate to future information.

The differences between two-stage and multistage robust optimization models were widely explored for the UC problem in [15], where the above multistage robust

UC model was proposed. In what follows, we will summarize modeling aspects and algorithms to find good solutions for the multistage robust UC problem and explore the advantages of this approach.

7.4.2 *Affine Policies to Confront Infinite Dimensionality*

The multistage robust UC problem (7.26) is an infinite-dimensional problem, since the decision space for dispatch decisions $\boldsymbol{p}(\cdot)$ is an arbitrary functional space over the uncountable uncertainty set $\mathscr{D}$. Given this, finding an optimal solution for the problem is intractable. However, if we restrict the functional space in some tractable way, we might be able to find good solutions for the problem. In particular, the concept of *affine policies* (also known as *linear decision rules*) has been proposed for multistage robust optimization problems [2]. For the multistage robust UC (7.26) this method works as follows. Instead of finding a solution over an arbitrary functional space, we restrict the search to those dispatch policies of the following affine form:

$$p_{it}(\boldsymbol{d}_{[t]}) = w_{it} + \sum_{j\in\mathscr{N}_d} \sum_{s\in[1:t]} W_{itjs}\, d_{js}, \tag{7.27}$$

for all i, t,for some $\boldsymbol{w}, \boldsymbol{W}$. We can also compactly write this restriction as $\boldsymbol{p}_t(\boldsymbol{d}_{[t]}) = \boldsymbol{w}_t + \boldsymbol{W}_t \boldsymbol{d}_{[t]}$, for all t, with which we obtain the affine multistage robust UC problem:

$$\min_{\boldsymbol{x},\boldsymbol{w},\boldsymbol{W}} \left\{ f(\boldsymbol{x}) + \max_{\boldsymbol{d}\in\mathscr{D}} \sum_{t\in\mathscr{T}} c_t \left(\boldsymbol{w}_t + \boldsymbol{W}_t \boldsymbol{d}_{[t]}\right) \right\} \tag{7.28a}$$

$$\text{s.t.}\ \ \boldsymbol{x} \in \mathscr{F} \tag{7.28b}$$

$$\boldsymbol{w}_t + \boldsymbol{W}_t \boldsymbol{d}_{[t]} \in \Omega_t \left(\boldsymbol{x}, \boldsymbol{d}_t, \boldsymbol{w}_{t-1} + \boldsymbol{W}_{t-1} \boldsymbol{d}_{[t-1]}\right) \quad \forall t \in \mathscr{T}, \boldsymbol{d} \in \mathscr{D}. \tag{7.28c}$$

The reason for making use of this restriction is that the solution space of the problem obtained is now finite-dimensional. With this, the problem is in the right format for directly trying known solution methods.

Affine policies have been recently employed for several power system operation problems. In [21] a stochastic optimization model for the economic dispatch problem with energy storage is presented, which was later extended to unit commitment decisions in [22]. In [10], the dispatch of automatic generation control units is addressed with affine policies under uncertainty in renewable energy outputs. [13] also presents a power dispatch model with automatic generation control units using affine policies, considering wind power uncertainty and the possibility to curtail wind power outputs. In [23] a real-time dispatch problem with affine policy is presented, considering the objective of maximizing the size of the set of wind power injections under which feasibility is maintained. [17] presents a capacity expansion planning problem that uses affine policies for capacity expansion decisions, power outputs and voltage angles.

In [15] the authors realize that it is very difficult to solve the affine multistage robust UC (7.28) under the full affine policy (7.27), due to the structure of the problem and the large number of W_{itjs} decision variables. In order to facilitate the solution of the problem, they further restrict the solution space to the following simplified affine policy form:

$$p_{it}(\boldsymbol{d}_{[t]}) = w_{it} + W_{it} \sum_{j \in \mathcal{N}_d} d_{jt}, \tag{7.29}$$

for all i, t, for some $\boldsymbol{w}, \boldsymbol{W}$. This dispatch policy choice allows exploiting the simpler structure of the problem and together with an innovative algorithmic approach, allows efficiently solving the problem, while still finding close-to-optimal solutions.

7.4.3 Two Paradigms to Solve the Problem

The affine multistage robust UC problem (7.28) can be written in the following form:

$$\min_{\boldsymbol{x},\boldsymbol{w},\boldsymbol{W},z} \quad f(\boldsymbol{x}) + z \tag{7.30a}$$

$$\text{s.t.} \quad \boldsymbol{x} \in \mathscr{F} \tag{7.30b}$$

$$\boldsymbol{a}_k(\boldsymbol{W})^\top \boldsymbol{d} \le b_k(\boldsymbol{x}, \boldsymbol{w}, z) \quad \forall k \in \{1, \dots, K\}, \boldsymbol{d} \in \mathscr{D}, \tag{7.30c}$$

where z corresponds to the worst-case dispatch cost, K is the number of robust constraints in the problem, and $\boldsymbol{a}_k(\boldsymbol{W})$ and $b_k(\boldsymbol{x}, \boldsymbol{w}, z)$ depend on each corresponding constraint. For example, to formulate

$$w_{it} + W_{it} \sum_{j \in \mathcal{N}_d} d_{jt} \le p_i^{\max} x_{it},$$

we need $a_{kjt}(\boldsymbol{W}) = W_{it}$ for all j, $a_{kjt'}(\boldsymbol{W}) = 0$ for all j and $t' \neq t$, and $b_k(\boldsymbol{x}, \boldsymbol{w}, z) = p_i^{\max} x_{it} - w_{it}$.

In order to solve (7.30) each of the K robust constraints can be reformulated using the duality-based approach discussed in Sect. 7.2. For a polyhedral uncertainty set $\mathscr{D} = \{\boldsymbol{d} : \boldsymbol{H}\boldsymbol{d} \le \boldsymbol{h}\}$, we have:

$$
\begin{aligned}
& \boldsymbol{a}^\top \boldsymbol{d} \leq b \quad \forall \boldsymbol{d} \in \mathscr{D} \\
\Leftrightarrow \quad & \max_{\{\boldsymbol{d}:\, H\boldsymbol{d} \leq h\}} \boldsymbol{a}^\top \boldsymbol{d} \leq b \\
\Leftrightarrow \quad & \min_{\{\boldsymbol{\pi}:\, \boldsymbol{\pi}^\top H = \boldsymbol{a},\, \boldsymbol{\pi} \geq 0\}} \boldsymbol{\pi}^\top h \leq b \\
\Leftrightarrow \quad & \exists \boldsymbol{\pi} : \boldsymbol{\pi}^\top H = \boldsymbol{a},\ \boldsymbol{\pi} \geq 0,\ \boldsymbol{\pi}^\top h \leq b.
\end{aligned}
$$

With this, the affine multistage robust UC problem (7.30) can be reformulated as

$$
\begin{aligned}
\min_{\boldsymbol{x},\boldsymbol{w},\boldsymbol{W},z,\boldsymbol{\pi}} \quad & f(\boldsymbol{x}) + z && (7.31a) \\
\text{s.t.} \quad & \boldsymbol{x} \in \mathscr{F} && (7.31b) \\
& \boldsymbol{\pi}_k^\top H = \boldsymbol{a}_k(\boldsymbol{W}) \quad \forall k \in \{1, \ldots, K\} && (7.31c) \\
& \boldsymbol{\pi}_k \geq 0 \quad \forall k \in \{1, \ldots, K\} && (7.31d) \\
& \boldsymbol{\pi}_k^\top h \leq b_k(\boldsymbol{x}, \boldsymbol{w}, z) \quad \forall k \in \{1, \ldots, K\}, && (7.31e)
\end{aligned}
$$

which is a mixed-integer optimization problem that can be directly coded to attempt solving with solvers such as CPLEX or GUROBI.

The duality-based approach presented above is the most traditional method for solving robust optimization problems and has been successfully employed in many different circumstances [3]. However, when attempting to employ this method for the affine multistage robust UC, due to the large dimensionality of the problem, a very large mixed-integer optimization problem can be obtained, which is intractable in practice. In [15], the authors report that for a real-world 2736-bus power system, even if using the simplified affine policy (7.29), this method requires the creation of more than 250 million $\boldsymbol{\pi}$-variables.

An alternative approach proposed in [15] is to use a constraint generation method. It can be easily shown that in problem (7.30) it is equivalent to replace the polyhedral uncertainty set $\mathscr{D}$ with its finite set of extreme points $\{\boldsymbol{d}_1, \ldots, \boldsymbol{d}_L\}$. This change does not help much directly, since the number of extreme points L can be exponential in terms of the input data of the problem, however, this suggest an algorithmic procedure in which we iteratively incorporate the relevant extreme points. To be precise, consider the following *master problem*:

$$
\begin{aligned}
\min_{\boldsymbol{x},\boldsymbol{w},\boldsymbol{W},z} \quad & f(\boldsymbol{x}) + z && (7.32a) \\
\text{s.t.} \quad & \boldsymbol{x} \in \mathscr{F} && (7.32b) \\
& \boldsymbol{a}_k(\boldsymbol{W})^\top \boldsymbol{d} \leq b_k(\boldsymbol{x}, \boldsymbol{w}, z) \quad \forall k \in \{1, \ldots, K\}, \boldsymbol{d} \in D_k, && (7.32c)
\end{aligned}
$$

where D_k is a set of extreme points of $\mathscr{D}$ considered as relevant for the k-th robust constraint. In the constraint generation method, we start with empty D_k's and iteratively solve the master problem and check for every k whether $\boldsymbol{a}_k(\boldsymbol{W})^\top \boldsymbol{d} \leq b_k(\boldsymbol{x}, \boldsymbol{w}, z)$ holds for all $\boldsymbol{d} \in \mathscr{D}$. If this does not hold, we add to D_k the extreme point $\boldsymbol{d}_k^*$ that achieves the maximum violation of the robust constraint. This procedure is repeated until the feasibility of all robust constraints is guaranteed. Formally, the constraint generation method is presented in Algorithm 1.

Algorithm 1 Constraint generation

1: **repeat**
2: Solve Master Problem (7.32)
3: **for all** $k \in \{1, ..., K\}$ **do**
4: $\boldsymbol{d}_k^* \leftarrow \operatorname{argmax} \left\{ \boldsymbol{a}_k(\boldsymbol{W})^\top \boldsymbol{d} \; : \; \boldsymbol{d} \in \mathscr{D} \right\}$
5: If $\boldsymbol{a}_k(\boldsymbol{W})^\top \boldsymbol{d}_k^* > b_k(\boldsymbol{x}, \boldsymbol{w}, z)$ let $D_k \leftarrow D_k \cup \{\boldsymbol{d}_k^*\}$
6: **end for**
7: **until** $\boldsymbol{a}_k(\boldsymbol{W})^\top \boldsymbol{d}_k^* \le b_k(\boldsymbol{x}, \boldsymbol{w}, z)$ for all $k \in \{1, \ldots, K\}$

The constraint generation method presented above iteratively enlarges the master problem, avoiding an extremely large formulation as that in (7.31). However, the drawback of this method is that it can require a very large number of iterations, since there is no guarantee that a small subset of extreme points will be sufficient to characterize each robust constraint. In [15] the authors propose to enhance the constraint generation method in several ways. The enhancements developed include using a very quick sorting method for the separation procedure in line 4 of Algorithm 1, using good starting extreme points in the D_k's, exploiting the structure of the problem and uncertainty set to reformulate output limit and ramping constraints in a direct and simple way, and fixing and releasing binary variables for generating multiple cuts to the master problem. The result of this is a very efficient solution method for the affine multistage robust UC (7.28), under the simplified affine policy (7.29).

7.4.4 Computational Experiments

We summarize here the main experimental results obtained in [15] for the affine multistage robust UC (7.28), under the simplified affine policy (7.29), using a real-world power system with 2736 buses, 289 generators, 2011 load buses (where uncertain net load occurs), and 100 transmission lines, over 24 h.

First, let's briefly go over the efficiency of the algorithm described above. If the enhanced constraint generation algorithm is employed, the problem can be solved within 3.6 h in a modest laptop, whereas the duality based approach runs out of memory, and the basic constraint generation method does not converge within a larger time limit. Further, for a much smaller 118-bus system the enhanced constraint generation algorithm solves the problem within 3 min, while the duality based approach still runs out of memory and the basic constraint generation method does not converge within 4 h.

Now, let's see if the simplified affine policy (7.29) achieves good-quality solutions for the original non-restricted multistage robust UC (7.26). For this purpose, a lower

Table 7.3 Guaranteed optimality gap of the simplified affine policy as a solution for multistage robust UC [15]

Γ	0.25	0.5	1	1.5	2	3
Opt. gap (%)	0.07	0.11	0.25	0.35	0.53	0.94

bound for this latter problem is obtained based on solving the two-stage robust UC (7.24), since it is a relaxation of the original multistage problem (non-anticipativity is relaxed in the two-stage problem). This allows bounding the quality of the solutions obtained. Table 7.3 presents the optimality gap. In this table, Γ is a parameter that determines the size of the uncertainty set. We can observe that even for a large uncertainty set the cost achieved by the simplified affine policy is within 1% of that achieved by the optimal solution. Surprisingly, the simplified affine policy achieves high-quality solutions.

From the above experiments, we can see that good quality solutions for the multistage robust UC problem can be obtained efficiently. So let's try now to see if there is any important practical differences with the two-stage robust UC. In other words, is it important to model non-anticipativity? For this purpose, in [15] two different comparisons are presented, one based on worst-case performance and another based on average performance under a probabilistic framework. In order to compare the worst-case performance of these two models, let's suppose that we solve the two-stage robust UC (7.24) to obtain a commitment solution $\boldsymbol{x}_{2S}$. Then, we take this solution and feed it to the multistage robust UC problem (7.28) to solve it forcing the constraint $\boldsymbol{x} = \boldsymbol{x}_{2S}$, so that we can see how well the solution from two-stage robust UC performs under the multistage robust UC framework. Then, we compare this to directly solving the multistage robust UC. Table 7.4 presents the results. In this table, the "Penalty" cost measures "how much" infeasibility was obtained by penalizing energy balance and transmission violations by \$5000/MWh. We can observe that under large uncertainty sets the two-stage robust UC can have a very limited

Table 7.4 Worst-case cost (US\$) of multistage robust dispatch under the two-stage and multistage UC solutions. Multistage models use the simplified affine policy [15]

	$\Gamma = 0.5$	$\Gamma = 1$	$\Gamma = 1.5$	$\Gamma = 2$	$\Gamma = 3$
Affine multistage UC solutions					
Total cost	9,445,069	9,596,788	9,746,685	9,905,527	10,234,459
Penalty	0	0	0	0	0
Two-stage UC solutions					
Total cost	9,505,651	9,745,889	10,183,433	10,975,403	12,864,719
Penalty	96,313	224,952	591,661	1,165,324	2,703,522
Rel Diff (%)	0.64	1.55	4.49	10.80	25.70

Table 7.5 Simulation performance for multistage and two-stage robust UC [15]

Γ	0.25	0.5	1	1.5	2	3
Affine multistage robust UC with policy-enforcement robust ED						
Cost Avg ($)	9,397,528	9,319,396	9,342,754	9,360,359	9,379,464	9,442,858
Cost Std ($)	113,725	15,970	12,828	12,509	12,363	12,092
Penalty cost Avg ($)	93,552	3497	727	61	5	0
Penalty Freq Avg (%)	10.00	1.47	0.40	0.01	0.00	0.00
Two-stage robust UC with look-ahead ED						
Cost Avg ($)	9,398,109	9,456,599	9,408,732	9,383,569	9,407,290	9,362,379
Cost Std ($)	93,470	195,774	173,884	144,698	162,469	45,584
Penalty cost Avg ($)	80,127	152,637	98,113	66,801	82,864	6,103
Penalty Freq Avg (%)	9.93	12.26	7.80	5.11	5.57	0.37

performance, achieving a very high penalty level. This means that modeling non-anticipativity can make a significant difference in practice.

Finally, we summarize the comparison presented in [15] based on average performance under a probabilistic framework. Consider the following simulation. Given a commitment solution $\boldsymbol{x}$ the net load vector $\boldsymbol{d}_1$ at time $t = 1$ is simulated using a normal distribution and then the dispatch decisions $\boldsymbol{p}_1$ are determined solving an ED problem, then the net load vector $\boldsymbol{d}_2$ at time $t = 2$ is simulated using a normal distribution and the dispatch decisions $\boldsymbol{p}_2$ are determined solving an ED problem, and so on, until time $t = 24$ is reached. This procedure is carried out under $N = 1000$ simulated net load trajectories (assuming independence over buses and time periods), for both models: (i) using the multistage robust UC solution with an ED method denoted as "policy-enforcement robust ED", which employs the policy given by $\boldsymbol{w}$, $\boldsymbol{W}$ identified when solving (7.28), and (ii) using the two-stage robust UC solution with a deterministic "look-ahead ED". The results are shown in Table 7.5. In this Table, "Cost Avg" denotes the average daily cost over the 1000 simulations, "Cost Std" the respective standard deviation, "Penalty Cost Avg" the average penalty cost, and "Penalty Freq Avg" the proportion of time periods where penalty occurred. We can observe that the best average cost is achieved at $\Gamma = 0.5$ for the multistage robust UC and at $\Gamma = 3$ for the two-stage robust UC, with the multistage model presenting a cost reduction of 0.46% as compared to the two-stage robust UC. For these solutions, we can further observe that the multistage robust UC achieves an improved system reliability, presenting a cost standard deviation which is 64.97% lower than that of the two-stage robust UC. Moreover, if Γ is sufficiently enlarged, the multistage robust UC can eliminate the occurrence of penalty, whereas the two-stage robust UC did not achieve this for any of the tested uncertainty set sizes. We can conclude that non-anticipativity is an important aspect that can lead to significant economic and reliability improvements in power system operations.

7.5 Robust Optimization as Decision Support for Determining Bidding Strategies

Robust optimization methods have also been employed for supporting bidding strategies of power producers and virtual power plants participating in power pools. The first such approach was developed in [1] where a robust profit maximization problem for a price-taker power production unit is proposed, considering power price uncertainty. This problem can be represented as:

$$\max_{p \in \Pi} \min_{\lambda \in \Lambda} \sum_{t=1}^{T} \left\{ \lambda_t p_t - c_t(p_t) \right\},$$

where p_t is the power output of the unit at time t, λ_t is power price at time t, $c_t(p_t)$ is a power production cost, and Λ is an uncertainty set for power prices over the T time periods. In this problem, the power production vector $\boldsymbol{p}$ is selected in such a way that worst-case profit is maximized. The uncertainty set considered by the authors is given by a *box* uncertainty set of the form

$$\Lambda = \left[\underline{\lambda}, \overline{\lambda}\right] = \left\{ \lambda : \lambda_t \in [\underline{\lambda}_t, \overline{\lambda}_t] \ \forall t \in \{1, \ldots, T\} \right\},$$

and the problem is solved using a traditional duality reformulation. In order to build a bidding curve with the support of this model, the authors propose to repeatedly solve the problem under different uncertainty sets Λ. In particular, suppose a bidding curve for prices ranging from λ^{min} to λ^{max} is desired. Then, in the bidding curve, the power production quantities offered under prices $\hat{\lambda}$ are selected as the solution to the robust profit maximization problem under $\Lambda = \left[\hat{\lambda}, \lambda^{max}\right]$. This problem is solved for multiple $\hat{\lambda}$ ranging from λ^{min} to λ^{max} in order to determine an entire bidding curve for each hour, where the offered production quantities are increasing functions of prices.

Another robust optimization approach for supporting bidding strategies was developed in [20], where a power production entity composed of a wind farm and an energy storage unit is considered. Uncertainty in both power prices and wind power are incorporated, using an uncertainty set related to the concept of conditional value at risk. Computational experiments show this approach can yield better returns than a deterministic model.

Reference [18] emphasizes the uncertain mismatch between day-ahead and real-time power prices. The authors consider an energy management system that controls price responsive demands, a wind farm, and an energy storage unit, composing a virtual power plant that can buy and sell power in the day-ahead and real-time markets. Two robust optimization models are proposed, one for the real-time market, in which uncertain real-time power prices are considered, and another for the day-ahead market, in which uncertainty is present in real-time and day-ahead power prices, as well as in available wind power. In order to model the relation between day-ahead and real-time prices, the latter are determined as day-ahead prices plus a "residual" term that is assumed to be independent of day-ahead prices, a choice justified based on a statistical analysis. In order to construct bidding curves for the day-ahead and real-time markets, the respective models are repeatedly solved under different corresponding uncertainty sets, as proposed in [1] (described above).

The above references build models based on maximizing a worst-case profit. In contrast to this objective function, and with the purpose of conservativeness reduction, [9] proposes a robust optimization model that minimizes the worst-case *regret*, with the purpose of designing bidding strategies for thermal generators under uncertain power prices. Their worst-case regret minimization problem can be represented as:

$$\min_{(\boldsymbol{x},\boldsymbol{p})\in X} \max_{\lambda\in\Lambda} \operatorname{Reg}(\boldsymbol{x},\boldsymbol{p},\lambda), \tag{7.33}$$

where $\boldsymbol{x}$ contains commitment decisions (on/off, start-up, shut-down), $\boldsymbol{p}$ contains dispatch decisions over the time horizon, λ contains uncertain power prices over the time horizon, Λ is the uncertainty set, and $\operatorname{Reg}(\boldsymbol{x},\boldsymbol{p},\lambda)$ is the regret, defined as

$$\operatorname{Reg}(\boldsymbol{x},\boldsymbol{p},\lambda) = \max_{(\hat{\boldsymbol{x}},\hat{\boldsymbol{p}})\in X} \left\{\lambda^{\top}\hat{\boldsymbol{p}} - c(\hat{\boldsymbol{x}},\hat{\boldsymbol{p}})\right\} - \left\{\lambda^{\top}\boldsymbol{p} - c(\boldsymbol{x},\boldsymbol{p})\right\},$$

where $c(\boldsymbol{x},\boldsymbol{p})$ includes fixed and variable costs, and $(\hat{\boldsymbol{x}},\hat{\boldsymbol{p}})$ are realized as the best possible commitment and dispatch decisions under a realization λ of power prices, so that the regret $\operatorname{Reg}(\boldsymbol{x},\boldsymbol{p},\lambda)$ represents the profit difference between optimal decisions $(\hat{\boldsymbol{x}},\hat{\boldsymbol{p}})$ given the knowledge of λ and the profit obtained under the current decisions $(\boldsymbol{x},\boldsymbol{p})$.

In order to solve problem (7.33), a reformulation method and Benders Decomposition algorithm are developed. Further, the model is used to obtain a bidding curve based on repeatedly solving the problem, in a way similar to the method proposed in [1] (described above), but updating a budget of uncertainty in the uncertainty set, and updating bounds on dispatch decisions. In computational experiments, the authors show that the proposed approach can increase profits as compared to a robust optimization approach with worst-case profit maximization as objective.

7.6 Distributionally Robust Optimization in Power System Operations

Consider the following two-stage distributionally robust optimization model:

$$\min_{x\in\mathscr{F}} \left\{ f^\top x + \max_{\mathbb{P}\in\mathscr{P}} \mathbb{E}_{\mathbb{P}}\left[Q(x,\tilde{\xi})\right] \right\}, \tag{7.34}$$

where for any given ξ

$$Q(x,\xi) = \min_{y\in\Omega(x,\xi)} c^\top y$$

with

$$\Omega(x,\xi) = \{y : By \geq M\xi - Ax + h\}.$$

In this problem, x encompasses first-stage decisions and y second-stage decisions, $\mathbb{P}$ denotes a probability distribution, $\mathscr{P}$ is an *ambiguity* set, $\mathbb{E}_{\mathbb{P}}[\cdot]$ denotes expectation under probability distribution $\mathbb{P}$, and $\tilde{\xi}$ is a vector of stochastic parameters that follows some probability distribution $\mathbb{P} \in \mathscr{P}$. Here, $Q(x,\xi)$ represents the optimized second-stage cost under first-stage decision x and realization ξ of stochastic vector $\tilde{\xi}$, so that $\mathbb{E}_{\mathbb{P}}\left[Q(x,\tilde{\xi})\right]$ represents the expected second-stage cost under probability distribution $\mathbb{P}$, and $\max_{\mathbb{P}\in\mathscr{P}} \mathbb{E}_{\mathbb{P}}\left[Q(x,\tilde{\xi})\right]$ represents the worst-case expected second-stage cost under first-stage decision x, obtained under some corresponding worst-case probability distribution $\mathbb{P} \in \mathscr{P}$.

In the above model, the ambiguity set $\mathscr{P}$ is a set of probability distributions. This set could be built encompassing all probability distributions with statistical properties (such as mean and covariance matrix) close in some sense to certain reference statistical properties estimated from historical data [8].

In [26] the authors propose a data-driven stochastic unit commitment model using the above framework, in which the first-stage decision x represents on/off, start-up and shut-down commitment decisions; the second-stage decision y represents dispatch levels, operating reserves and spinning reserves; and stochastic vector $\tilde{\xi}$ determines stochastic wind power outputs.

In their approach, $\tilde{\xi}$ is assumed to have a finite support composed of K scenarios $\xi^1, \ldots, \xi^K$, and any probability distribution $\mathbb{P}$ determines their respective probabilities $p_1, \ldots, p_K$, so that

$$\mathbb{E}_{\mathbb{P}}\left[Q(x,\tilde{\xi})\right] = \sum_{k=1}^{K} p_k\, Q(x,\xi^k)$$

for some vector p associated to $\mathbb{P}$. Given this critical assumption, the ambiguity set $\mathscr{P}$ can be represented through a set for the K-dimensional vector p. The authors propose two such sets, based on constraining the distance of $\boldsymbol{p}$ to an empiric distribution $\boldsymbol{p}^0$ obtained from historical data, employing the L_1 and L_∞ norms to measure distance:

$$\mathscr{P}_1 = \left\{ p \in \mathbb{R}_+^K : \; ||p - p^0||_1 \leq \theta \right\}$$
$$\mathscr{P}_\infty = \left\{ p \in \mathbb{R}_+^K : \; ||p - p^0||_\infty \leq \theta \right\},$$

where θ is a distance tolerance that is chosen according to the amount of data available, according to a confidence level criteria. The authors also show that the conservativeness of their approach vanishes as the amount of data available increases to infinity, under both ambiguity sets.

In order to solve the resulting problem (7.34) under this setting, the authors develop a Benders Decomposition algorithm that converges to a global optimal solution. See [26] for more details.

7.7 Chance-Constrained Optimal Power Flow

7.7.1 Modeling Approach

In robust optimization models constraints are required to hold under any realization of uncertain parameters in a given uncertainty set. In these models there is no need for probabilistic distributions for uncertain parameters. However, if one were to have good probabilistic models for the uncertain parameters in the problem, then one can also formulate *chance constraints*, namely, constraints that state that the probability of certain event has to be high (or low). For example, one could formulate the following chance constraint for the maximum flow capacity of a transmission line (i,j):

$$\mathbb{P}\left(\tilde{f}_{ij} > f_{ij}^{max} \right) < \varepsilon.$$

This constraint states that the probability that the power flow $\tilde{f}_{ij}$ is greater than the capacity f_{ij}^{max} on has to be smaller than ε. Similarly, one could also incorporate the analogous constraint for the backward flow on this line:

$$\mathbb{P}\left(\tilde{f}_{ij} < -f_{ij}^{max} \right) < \varepsilon.$$

Further, another alternative is to enforce the more conservative joint chance constraint that implies both of the above:

$$\mathbb{P}\left(|\tilde{f}_{ij}| > f_{ij}^{max}\right) < \varepsilon.$$

In an optimization problem, one could impose the above chance constraints for all transmission lines, or one could also enforce a much more conservative joint chance constraint for all transmission lines:

$$\mathbb{P}\left(\exists (i,j) \text{ s.t.} |\tilde{f}_{ij}| > f_{ij}^{max}\right) < \varepsilon.$$

The above discussion is developed in detail in [6], and in this paper, the following chance-constrained optimal power flow problem with stochastic wind power is proposed:

$$\min_{\tilde{p},\tilde{\theta},\boldsymbol{\alpha},\overline{\boldsymbol{p}}} \quad \mathbb{E}\left[c(\tilde{\boldsymbol{p}})\right] \tag{7.35a}$$

$$\text{s.t. } \tilde{p}_i = \overline{p}_i - \alpha_i \sum_{j\in\mathscr{W}} \tilde{\omega}_j \quad \forall i \in \mathscr{G} \tag{7.35b}$$

$$\tilde{p}_i = \mu_i + \tilde{\omega}_i \quad \forall i \in \mathscr{W} \tag{7.35c}$$

$$\tilde{p}_i = 0 \quad \forall i \in \mathscr{V} - (\mathscr{G} \cup \mathscr{W}) \tag{7.35d}$$

$$\boldsymbol{B}\tilde{\boldsymbol{\theta}} = \tilde{\boldsymbol{p}} - \boldsymbol{d} \tag{7.35e}$$

$$\tilde{f}_{ij} = \beta_{ij}\left(\tilde{\theta}_i - \tilde{\theta}_j\right) \quad \forall \{i,j\} \in \mathscr{E} \tag{7.35f}$$

$$\mathbb{P}\left(\tilde{p}_i > p_i^{max}\right) < \varepsilon_i \quad \forall i \in \mathscr{G} \tag{7.35g}$$

$$\mathbb{P}\left(\tilde{p}_i < p_i^{min}\right) < \varepsilon_i \quad \forall i \in \mathscr{G} \tag{7.35h}$$

$$\mathbb{P}\left(\tilde{f}_{ij} > f_{ij}^{max}\right) < \varepsilon_{ij} \quad \forall \{i,j\} \in \mathscr{E} \tag{7.35i}$$

$$\mathbb{P}\left(\tilde{f}_{ij} < -f_{ij}^{max}\right) < \varepsilon_{ij} \quad \forall \{i,j\} \in \mathscr{E} \tag{7.35j}$$

$$\boldsymbol{\alpha} \geq 0, \tag{7.35k}$$

where $\mathscr{V} = \{1, \dots, n\}, \mathscr{G}, \mathscr{W}, \mathscr{E}$ are the sets of buses, generators, wind farms and transmission lines, respectively, with $\mathscr{G}, \mathscr{W}$ disjoint and $\mathscr{G}, \mathscr{W} \subset \mathscr{V}$; c is a cost function defined as $c(\boldsymbol{p}) = \sum_{i\in\mathscr{G}} c_i(p_i)$, where each c_i is convex quadratic; $\tilde{p}_i$ is power output at generator i; $\overline{p}_i$ and α_i are decision variables that determine how $\tilde{p}_i$ adapts to wind power output variations, for generator i; $\tilde{\theta}_i$ is the voltage angle at bus i; μ_i is the expected wind power output at wind farm i; $\tilde{\omega}_i$ is a stochastic component that determines the wind power output deviation from its mean at wind farm i; $\boldsymbol{d}$ is a vector of electricity loads at all buses; β_{ij} and $\tilde{f}_{ij}$ are the susceptance and power flow on transmission line $\{i,j\}$, respectively; $\boldsymbol{B}$ is a matrix determined by the susceptance values; and ε_i and ε_{ij} are tolerance levels on the respective chance constraints.

The objective of this chance-constrained optimal power flow problem is to minimize the expected dispatch cost. Constraints (7.35b) determine how generator dispatch decisions adapt to wind power output variations. Constraints (7.35c) determine the power output at wind farms. Constraints (7.35d) enforce no power output at those buses without generators or wind farms. Constraint (7.35e) represents the energy balance at all buses through the DC power flow equations. Constraints (7.35f) determine power flows on transmission lines. Constraints (7.35g) determine chance constraints that enforce a low probability of power output at generators exceeding their maximum power output levels. Similarly, constraints (7.35h) impose this idea for the minimum power output levels. Constraints (7.35i) enforce a low probability of transmission line flows exceeding their maximum flow levels in the forward direction. Similarly, constraints (7.35j) enforce this idea for flows in the backward direction. Constraint (7.35k) imposes nonnegativity on the α_i's.

In this problem, the wind power fluctuation $\tilde{\omega}_i$ is stochastic, with mean 0 and standard deviation σ_i^2. Based on these fluctuations, power output at generators is determined as an affine function of the aggregated fluctuation $\sum_{j\in\mathcal{W}} \tilde{\omega}_j$. Based on this dependence structure, chance constraints are formulated on the maximum and minimum power output levels and transmission line flow capacities.

7.7.2 Reformulation as a Second-Order Cone Program and Solution Method

Under the assumption that the $\tilde{\omega}_i$ fluctuations are normally distributed and independent, the authors show that the chance-constrained optimal power flow (7.35) can be reformulated as the following second-order cone program:

$$\min_{\alpha,\overline{p},\overline{\theta},\delta,s} \quad \sum_{i\in\mathcal{G}} \left\{ c_{i2}\overline{p}_i^2 + c_{i2}\alpha_i^2 \left(\sum_{k\in\mathcal{W}} \sigma_k^2 \right) + c_{i1}\overline{p}_i + c_{i0} \right\} \tag{7.36a}$$

$$\text{s.t.} \quad \sum_{j=1}^{n-1} \hat{B}_{ij}\delta_j = \alpha_i \quad \forall i \in \{1,\dots,n-1\} \tag{7.36b}$$

$$\sum_{j=1}^{n-1} \hat{B}_{ij}\overline{\theta}_j - \overline{p}_i = \mu_i - d_i \quad \forall i \in \{1,\dots,n-1\} \tag{7.36c}$$

$$\sum_{i\in\mathcal{G}} \alpha_i = 1, \ \boldsymbol{\alpha} \geq 0, \ \overline{\boldsymbol{p}} \geq 0, \tag{7.36d}$$

$$\overline{p}_n = \alpha_n = \delta_n = \overline{\theta}_n = 0, \tag{7.36e}$$

$$\beta_{ij}|\overline{\theta}_i - \overline{\theta}_j| + \beta_{ij}\eta(\varepsilon_{ij})s_{ij} \leq f_{ij}^{max} \quad \forall \{i,j\} \in \mathcal{E} \tag{7.36f}$$

$$\left[\sum_{k\in\mathcal{W}} \sigma_k^2(\pi_{ik} - \pi_{jk} - \delta_i + \delta_j)^2 \right]^{1/2} - s_{ij} \leq 0 \quad \forall \{i,j\} \in \mathcal{E} \tag{7.36g}$$

$$-\overline{p}_i + \eta(\varepsilon_i)\left(\sum_{k\in\mathscr{W}} \sigma_k^2\right)^{1/2} \leq -p_i^{min} \quad \forall i \in \mathscr{G} \tag{7.36h}$$

$$\overline{p}_i + \eta(\varepsilon_i)\left(\sum_{k\in\mathscr{W}} \sigma_k^2\right)^{1/2} \leq p_i^{max} \quad \forall i \in \mathscr{G} \tag{7.36i}$$

where c_{i2}, c_{i1}, c_{i0} are the parameters that determine the convex quadratic function c_i; $\hat{B}_{ij}$ and π_{ij} are determined by the susceptance matrix of the network; $\overline{\boldsymbol{\theta}}, \boldsymbol{\delta}, \boldsymbol{s}$ are new decision variables; and η is a function defined as $\eta(r) = \Phi^{-1}(1-r)$, where Φ is the cumulative distribution function of the standard normal distribution.

In order to obtain this reformulation, several properties and observations are employed. In particular, the properties of the DC power flow equations (7.35e) allow eliminating the power flow variables $\tilde{f}_{ij}$ in (7.35) and reformulate the affine policy (7.35b) into simpler direct expressions for $\boldsymbol{\alpha}$ and $\overline{\boldsymbol{p}}$. Moreover, the key element to reformulate chance constraints (7.35g)–(7.35j) is to notice that under the assumptions specified and the conditions in the problem, $\tilde{f}_{ij}$ can be written as an affine function of the ω_i's and is thus normally distributed itself, so that

$$\mathbb{P}\left(\tilde{f}_{ij} > f_{ij}^{max}\right) < \varepsilon_{ij}$$

holds if and only if

$$\mathbb{E}[\tilde{f}_{ij}] \;+\; \eta(\varepsilon_{ij}) \;\mathrm{var}(\tilde{f}_{ij}) \leq f_{ij}^{max},$$

where $\mathrm{var}(\tilde{f}_{ij})$ is the variance of $\tilde{f}_{ij}$. See [6] for details.

The above second-order cone formulation (7.36) of the chance-constrained optimal power flow problem can in theory be solved directly with efficient specialized algorithms. However, the authors report that numerical issues led to significant difficulties of optimization solvers when trying to solve the problem for large-scale instances. Given this, they implemented an effective cutting plane algorithm by reformulating second-order cone constraints through an infinite set of linear inequalities parameterized by a vector $\hat{\delta} \in \mathbb{R}^n$. Based on this reformulation, the algorithm consists of iteratively solving a master problem with a partial set of these linear constraints, and separating over the conic constraints in order to identify new of these linear constraints parameterized by certain new $\hat{\delta}$'s, adding then these new linear constraints to the master problem, repeating this procedure until all conic and chance constraints are satisfied up to a numerical tolerance. See [6] for details.

7.7.3 Data-Robust Chance-Constrained Optimal Power Flow

In [6] the authors also formulate a data-robust version of the chance-constrained optimal power flow, in which the wind power output mean μ_i and variance σ_i^2 are not assumed to be known exactly. In fact, an uncertainty set representation is used for these parameters by setting $\mu_i = \overline{\mu}_i + r_i$ and $\sigma_i^2 = \overline{\sigma}_i^2 + v_i$, where $\boldsymbol{r} \in \mathscr{M}$ and $\boldsymbol{v} \in \mathscr{S}$, with certain uncertainty sets $\mathscr{M}$ and $\mathscr{S}$ that can be either polyhedral or ellipsoidal. Based on this, the *nominal* chance constraint

$$\mathbb{P}\left(\tilde{f}_{ij} > f_{ij}^{max}\right) < \varepsilon_{ij}.$$

is replaced by the *robust* chance constraint

$$\max_{\{\boldsymbol{r} \in \mathscr{M}, \boldsymbol{v} \in \mathscr{S}\}} P_{\boldsymbol{r},\boldsymbol{v}}\left(\tilde{f}_{ij} > f_{ij}^{max}\right) < \varepsilon_{ij},$$

where $P_{\boldsymbol{r},\boldsymbol{v}}$ denotes the probability function when the wind power output mean μ_i and variance σ_i^2 are determined as $\mu_i = \overline{\mu}_i + r_i$ and $\sigma_i^2 = \overline{\sigma}_i^2 + v_i$. Similarly, other nominal chance constraints in the problem are replaced by their robust counterparts to obtain a data-robust chance-constrained optimal power flow problem. Further, to solve the resulting data-robust problem, the authors develop another cutting-plane algorithm. See [6] for details.

Finally, [16] presents another robust chance-constrained optimal power flow problem that builds upon the work in [6].

7.8 Concluding Remarks

We have reviewed the robust optimization approach for power system operations. After an introduction to the major tools of robust optimization, we presented promising recent two-stage adaptive robust models for day-ahead and real-time power system operations, including a description of dynamic uncertainty sets for capturing temporal and spatial correlations in wind power. After this, we discussed the concept of multistage robust optimization, emphasizing the employment of affine policies and the development of an innovative algorithm for multistage robust unit commitment. Then, we further went on to show how robust optimization can be used to support bidding strategies of power producers. Finally, we presented two promising areas of optimization under uncertainty approaches that are closely related to robust optimization: distributionally robust optimization and chance constraints, showing applications on unit commitment and optimal power flow, respectively.

Many important questions on robust optimization for power system operations remain open. For example, can robust optimization support the selection of adequate day-ahead and real-time locational prices? Can more efficient algorithms be devised to solve large-scale robust unit commitment problems much faster? In the future, we expect many exciting further developments on this vibrant research area.

References

1. L. Baringo, A.J. Conejo, Offering strategy via robust optimization. IEEE Trans. Power Syst. **26**(3), 1418–1425 (2011)
2. A. Ben-Tal, A. Goryashko, E. Guslitzer, A. Nemirovski, Adjustable robust solutions of uncertain linear programs. Math. Program. Ser. A **99**, 351–376 (2004)
3. D. Bertsimas, D.B. Brown, C. Caramanis, Theory and applications of robust optimization. SIAM Rev. **53**(3), 464–501 (2011)
4. D. Bertsimas, E. Litvinov, X.A. Sun, J. Zhao, T. Zheng, Adaptive robust optimization for the security constrained unit commitment problem. IEEE Trans. Power Syst. **28**(1), 52–63 (2013)
5. D. Bertsimas, M. Sim, The price of robustness. Oper. Res. **52**(1), 35–53 (2004)
6. D. Bienstock, M. Chertkov, S. Harnett, Chance-constrained optimal power flow: risk-aware network control under uncertainty. SIAM Rev. **56**(3), 461–495 (2014)
7. E. Delage, D. Iancu, Robust multi-stage decision making, in *In INFORMS Tutorials in Operations Research* (2015), pp. 20–46
8. E. Delage, Y. Ye, Distributionally robust optimization under moment uncertainty with application to data-driven problems. Oper. Res. **58**(3), 595–612 (2010)
9. L. Fan, J. Wang, R. Jiang, Y. Guan, Min-max regret bidding strategy for thermal generator considering price uncertainty. IEEE Trans. Power Syst. **29**(5), 2169–2179 (2014)
10. R.A. Jabr, Adjustable robust OPF with renewable energy sources. IEEE Trans. Power Syst. **28**(4), 4742–4751 (2013)
11. R. Jiang, J. Wang, Y. Guan, Robust unit commitment with wind power and pumped storage hydro. IEEE Trans. Power Syst. **27**(2), 800–810 (2012)
12. R. Jiang, M. Zhang, G. Li, Y. Guan, Two-stage network constrained robust unit commitment problem. Euro. J. Oper. Res. **234**(3), 751–762 (2014)
13. Z. Li, W. Wu, B. Zhang, B. Wang, Adjustable robust real-time power dispatch with large-scale wind power integration. IEEE Trans. Sustain. Energ. **6**(2), 357–368 (2015)
14. A. Lorca, X.A. Sun, Adaptive robust optimization with dynamic uncertainty sets for multi-period economic dispatch under significant wind. IEEE Trans. Power Syst. **30**(4), 1702–1713 (2015)
15. A. Lorca, X.A. Sun, E. Litvinov and T. Zheng, Multistage adaptive robust optimization for the unit commitment problem. Oper. Res. (2016)
16. M. Lubin, Y. Dvorkin, S. Backhaus, A robust approach to chance constrained optimal power flow with renewable generation. IEEE Trans. Power Syst. (2015)
17. D. Mejia-Giraldo, J. McCalley, Adjustable decisions for reducing the price of robustness of capacity expansion planning. IEEE Trans. Power Syst. **29**(4), 1573–1582 (2014)
18. M. Rahimiyan and L. Baringo, Strategic bidding for a virtual power plant in the day-ahead and real-time markets: a price-taker robust optimization approach. IEEE Trans. Power Syst.
19. X.A. Sun and A. Lorca, Adaptive robust optimization for daily power system operation, in *Power Systems Computation Conference (PSCC)*. (IEEE, 2014), pp. 1–9
20. A. Thatte, L. Xie, D.E. Viassolo, S. Singh et al., Risk measure based robust bidding strategy for arbitrage using a wind farm and energy storage. IEEE Trans. Smart Grid **4**(4), 2191–2199 (2013)
21. J. Warrington, P.J. Goulart, S. Mariéthoz, M. Morari, Policy-based reserves for power systems. IEEE Trans. Power Syst. **28**(4), 4427–4437 (2013)
22. J. Warrington, C. Hohl, P.J. Goulart, and M. Morari, Rolling unit commitment and dispatch with multi-stage recourse policies for heterogeneous devices. IEEE Trans. Power Syst. (2015)

23. W. Wei, J. Wang, S. Mei, Dispatchability maximization for co-optimized energy and reserve dispatch with explicit reliability guarantee. IEEE Trans. Power Syst. (2015)
24. L. Xie, P. Carvalho, L. Ferreira, J. Liu, B. Krogh, N. Popli, M. Ilic, Wind integration in power systems: operational challenges and possible solutions. Proc. IEEE. **99**(1), 214–232 (2011)
25. B. Zeng, L. Zhao, Solving two-stage robust optimization problems using a column-and-constraint generation method. Oper. Res. Lett. **41**(5), 457–461 (2013)
26. C. Zhao, Y. Guan, Data-driven stochastic unit commitment for integrating wind generation. IEEE Trans. Power Syst. (2015)
27. L. Zhao, B. Zeng, Robust unit commitment problem with demand response and wind energy. IEEE Power Energ. Soc. Gen. Meet. (2012)

Chapter 8
Planning of Large Scale Renewable Energy for Bulk Power Systems

José Conto

Preface

This chapter presents the challenging problems and potential solutions that large interconnection network systems face due to the high penetration of renewable power plants using asynchronous generators. The penetration of renewable power sources into wide-area power system has been increase steadily over the last years all over the world. As an example, wind power at the Electric Reliability Council of Texas, (ERCOT) has growth to over 12 GW capacity installed (2015) and it is expected that an additional 7–10 GW can be added in the next 2–4 years. A similar wave of new generation interconnections is expected for utility-scale solar PV plants.

This high penetration of asynchronous generation systems comes with new challenges and potential new operational practices to continuously operate reliably the grid. Under an economy-driven restructured electric market, old conventional generation are being replaced by cheaper generation, mostly wind power. This new generation fleet provide less dynamic reactive capability, inertia and synchronism support during disturbance events due to performance limitations of its technology or legendary interconnection requirements. To bring all these renewable energy from remote locations to load centers, new long transmission lines are designed with series compensation together with relative large amount of reactor shunts and SVC-like devices to maximize power transfers, which could lead to potential sub-synchronous resonance.

Sub-synchronous resonance can manifest itself when asynchronous power sources' controls interacts with the dynamics of the grid, resulting in voltage oscillations due to too-fast uncoordinated reactive control as it had occurred in practice. Networks with very low short-circuit ratios can induce unexpected oscillations due to plant controls operating outside their normal range. Regulatory

J. Conto (✉)
ERCOT's System Planning, Dynamic Studies Group, Taylor, TX 76574, USA
e-mail: jconto@ieee.org; Jose.Conto@ercot.com

P. Du et al. (eds.), *Integration of Large-Scale Renewable Energy into Bulk Power Systems*, Power Electronics and Power Systems,
DOI 10.1007/978-3-319-55581-2_8

agencies (FERC, NERC) have imposed more restrict reliability standards, asking for more demanding and comprehensive system studies. Next generation planners and operators should work together to provide sound solutions to each and all of these challenges. All of these challenges can be solved by coordinated system planning as well as emerging techniques, like parallel processing, for these computation-intensive studies.

8.1 Introduction

Large electric interconnections like ERCOT, has adjusted planning and operation processes to accommodate a high penetration levels of wind power, over 12 GW of installed capacity and providing, on high wind times, up to 25% of total system load [1]. Even though a lower marginal cost energy source like wind energy is preferred from an economic standpoint, there are several technical problems that large amount of wind power, due to technology limitations, network location and effect on conventional generation, have resulted in new challenges to operate the grid in a reliably manner. The trend for new wind plants is toward using double-fed or full-conversion technology (Type III or Type IV wind turbine technology.)

Moving renewable energy located very far from the load centers would require new transmission lines and depending on economics, be series-capacitor compensatedcompensated to increase their transfer capability—see Fig. 8.1.

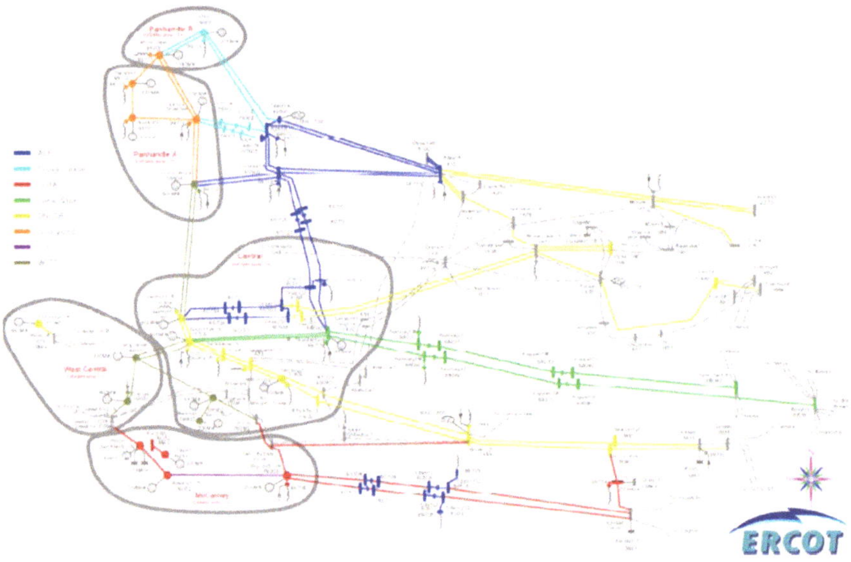

Fig. 8.1 ERCOT West Texas network—design proposal

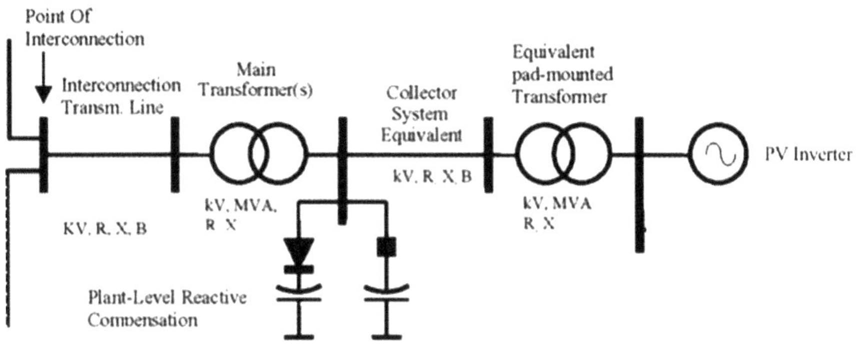

Fig. 8.2 Solar PV plant modeled as single machine equivalent [3]

Technical challenges referenced here, can occur due to the intermittency of the energy source, fluctuating levels of wind or solar irradiation, or to unique network characteristics like network having very low short circuit ratio values. For example, the short-circuit ratio (typically in the West Texas region) can get reduced to lower values under N-1 line outage. Appropriate modeling, not typically used in transmission studies, would be required to study phenomena like signal stability or sub-synchronous resonance.

8.2 Renewable Modeling for Steady State Studies

Simple models representing power output (MW and MVARS) has been used to represent renewable plants at the transmission model [2]. Simulations tools allows for fixed power factor generation as well as voltage control model, similar to a synchronous machine. Equivalent models representing tens or hundreds units in a wind plant are used in steady state model to evaluate load flows—see Fig. 8.2.

8.3 Wind Modeling for Dynamic Studies

The modeling of wind power plant (WPP) was initially done by using induction generator models to represent the dynamics of the wind turbine (in the early years, before 2003, wind turbine Type I and Type II were highly popular), but due to deficiencies of the model in reproduce the dynamics of the wind plant, a new set of manufacturer-driven models were soon adopted by the industry.[1] Along the years,

[1]ERCOT developed a first generation wind models in 2003. PSS/e user-defined models were prepared for manufacturer specific wind turbines, representing the existing fleet of type I, II, III and IV wind turbine machines.

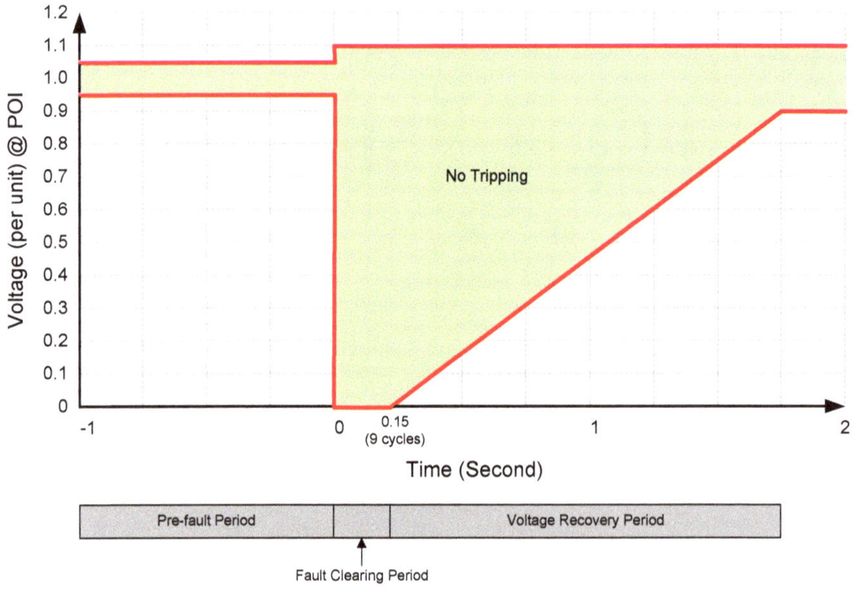

Fig. 8.3 ERCOT VRT requirements for intermittent renewable resources

new wind turbine machines were deployed and their relative impact to voltage response under transient events were recognized resulting in an update of manufacturer's wind models with Voltage Ride Through (VRT) requirements—see Fig. 8.3. Better methodology were developed to aggregate full detail wind plant sub-network into a few equivalent machines, retaining its steady state, short circuit and transient stability behavior [4].

Tests and validation of wind models in transient simulations led to the conclusion that the existing models provide acceptable response to voltage transient tests but were not suitable for small signal stability study, large frequency excursion or when the network is weak (measured as a low short-circuit ratio[2]). For these situations, detail modeling of wind plants is usually done in PSCAD, a three-phase network EMTP-like tool, with a further requirement to obtain compatible models for each wind turbine model to be studied.

Recognizing the advantages of generic wind turbine models, several organizations like IEEE, NERC, WECC, participate in wind turbine generic model development. Operational tools to perform quasi real-time transient studies that include wind models have been deployed at several EMS centers (ERCOT, PJM, CAISO, WECC).

[2]Short-circuit ratio (SCR) is defined as the ratio of the system three phase short circuit MVA at WPP's point of interconnection (POI) to the WPP's rated capacity in MW. In practice, low SCR values are below 3.

8.4 Solar PV Modeling for Dynamic Studies

Solar PV plants are also modeled using dynamic models available in the simulation tools or by using the proprietary user-defined model provided by the manufacturer of the inverter [3]. Type IV wind modelingtechniques are applicable to solar PV models.

8.5 Frequency Support

Conventional synchronous generation contribute not only MW or MVAR to the grid but also provides power synchronism, inertia, damping and through its controls, capability to react in a positive way to transient disturbances. Most new wind power being proposed through the interconnection process are type III or IV, and solar PV plants can be considered to behave similar to a wind type IV. Both plant types are known for their inability to provide inertia support (instantaneous reaction to counteract the sudden change in frequency) as well as lack of governor function to help restore frequency unless their power production is constrained.

The effect of having higher penetration of renewable power relative to conventional generation is a reduction in dynamic frequency response of the system, not only loss of inertia but also less governor response. As an illustration, Fig. 8.4 shows the system response to comparable frequency disturbances for two different levels of wind power penetration [5]. The curve in red belong to a July 2009 power system condition where a unit generating at 890 MW tripped when total system load was 49,209 MW with 675 MW wind generation in service, while the dotted

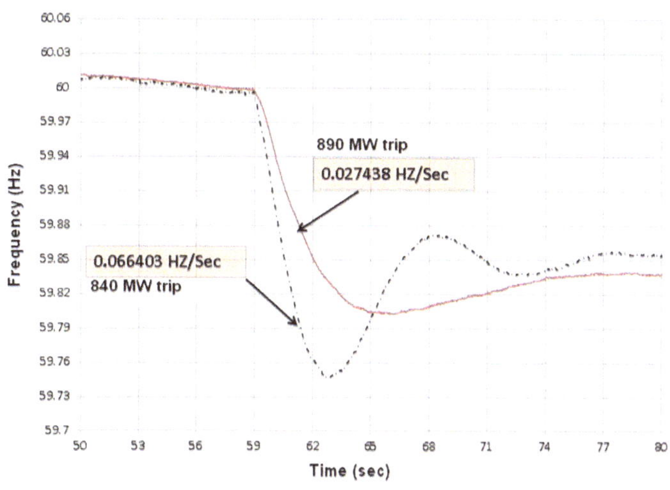

Fig. 8.4 Frequency response comparison

curve correspond to a March 2010 power system condition where a unit generating at 837 MW tripped when total system load was 23,655 MW with 4,300 MW wind generation in service. The frequency response for the 2010 power system conditions is characterized by the high rate of change of frequency (ROCOF) immediately following the disturbance and the lower nadir point. The power system operating condition had a combination of high-wind and low-load conditions, which mainly caused the deteriorated frequency response.

Large electric interconnections like ERCOT require all new renewable generation to have primary frequency support, similar in response to an equivalent size conventional generation. Renewable plant's lack of inertia and governor action during frequency events could impact the amount of appropriate system spinning reserve required to sustain the loss of the largest generator. From an Operational perspective, the total amount of spinning reserve should be sufficient to arrest frequency decay below critical limits (NERC Standard PRC-006) under credible frequency events. Load resources can also participate in frequency response program to supplement total spinning reserve (at ERCOT, load resources participate in a high-frequency load shed program up to 50% of total responsive reserve service, currently up to 1400 MW).

8.6 Voltage Oscillation

Certain remote grid network can experience voltage oscillation due to poor voltage control coordination and weak grid conditions. As an example, the ERCOT West Texas region has new 345 kV lines where up to 13 of these lines will be 50% series compensated [6]. In times when the West Texas region is operating in low-load conditions, the planning network could experience transient over-voltages above 1.2 p.u. even though the project does include a large amount of capacitive and reactor devices to manage the bus voltage under N-1 contingency conditions. In some network design scenarios, SVCs of very large size were located near to large wind power clusters. Dynamic simulations have shown that this arrangement produces not only high bus voltage but also almost un-damped voltage oscillations.

Proprietary dynamic models of renewable plants do not provide insight of the control blocks, usually being modeled as a user-defined model with few parameters for its tuning and adding to the injury, lack of technical information. Generic wind plant models with sufficient block control representation (PSCAD models) were used to test control strategies to slow the reactive compensation response of a wind unit. Figure 8.5 shows that reduction in voltage oscillation and also a lower voltage transient post-fault for different settings of the voltage controller gain. As an alternative, similar strategies were successfully tested on a generic SVC control block to slow down the reactive contribution during post-fault conditions. It was concluded that a better coordination of dynamic response to voltage support during transient is the solution to this issue. The problem of voltage oscillations in weak grid with high penetration of wind power is discussed in more detail at [7].

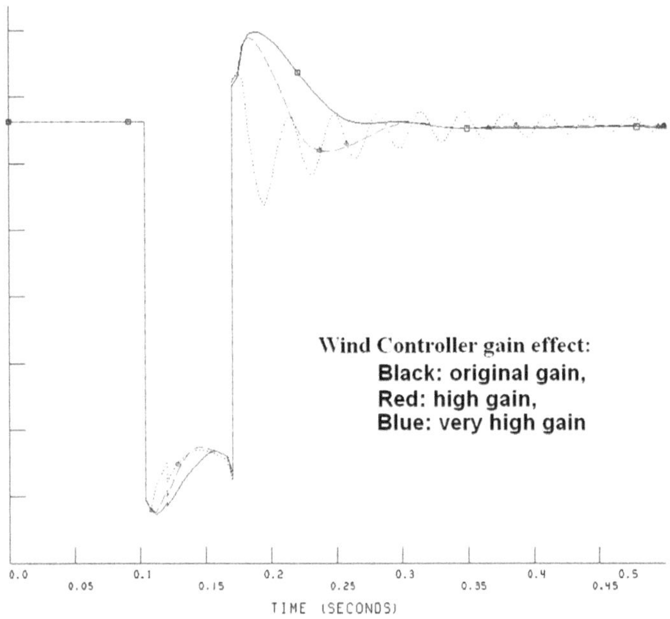

Fig. 8.5 Voltage oscillatory response

8.7 Sub-synchronous Resonance

Sub-synchronous resonance (SSR) is known to occur in series compensated networks affecting mostly nearby thermal generation. A similar effect, named sub-synchronous control instability[3] (SSCI) can occur due to control interactions within a wind plant (type III) when left in series with a compensated line.

A wind plant in Texas experienced this ordeal when the clearing of a fault tripped a primary line and put the wind plant in series with a 50% compensated 345 kV line [8]. SSCI induced oscillations were severe enough to tripped a wind plant causing damages to some wind units and to the series capacitor in a nearby 345 kV transmission line. Figure 8.6 shows the voltage and current recording during the SSCI event.

Networks with large penetration of renewable plants with power electronics inverters are prompt to SSCI effects in weak grids. In the case of Texas, ERCOT has implemented additional interconnection processes to request that SSCI studies be performed using appropriate models to assess if new wind project planned near series compensated lines would be susceptible to SSCI effects.

[3]SSCI = sub-synchronous control instability, undesirable interactions between series compensation and power electronic devices such as wind turbines, HVDC terminals, or static var compensators.

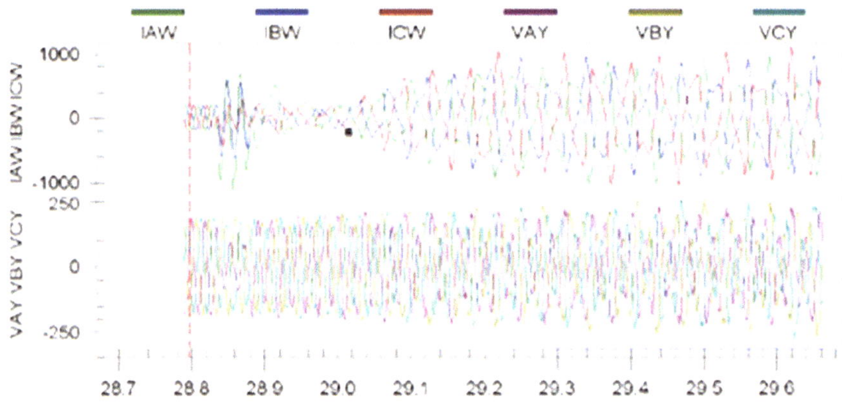

Fig. 8.6 V and I recordings during SSCI event. *Source* [8]

8.8 Performing Dynamic Studies in Parallel

In practice, all dynamic models, including renewable plant models, are aggregated together with a network base case to build a dynamic data set ready for dynamic simulations. This dynamic data set will be user over and over for the different events being tested. Runs of 10 s simulations are typical for voltage transient study while 20 s simulation time or more are recommended for frequency disturbance tests. Dynamic studies can be executed in parallel since the evaluation of a single

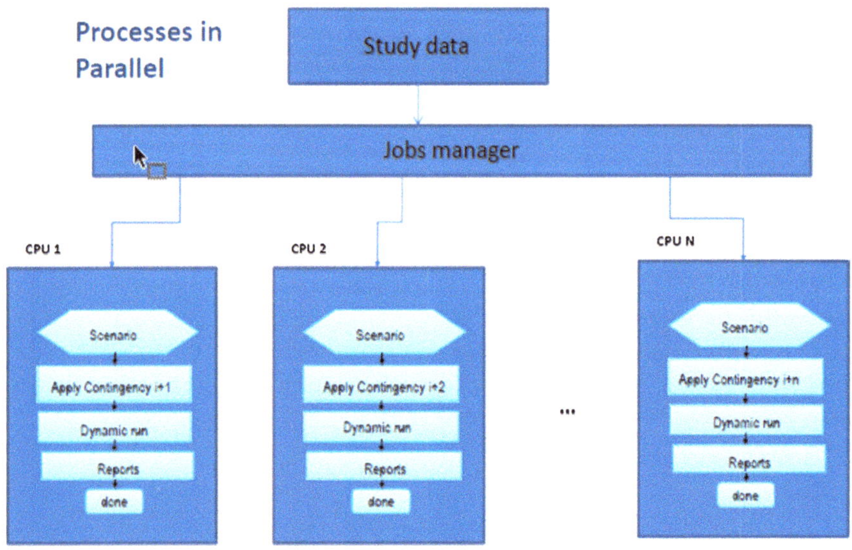

Fig. 8.7 Dynamic runs in parallel

fault event is totally independent of another event (see Fig. 8.7). Keep in mind that data preparation and miscellaneous adjustment will still be using existing processes.

Using python scripts to control PSS/e, multiple dynamic runs can be executed up to the number of CPU available in one's personal computer [9]. Parallel processing is achieved with the python multiprocessing library by calling PSS/e as another computing process. In a single-loop setup, a new instance of PSS/e is activated "on its own CPU" at each iteration with a single loop variable being changed (like the event name, identifying the text file containing an event definition in PSS/s format). A double-loop setup can be used such two loop variables can be used like changing the base case while varying the event name.

8.9 Conclusions

Planning and operation engineers should constantly work to identify the unique problems that high penetration levels of asynchronous generation impose to electric networks by performing simulations of the problems, enabling real-time calculation engines, acquiring the right amount of modeling, developing policy and processes to mitigate the effect of large frequency events or sub-synchronous resonance; and contacting manufacturers on control strategies suitable of their equipment for weak grid conditions (i.e. wind turbine manufacturers).

High penetration of renewable power brings new operational challenges when combined with weak systems having a short-circuit ratio. The effect that renewable power impose on the grid after displacing conventional generation include constraints like economics, environmental but also reduction in the amount of dynamic reactive support, governor action, synchronism torque and inertia available to the grid operator to cope with network disturbances. Large Interconnections like ERCOT [10] had experienced several of these issues and adjusted planning and operation processes to deal with them.

Selection of appropriate simulation tools, modeling and data will impact the result of studies.

Parallel processing is a solution to the increase number of dynamic simulations in the planning and operation of a large interconnected system. Now planners can maximizing the use of multiple cores available in personal computers using python script to control PSS/e, to execute dynamic runs in parallel. Distributed computer or Cloud computing would render bigger benefits in dynamic run applications.

References

1. System Planning Report to ROS, ERCOT, October 2011
2. MRO Model Building Manual—Addendum Wind Generation, November 2008

3. WECC Guide for Representation of Photovoltaic Systems in Large-Scale Load Flow Simulations, WECC Renewable Energy Modeling Task Force, August 2010
4. Y. Cheng, M. Sahni, J. Conto, S.-H. Huang, J. Schmall, Voltage-profile-based approach for developing collection system aggregated models for wind generation resources for grid voltage ride-through studies. IET Renew. Power Gener. (2011)
5. S. Sharma, S.-H. Huang, N.D.R. Sarma, System inertial frequency response estimation and impact of renewable resources in ERCOT interconnection, in *IEEE PES Meeting*, July 2011
6. ERCOT CREZ Reactive Power Compensation Study, ERCOT, 2010
7. S.-H. Huang, J. Schmall, J. Conto, J. Adams, Y. Zhang, C. Carter, Voltage stability concerns on weak grids with high penetration of wind generation: ERCOT experience, in *IEEE PES Meeting*, July 2012
8. P. Belkin, Event of 10/22/09, in *AEP, ERCOT Technical Conference on CREZ System Design and Operation*, 26 Jan 2010
9. J. Conto, MPjobs—running PSSe in parallel, Siemens PTI UGM, 2015
10. ERCOT Quick Facts—Oct 2011, ERCOT, October 2011

Chapter 9
Voltage Control for Wind Power Integration Areas

Qinglai Guo and Hongbin Sun

9.1 Introduction of Hierarchical AVC for Integration of Large-Scale Wind Power

9.1.1 Background and Challenges

Accommodating greater levels of penetration of renewable energy sources is an important feature of smart grids. Currently, wind energy is one of the most popular and promise renewable energy sources, and significant wind power integration has already been achieved in electricity grids in a number of countries across the globe. Wind power can be integrated into distribution grids as distributed generators (DGs), or aggregated as large wind farms and connected to transmission grids. The latter approach accounts for the largest portion of the wind power generation in China, where the wind power capacity reached 145.1 GW at the end of 2015, which represents the largest scale integration of wind power in the world [1]. In China, there is a long-term plan to build seven or eight wind power bases with a minimum capacity of 10 GW each by 2020. However, many of the regions that this large-scale wind power integration is intended supply have relatively weak power grids. For example, in Zhangbei Wind Base, 18 wind farms are connected to the aggregation substation via a single 220 kV transmission line, where the power flow has reached 900 MW, approximately four times the surge impedance loading of the line. The short-circuit capacity in this area is also very small, so the voltage may change greatly in response to variations in the reactive power. During seasons when the wind is strong, switching a single 10 Mvar capacitor may lead to a voltage increase of more than 10 kV. Such an operation scheme, with large-scale wind

Q. Guo (✉) · H. Sun
Tsinghua University, Beijing, China
e-mail: guoqinglai@tsinghua.edu.cn

P. Du et al. (eds.), *Integration of Large-Scale Renewable Energy into Bulk Power Systems*, Power Electronics and Power Systems,
DOI 10.1007/978-3-319-55581-2_9

power integration into a weak power grid, brings about considerable challenges for power system engineers.

The first challenge is related to the voltage fluctuations caused by the intermittent output of the wind turbine generators (WTGs) and the relatively weak grid structure. Historical data show that the voltage from the wind farms varies dramatically compared with conventional power generation. The range of fluctuation was more than 10% per day; for comparison, conventional substations typically fluctuate by less than 5% in a day. The rate of change in the voltage during these fluctuations may also be very rapid. In some wind farms, a fluctuation of more than 6 kV may occur within 10 s, and voltage variations of up to 5 kV have been observed to happen over only 2 s. Moreover, these fluctuations are determined by the intermittent nature of the wind, rendering them inherently irregular and unpredictable. For these reasons, conventional automatic voltage control (AVC) technology is not suitable for large-scale wind power integration, and so a modified approach is required.

The second challenge is how to improve the wind pool area's stability under disturbances. There have been a number of cascading trip-off failures in Northern and Northwest China involving thousands of WTGs. Investigations of these cascading faults have uncovered a commonality in the process [2–4]. First, all of these cascading faults occurred when the WTGs were operating at close to full capacity. The long transmission lines to transfer heavy wind power led to low voltage profiles in the area, which resulted in the capacitors in the wind farms and nearby substations being switched on to support the voltage. The cascading faults were triggered by short-circuit faults in one wind farm or substation, which caused the very low voltage. Unfortunately, most of the WTGs in China were not equipped with effective low-voltage ride through (LVRT) control that time, so these WTGs were shut down. Following the tripping of a large number of WTGs, the transmission line connected to the wind farm where the fault occurred changed from being heavily loaded to carrying a light load. Combined with the capacitors that were not switched off in time, this led to a sudden large amount of redundant reactive power. The wind power pool area was connected with a relatively weak power grid, which resulted in the voltage of the 220-kV transmission line reaching 245–255 kV. As a consequence, WTGs in other wind farms were tripped by the high-voltage protection. As more and more WTGs tripped, the voltage in this area became higher and higher, which resulted in even more WTGs being tripped off, i.e., a cascading failure occurred.

The cascading process can be divided into two phases. The first is the trigger phase, which starts when the fault occurs (just before 0.2 s) and ends when very low voltage occurs (just after 0.2 s). During this process, LVRT may improve the stability of the WTGs and provide reactive power support. Positive LVRT may avert a cascading failure before it starts. The second phase starts from the first tripping of the WTGs in the wind farm and ends with all the wind farms that are involved in the failure being tripped off. This is termed the spreading phase, during which the voltage increased considerably over a period of 2–3 s. During this phase, in addition to high voltage ride through (HVRT) technology to mitigate the voltage increase, a well-designed AVC technology, which will be discussed in this paper, is expected to be very useful.

AVC is one of the most important control systems employed in electrical power grids [5–7]. Many successful applications of this technology have been reported, with generally positive results [8–12]. In the emerging smart transmission grid [13], an AVC system supporting large-scale wind power integration (referred to hereafter as Wind-AVC) is urgently required to address the new challenges described above. Some researches [14–17] have been reported on this topic recently, covering WTG control, wind farm primary control and area secondary control. However, most of the current methods are still not satisfying the requirements of Chinese large wind bases with critical voltage fluctuations and high cascading trip risks. In this chapter, Wind-AVC-related technologies will be discussed, and based on these technologies, a Wind-AVC system and its implementation in a large-scale wind pool area will be described.

9.1.2 Requirements for Wind-AVC

The available control devices for Wind-AVC are also much more complicated than those of traditional AVC. For each wind farm, there are dozens of WTGs whose reactive powers can be online regulated. Besides, there may be shunt capacitors (or reactors) in the wind farms or integration substations. Furthermore, to prevent cascading failures, guidelines now require that most wind farms in China be equipped with SVCs (Static Var Compensator) or SVGs (Static Var Generator), which can provide dynamic reactive power support to compensate for disturbances. However, these fast compensators are not well coordinated with the relatively slower ones (such as WTGs and shunt capacitors/reactors), in some cases, the fast-response Var is firstly exhausted during normal operation and therefore loses the dynamic response ability when disturbance occurs. The coordination of these reactive power regulators with different temporal response characteristics is another problem to be addressed.

The third requirement is the coordination of spatially distributed wind farms and substations. Many wind power pool areas are with relatively weak power grid, so the wind farms and substations that are connected together are strongly coupled. Disturbances (such WTGs being tripped) or control actions (such as switching a capacitor) in one wind farm will therefore significantly influence the voltage at other farms in the same region. This has been proved to be a major factor leading to WTGs cascading trip-off spreading from the first wind farm to others. So there should be a system-wide coordinator for the whole wind pool area to optimize all the wind farms and substations.

In a wind power pool area, there may be several dozens of wind farms with thousands of WTGs and hundreds of SVCs, SVGs, capacitors and/or reactors. It is unimaginable to find an area with such large amounts of control devices coupled together in a conventional power grid. Automatic voltage control to support large-scale wind power integration is therefore a new challenge for the emerging smart transmission grid.

9.1.3 Architecture of Hierarchical AVC

An autonomous-synergic Wind-AVC system is designed and implemented in China to meet the requirements proposed above. Figure 9.1 shows the architecture of the Wind-AVC system, which has two-layer hierarchical structure. At the top level, a system-wide voltage controller is deployed in the control center, which is termed the control center voltage controller (CCVC). At the lower level, in addition to the conventional voltage controllers installed in conventional power plants and substations, a wind farm voltage controller (WFVC) is deployed in each wind farm.

The WFVC should coordinate the control devices (WTGs, SVCs, SVGs, capacitors and reactors) within a wind farm. To consider the detailed voltage

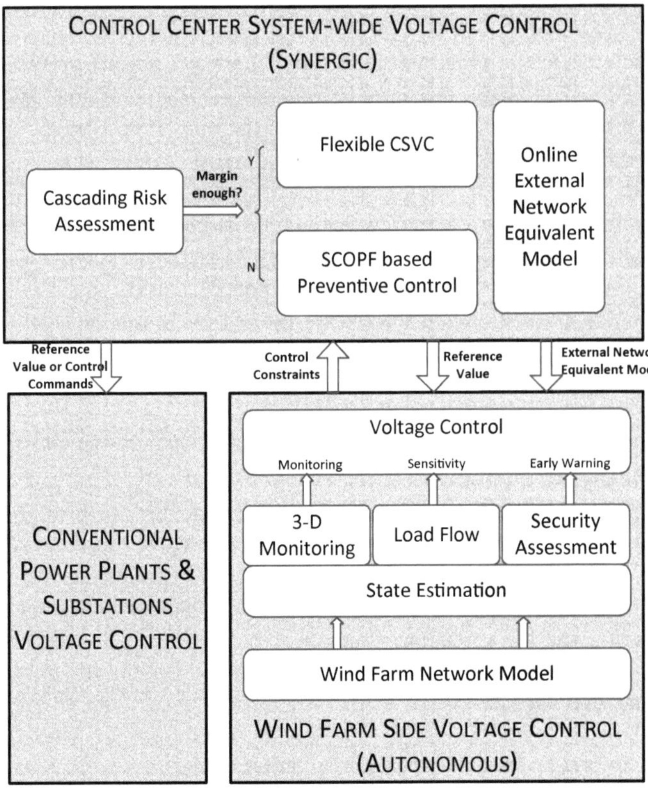

Fig. 9.1 Architecture of the implemented Wind-AVC system

profiles and Var distribution inside the wind farm, the WFVC was designed as a grid-model-based local voltage controller covering the whole wind farm grid. Such a system differs significantly from the conventional thermal or hydro power plant voltage controller. The accuracy of the WFCV calculation depends on the grid-model, including the internal radial grid inside the farm, which can be directly modeled by the WFVC, as well as the external network, which may vary in accordance with the topology and operational scheme of the whole power grid. This time-varying external network model, which describes the influence of the other parts of the power system when the WFVC makes local decision, is essential to ensure the autonomous feature of WFVC and is periodically generated and released online by the control center [18]. Based on the complete and independent network model that combines the internal and external ones, a set of advanced analysis modules is implemented in WFVC, including SE (state estimation), 3-D monitoring, LF (Load Flow), SA (Security Assessment) and VC (Voltage Control). In a sense, a WFVC, which covers modeling, analysis, assessment and close-loop control, can be considered a distributed Energy Management System (EMS) [19] for a single wind farm, which is why we define it as autonomous. Some details of the systems can be found in the preliminary paper [20], and the voltage control decision part will be proposed further in Sect. 9.1.2.

On the control center side, CCVC is implemented to coordinate all the autonomous WFVCs in the region as well as the local controllers within the conventional power plants and substations. According to the current controllable capacity of the SVGs, SVCs, WTGs and shunt capacitors, the WFVC online calculates the Var regulating ability and sends it to the control center as constraints. In CCVC, all the Var control devices within a wind farm are aggregated into a single-machine model with the regulating constraints determined by the farm itself. The control objective of the CCVC is to achieve optimal voltage and Var distribution satisfying the operation constraints, and its outputs include the reference values of the high-side voltage for WFVCs as well as conventional power plants, and control commands to directly switch on/off shunt capacitors/reactors in the conventional substations. The risk of cascading failures is assessed periodically to check if the security margin is sufficient. If it is, the CCVC is set to a CSVC (Coordinated Secondary Voltage Control) mode, the details of which can be found in reference [11]. Otherwise, CCVC will be switched to a preventive control mode, which is based on a SCOPF (Security-Constrained Optimal Power Flow) model. In CSVC mode, only the current operation constraints are considered, while in SCOPF-based preventive control mode, N-1 contingency constraints are also included to ensure that the voltage of the wind farms satisfy the operational limits not only in base-case but also in all the possible N-1 scenarios. Preventive control is to get synergic effects to keep the wind pool area safe even under potential cascading faults, and the details will be discussed in Sect. 9.1.3.

9.2 Wind Farm Side Voltage Control

9.2.1 *Distributed Wind Farm AVC Functions*

The distributed wind farm AVC system is organized as the modules shown in Fig. 9.2 and introduced as follows.

9.2.1.1 Network Model

The network model includes the parameters as well as the topology connections of WTGs, feeder lines, transformers, SVCs/SVGs and shunt compensators. A graphic tool, shown as Fig. 9.3, is developed to build the network model, which integrating diagram, model and database together. The operators draw the diagram directly and inputting all necessary parameters on the diagram. As soon as finished, the topology of the whole wind farm will be created automatically and all the models will be stored into the database. The detailed network model is the basis for all the following functions.

9.2.1.2 State Estimation

In wind farms, a typical measurements deployment for a feeder includes a voltage measurement on the root and a pair of P/Q measurements for each WTG connecting point. Generally, not all the terminal voltage of WTGs can be directly measured. So state estimator is necessary to compute the terminal voltage amplitude as well as angle for each WTG with the available measurements [22, 23]. As our experience, the measurements are enough to ensure the observability, but not enough to identify

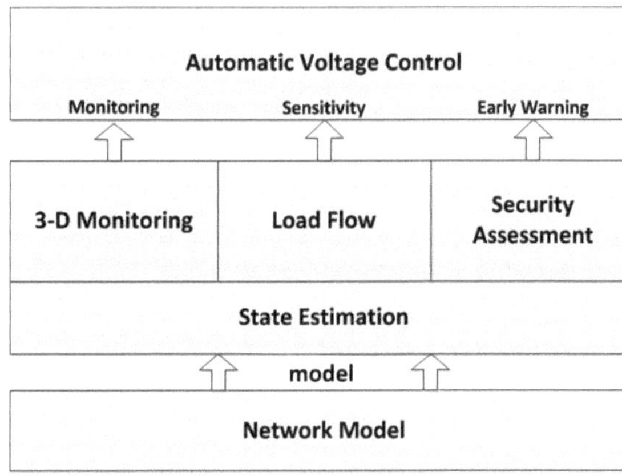

Fig. 9.2 Distributed wind farm AVC functions

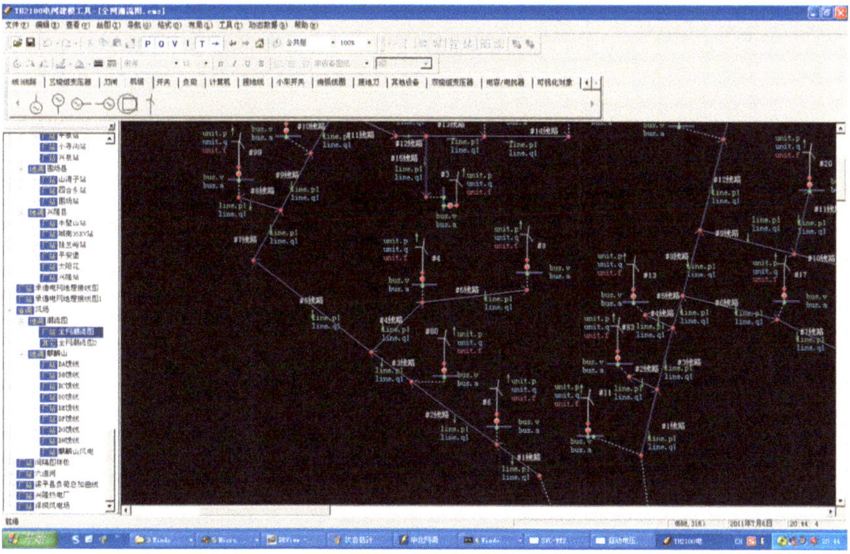

Fig. 9.3 Graphic model tool

bad data. So it is important to improve the measurement redundancy in the future, especially on the WTG terminal voltage.

9.2.1.3 3-D Visualization Based Monitoring

It is really a hard work to monitor all the details in a wind farm since there may be hundreds of WTGs. To help the operators find key information from massive data, a 3-D visualization technology is applied on the monitoring module, including contour views with voltage distribution, pipe views or flow arrow views with P/Q distribution and so on [24–27]. The operators may quickly know where is the dangerous spot with voltage too high or too low, or which branch is overloaded. Figures 9.4 and 9.5 are some examples.

9.2.1.4 Load Flow and Sensitivity Calculation

Load flow is indispensable if we want to know what will happen when a capacitor switched on or a WTG tripped off. Both the sensitivities in the control decision and the security assessment are also based on load flow model. The network in a wind farm is a typical radial grid, thus a loop based distribution power flow algorithm is adopted as a fast and practical method. Furthermore, quasi-steady-state sensitivities considering the Var sources regulating features are carried out and used in the coordinated voltage control module.

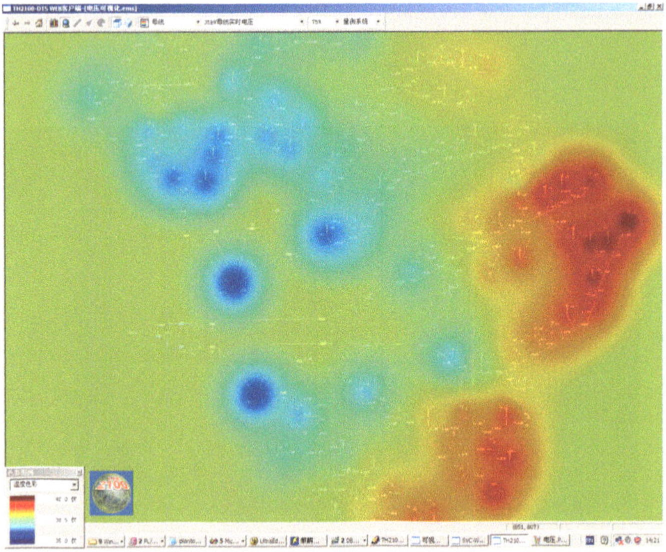

Fig. 9.4 Contour view with voltage distribution

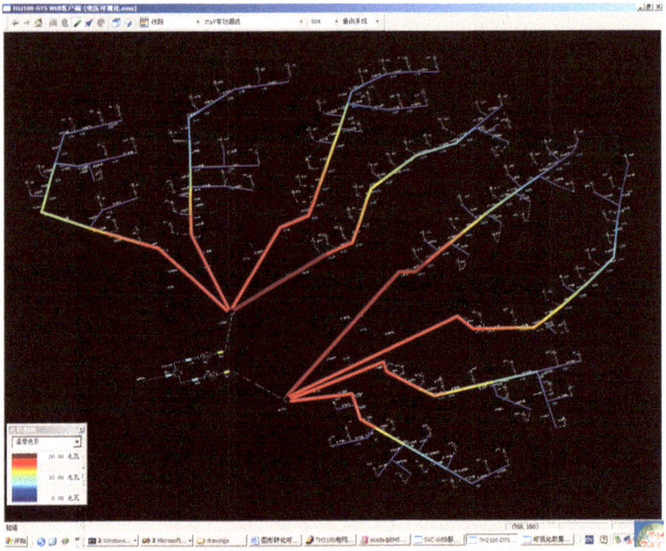

Fig. 9.5 Pipe view with power flow distribution

9.2.1.5 Security Assessment and Early Warning

Two kinds of N-1 security assessment are online scanned automatically in period. The first is to predict all the terminal voltages when a specified WTG tripped, and the other is to assess the voltage distribution in the whole wind farm when a feeder line out of operation. If any potential risk of cascading trip found, operators will receive the warning message. All these assessments will also be carried out based on the predicted snapshot 15 min later considering wind power prediction results.

9.2.1.6 Voltage control Strategy Calculation

Voltage control strategy calculation is the kernel of the whole system, which outputs the regulation commands for all the WTGs, SVC/SVGs and shunt capacitors. We will discuss two different methods in this chapter, the first is based on multi control modes, which will be introduced in Sect. 9.2.2, and the second is based on model preventive control (MPC), which will be introduced in Sect. 9.2.3.

9.2.2 Wind Farm Side Voltage Control Based on Multi Control Modes

Three control modes were designed for different operating requirements ranked by the priority:

(a) corrective control mode which aims to maintain all the WTGs' terminal voltage within limits.
(b) coordinated control mode, which aims to follow a reference value sent from control center and mitigate the voltage fluctuations considering all necessary operation constraints.
(c) preventive control mode, which substitutes the dynamic reactive power reserve with other slower Var sources, on the premise of keeping both the WTGs' terminal voltages and the high-side voltage within the required threshold. The detailed models of the three modes will be presented in the following sections (Fig. 9.6).

The outputs are the set-points and action commands for the reactive power regulation devices. Series of improvements on the Var regulators in the wind farm have been carried out according to the investigation suggestions for the cascading issues. First of all, the WTG, which once did not support online regulation of its Var outputs, has now been able to track the Var set-point by upgrading the converters with constant-Q control loop. It is essential for the voltage controlbecause the number of WTGs is considerable. Furthermore, the WTGs are distributed along the feeders so it becomes possible to control the voltages on different nodes all over the wind farm grid, rather than merely changing the root voltage by controlling the capacitors or SVCs in the past.

Secondly, for SVCs/SVGs, both constant-V and constant-Q control are optional. We should choose constant-V control if we suppose SVCs/SVGs to offer dynamic Var support during disturbances. However, considering that the WTGs are working with constant-Q mode, the SVCs/SVGs, if adopting constant-Q mode, will be much

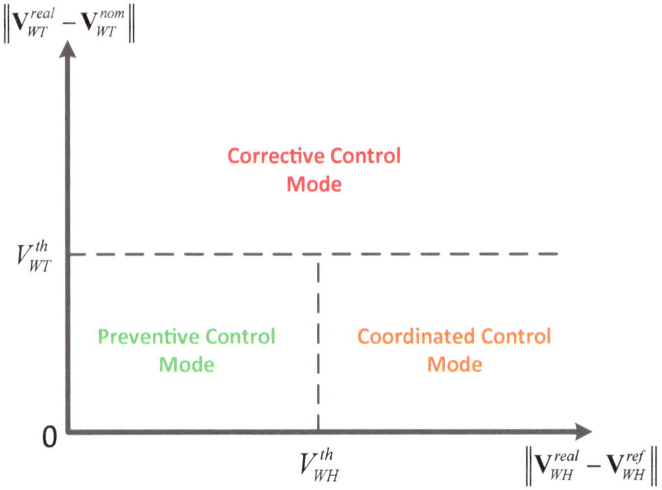

Fig. 9.6 Three voltage control modes for WFVC. The x-axis is the bias of the high-side voltage, and the y-axis is the difference between the terminal voltage of the WTGs and the nominal value

easier to be coordinated with WTGs. So a new control algorithm combining the constant-V and constant-Q modes has been designed and developed for SVCs/SVGs. For the SVC i (SVG is just the same), the WFVC will output a pair of set-points as $\left(\left[\underline{V}_{svc_i}^{ref},\bar{V}_{svc_i}^{ref}\right],Q_{svc_i}^{ref}\right)$. Here, $\left[\underline{V}_{svc_i}^{ref},\bar{V}_{svc_i}^{ref}\right]$ is a voltage limits within which the SVC must maintain its terminal voltage. If the SVC's terminal voltage violates this range, it will regulate the Var output to pull the voltage back in several milliseconds. So the SVCs will still retain the dynamic voltage support capability to against disturbances. When $\left[\underline{V}_{svc_i}^{ref},\bar{V}_{svc_i}^{ref}\right]$ is satisfied, the SVC will enter the constant-Q mode and track $Q_{svc_i}^{ref}$. In this way the SVC can be coordinated with the WTGs effectively.

As discussed in the previous section, the capacitors are proved to have taken negative effect during the cascading failure. Therefore, in WFVC, the capacitors are the last considered devices that will only be switched on when all the WTGs and SVCs/SVGs have reached the upper limit while the voltage is still violating the lower limit.

The detailed model for the three modes will be discussed in the following subsections.

9.2.2.1 Corrective Voltage Control Mode

If $\left\|\mathbf{V}_{WT}^{real}-\mathbf{V}_{WT}^{nom}\right\|\geq V_{WT}^{th}$, the WFVC will switch to the corrective control mode. Here, $\mathbf{V}_{WT}^{real}=\left[V_{wt,1}^{real},\ldots,V_{wt,n_w}^{real}\right]^T$, where $V_{wt,i}^{real}$ is the real-time magnitude value of the terminal voltage of WTG i and n_w is the number of WTGs in the wind farm. $\mathbf{V}_{WT}^{nom}$ is a vector describing the nominal terminal voltage for each WTG (typically 1.0 p.u.). V_{WT}^{th} refers to the threshold value, which can be selected as 0.1 p.u. as the protection configuration usually is [0.9 p.u., 1.1 p.u.]. However, to ensure sufficient operation margins, V_{WT}^{th} is usually configured as 0.07–0.08 p.u. in the real-life implementation.

The corrective control mode model is as follows:

$$
\begin{aligned}
&\min_{\Delta Q_{WT},\Delta Q_S} W_T\left\|V_{WT}^{real}-V_{WT}^{nom}+C_{TW}\Delta Q_{WT}+C_{TS}\Delta Q_S\right\|^2+W_W\left\|\Delta Q_{WT}\right\|^2+W_S\left\|\Delta Q_S\right\|^2\\
&s.t.\ V_{WT}^{\min}\le V_{WT}^{real}+C_{TW}\Delta Q_{WT}+C_{TS}\Delta Q_S\le V_{WT}^{\max}\\
&\quad V_S^{\min}\le V_S^{real}+C_{SW}\Delta Q_{WT}+C_{SS}\Delta Q_S\le V_S^{\max}\\
&\quad V_{WH}^{\min}\le V_{WH}^{real}+C_{HW}\Delta Q_{WT}+C_{HS}\Delta Q_S\le V_{WH}^{\max}\\
&\quad Q_{WT}^{\min}\le Q_{WT}+\Delta Q_{WT}\le Q_{WT}^{\max}\\
&\quad Q_S^{\min}\le Q_S+\Delta Q_S\le Q_S^{\max}
\end{aligned}
\tag{9.1}
$$

where ΔQ_{WT} and ΔQ_S are the Var regulation vector of the WTGs and SVCs/SVGs respectively, which are to be optimized. C_{TW} and C_{TS} are the sensitivity matrices for the WTGs' terminal voltage V_{WT} w.r.t ΔQ_{WT} and ΔQ_S. C_{SW} and $\mathbf{C}_{SS}$ denote the sensitivity matrix for the SVCs/SVGs' terminal voltage V_S w.r.t ΔQ_{WT} and ΔQ_S. C_{HW} and C_{HS} denote the sensitivity matrices for the wind farm's high-side voltage V_{WH} w.r.t $\Delta \mathbf{Q}_{WT}$ and ΔQ_S. The superscript real refers to the current real-time value sampled by the controller. $V_{WT}^{\min}$, $V_{WT}^{\max}$, $V_S^{\min}$, $V_S^{\max}$, $V_{WH}^{\min}$, $V_{WH}^{\max}$, $Q_{WT}^{\min}$, $Q_{WT}^{\max}$, $Q_S^{\min}$ and $Q_S^{\max}$ are the operating limits for V_{WT}, V_S, V_{WH}, Q_{WT} and Q_S, respectively, and we have $V_{WT,i}^{\min}=V_{WT,i}^{nom}-V_{WT}^{th}$ and $V_{WT,i}^{\max}=V_{WT,i}^{nom}+V_{WT}^{th}$. W_T, W_W and W_S are weighting coefficients. Corrective control is to find optimal solutions ΔQ_{WT}^* and ΔQ_S^* to ensure that all the terminal voltages remains within the limitations, as well as satisfying other operating constraints.

9.2.2.2 Coordinated Voltage Control Mode

If $\left\|\mathbf{V}_{WT}^{real}-\mathbf{V}_{WT}^{nom}\right\|<V_{WT}^{th}$ and $\left\|\mathbf{V}_{WH}^{real}-\mathbf{V}_{WH}^{ref}\right\|\ge V_{WH}^{th}$, the WFVC will switch to the coordinated control mode. In this situation, all the terminal voltages of WTGs satisfy the operating constraints, but the bias of the high-side voltage $\mathbf{V}_{WH}^{real}$ and its reference value, $\mathbf{V}_{WH}^{ref}$, which is got from the control center periodically, violates the allowed dead band V_{WH}^{th}. In general, if the high-side voltage of the wind farm is 220 kV, then V_{WH}^{th} is usually set as 1.0 kV.

Coordinated voltage control is designed to track the reference value updated by CCVC satisfying the operating constrains. Its model is as the follows:

$$\begin{aligned}&\min_{\Delta Q_{WT},\Delta Q_S} W_H\left\|V_{WH}^{real}-V_{WH}^{ref}+C_{HW}\Delta Q_{WT}+C_{HS}\Delta Q_S\right\|^2+W_W\|\Delta Q_{WT}\|^2+W_S\|\Delta Q_S\|^2\\&s.t.\ V_{WT}^{\min}\le V_{WT}^{real}+C_{TW}\Delta Q_{WT}+C_{TS}\Delta Q_S\le V_{WT}^{\max}\\&V_S^{\min}\le V_S^{real}+C_{SW}\Delta Q_{WT}+C_{SS}\Delta Q_S\le V_S^{\max}\\&V_{WH}^{\min}\le V_{WH}^{real}+C_{HW}\Delta Q_{WT}+C_{HS}\Delta Q_S\le V_{WH}^{\max}\\&Q_{WT}^{\min}\le Q_{WT}+\Delta Q_{WT}\le Q_{WT}^{\max}\\&Q_S^{\min}\le Q_S+\Delta Q_S\le Q_S^{\max}\end{aligned}\tag{9.2}$$

The constraints in (9.2) are just the same as the corrective control model (9.1), while the objective function here is designed to follow the reference value V_{WH}^{ref} as well as to minimize the control cost. W_H, W_W and W_S are weighting coefficients.

9.2.2.3 Preventive Voltage Control Mode

When $\|\mathbf{V}_{WT}^{real}-\mathbf{V}_{WT}^{nom}\|<V_{WT}^{th}$ and $\left\|\mathbf{V}_{WH}^{real}-\mathbf{V}_{WH}^{ref}\right\|<V_{WH}^{th}$, the preventive control as (9.3) will be carried out to maximize the dynamic Var reserves.

$$\begin{aligned}&\min_{\Delta Q_{WT}} W_S'\|Q_S+\Delta Q_S-\tfrac{1}{2}(Q_S^{\max}-Q_S^{\min})\|^2+W_W'\|\Delta Q_{WT}\|^2\\&s.t.\ V_{WT}^{\min}\le V_{WT}^{real}+C_{TW}\Delta Q_{WT}+C_{TS}\Delta Q_S\le V_{WT}^{\max}\\&\underline{V}_{WH}^{ref}\le V_{WH}^{real}+C_{HW}\Delta Q_{WT}+C_{HS}\Delta Q_S\le\overline{V}_{WH}^{ref}\\&Q_{WT}^{\min}\le Q_{WT}+\Delta Q_{WT}\le Q_{WT}^{\max}\\&Q_S^{\min}\le Q_S+\Delta Q_S\le Q_S^{\max}\\&C_{SS}\Delta Q_S+C_{SW}\Delta Q_{WT}=0\end{aligned}\tag{9.3}$$

In corrective or coordinated control modes, the reference values of the SVCs/SVGs updated by the WFVC are as $\left(\left[V_{S,i}^{\min},V_{S,i}^{\max}\right],Q_{S,i}^{ref}\right)$. Since the terminal voltage limitation of the SVC/SVG is adopted as in its reference, unless some disturbance occurs, or the SVC/SVG will simply follow the Var reference $Q_{S,i}^{ref}$. So the SVCs/SVGs are mainly working with constant-Q mode. However, in preventive control mode, the voltage reference of the ith SVC/SVG from the WFVC is $\left[V_{S,i}^{real}-\xi,V_{S,i}^{real}+\xi\right]$, where ξ is a small deviation. So the SVC/SVG is supposed to work with constant-V mode and maintain the voltage at $V_{S,i}^{real}$. In Eq. (9.3), the control variable is ΔQ_{WT} only, and a new constraint $C_{SS}\Delta Q_S+C_{SW}\Delta Q_{WT}=0$ is added to describe the quasi-steady-state response of SVCs/SVGs' constant-V control, from which the ΔQ_S can be estimated by assuming that V_S remains constant. Another difference in the constraints is that the high-side voltage should be maintained in the threshold $\left[\underline{V}_{WH}^{ref},\overline{V}_{WH}^{ref}\right]$ (i.e., around V_{WH}^{ref}), rather than the original operation limits $\left[V_{WH}^{\min}, V_{WH}^{\max}\right]$. These constraints require that preventive control should be carried out without affecting the high-side voltage.

The objective of (9.3) is to maximize the dynamic Var reserve of SVCs/SVGs by driving their Var output to the middle, so there will be the largest capacities for both upward and downward regulation. Here, W_S' and W_W' are weighting coefficients, and the former has the higher priority.

9.2.3 *Wind Farm Side Voltage Control Based on MPC*

For the convenience of programming, the MPC problem could be formulated in the context of a discrete-time, nonlinear system. The MPC algorithm is minimizing the objective function:

$$J = \sum_{i=1}^{N_p} \|y_{k+i} - y_s\|_Q^2 + \sum_{i=0}^{N_c-1} \|u_{k+i} - u_{k+i-1}\|_S^2. \tag{9.4}$$

Subject to the constraints:

$$\begin{cases} x_{k+i} = f(x_{k+i-1}, u_{k+i-1}) \\ y_{k+i} = g(x_{k+i}) \\ \underline{y} \leq y_{k+i} \leq \overline{y} \end{cases} \qquad i = 1, \ldots, N_p \tag{9.5}$$

$$\begin{cases} \underline{u} \leq u_{k+i} \leq \overline{u} \\ \underline{\delta} \leq u_{k+i+1} - u_{k+i} \leq \overline{\delta} \end{cases} \qquad i = 0, \ldots, N_c - 1 \tag{9.6}$$

Above x_k is a vector of state variables which mean active and reactive power of WTG and SVG. u_k is a vector of controlled variables which mean set points of WTG reactive power and SVG voltage. y_k is a vector of outputs which mean voltage of PCC and other places. Q and S are weighting matrices, N_p and N_c are prediction and control points respectively which are made the same in this paper. With time passing, real-time data are substituted for x_k, u_k and y_k representing feedback and the ongoing of receding-horizon optimization. y_s stands for both the system steady state or reference trajectory, while formulas (9.5) represent the predictive model. Following are specific forms of objectives and predictive model.

9.2.3.1 Objectives

Commonly the AVC control center in wind integration region is not acquainted with complicated model of each wind farm, and does not directly control every component of wind farms. Instead of that, reference values of PCC voltage are sent to AVC substations in each wind farm. Accordingly, an objective of the wind farm voltage controller is to minimize the deviations between set point and predicted value of PCC voltage, which could be expressed as below:

$$f(k) = \left[V_{\mathrm{PCC}}^{\mathrm{pre}}(k) - V_{\mathrm{PCC}}^{\mathrm{ref}}(k)\right]^2. \tag{9.7}$$

We assume that the control interval is t_{itv}, then k represents for kt_{itv} after present time. $V_{\mathrm{PCC}}^{\mathrm{ref}}$ is the PCC voltage set point, while $V_{\mathrm{PCC}}^{\mathrm{pre}}$ is the predicted value.

SVG is a kind of fast var generator, whose reactive power can reach its set value in less than 100 ms. Consequently, SVG is always set to track the reference voltage of its connecting bus. In this way, SVGs can guarantee voltage security more effectively under emergency circumstance than the AVC substation directly control SVGs' reactive power, since wind farm AVC control is much slower than devices' local control. The local voltage control of the SVG can be summarized as follows:

$$g(k) = \left[V_{\mathrm{SVG}}^{\mathrm{pre}}(k) - V_{\mathrm{SVG}}^{\mathrm{set}}(k-1)\right]^2. \tag{9.8}$$

Above $V_{\mathrm{SVG}}^{\mathrm{set}}$ is the SVG voltage set point, while $V_{\mathrm{SVG}}^{\mathrm{pre}}$ is the predicted value. Usually control variables don't appear in objective functions, here is just a method to simplify the modeling of SVG predictive model.

It has been discussed that SVG's reactive power is crucial to voltage security under emergency circumstance. Moreover, when wind power fluctuates swiftly, WTGs could hardly regulate reactive power in time to smooth voltage fluctuation. Thus sufficient fast reactive power of SVGs should be reserved beforehand, and another objective is obtained:

$$h(k) = \left[Q_{\mathrm{SVG}}^{\mathrm{pre}}(k) - Q_{\mathrm{SVG}}^{\mathrm{mid}}\right]^2. \tag{9.9}$$

$Q_{\mathrm{SVG}}^{\mathrm{pre}}$ is the predicted value of SVG reactive power, while $Q_{\mathrm{SVG}}^{\mathrm{mid}}$ is the middle output of reactive power. MPC is a receding-horizon optimization problem over a finite window in time. Multiple objectives have different weights; meanwhile objectives of different time (or snapshots) have weights as well. As we all know, when $k = 0$ the so-called predictive values of voltages or other variables are actually present real value. With k increasing, the uncertainty of predictive values likewise increases. Therefore, the weights of objectives with larger k should be smaller. After all, the entire control objective we got in this paper is shown below:

$$\min \sum_{k=0}^{N} \rho^k [\alpha \cdot f(k) + \beta \cdot g(k) + \gamma \cdot h(k)]. \tag{9.10}$$

It could be seen that the width of MPC's window in time is Nt_{itv}. α, β and γ are weights of PCC voltage, SVG voltage and SVG reactive power respectively, while ρ is a number less than 1. When ρ becomes close to 0, the controller will mainly take consider of only present condition of system, just like most of the voltage controllers in use in wind farms.

9.2.3.2 Predictive Model

Controller proposed in this paper mostly concerns voltage problems in tens of seconds. Under this time scale, the active power change of WTGs is always processed into stationary process. Hence, we can use ARMA model to forecast WTG active power [24]:

$$P_{\text{pre}}^{\text{WTG}}(k) = \sum_{i=1}^{N_a} \phi_k P_{\text{pre}}^{\text{WTG}}(k-i) + \varepsilon(k) - \sum_{i=1}^{N_m} \theta_k \varepsilon(k-i). \tag{9.11}$$

Here $P_{\text{pre}}^{\text{WTG}}$ is the predicted value of WTG active power, while ε is a random variable. N_a and N_m are respectively the rank of AR and MA model, while ϕ_k and θ_k are weights. Every time before startup of the forecast process, both ranks and weights would be reassessed based on the latest historic data of WTG active power.

Comparing to the control interval, interval of forecast could be much shorter with the help of PMU data collected from PCC. Active power of each WTG could be divided proportionately based on WTG's SCADA measurements. By the way, WTG monitor system is a system used to acquire operating data of all WTGs in a wind farm, and the technique now allow getting active power, wind speed and other information within a time interval less than 5 s. Then, the forecasts of WTG active power by wind power information in the future 10–30 s also come to be possible.

The situation of another state variable, reactive power, is different. In China, local control of WTGs always makes them track the set value of reactive power since it's much more convenient to manage the whole reactive power output in a wind farm and problems such as circle flow of reactive power are easier to solve than that when WTGs track voltage set points. Thus we assume that the WTG predictive reactive power could approach its set value at former time step.

$$Q_{\text{WTG}}^{\text{pre}}(k) = Q_{\text{WTG}}^{\text{set}}(k-1). \tag{9.12}$$

$Q_{\text{WTG}}^{\text{set}}$ is the WTG reactive power set point, while $Q_{\text{WTG}}^{\text{pre}}$ is the predicted value. In addition, WTG and SVG's reactive power must be in definite region, and not change too fast. These constraints are described in (9.10), $Q_{\text{WTG}}^{\min}$, $Q_{\text{WTG}}^{\max}$, $Q_{\text{SVG}}^{\min}$, $Q_{\text{SVG}}^{\max}$, $\Delta Q_{\text{WTG}}^{\min}$, $\Delta Q_{\text{WTG}}^{\max}$, $\Delta Q_{\text{SVG}}^{\min}$ and $\Delta Q_{\text{SVG}}^{\max}$ are bounds of variables discussed above:

$$\begin{cases} Q_{\text{WTG}}^{\min} \le Q_{\text{WTG}}^{\text{pre}}(k) \le Q_{\text{WTG}}^{\max} \\ Q_{\text{SVG}}^{\min} \le Q_{\text{SVG}}^{\text{pre}}(k) \le Q_{\text{SVG}}^{\max} \\ \Delta Q_{\text{WTG}}^{\min} \le Q_{\text{WTG}}^{\text{pre}}(k) - Q_{\text{WTG}}^{\text{pre}}(k-1) \le \Delta Q_{\text{WTG}}^{\max} \\ \Delta Q_{\text{SVG}}^{\min} \le Q_{\text{SVG}}^{\text{pre}}(k) - Q_{\text{SVG}}^{\text{pre}}(k-1) \le \Delta Q_{\text{SVG}}^{\max} \end{cases}. \tag{9.13}$$

Commonly it is thought that both SVGs and DFIGs change reactive power through power electronic devices which would not take much time. This concept meets the reality of SVGs whose response time is usually less than tenths of seconds. But when it comes to WTGs, test results got from the field reveal that the time it takes for a WTG to generate reactive power from zero to full load various from 2 to 30 s which depends on manufacturers. In this paper, WTGs are considered able to adjust their reactive power from its minimum to maximum value in 10 s.

It has been discussed before that forecast interval of active power could be much shorter than control interval. Concerning the delay of reactive power, voltage at the first point of optimization is affected mostly by former control strategies but present ones. Furthermore, present control strategies will be replaced after the second control point of MPC optimization. It is naturally to think that more detailed information of more snapshots within this control period should be acquired. WTG active power could be obtained by linear interpolation. When it comes to reactive power, an exponential function is used to draw the trends of reactive power (T_s is the time constant of WTG):

$$\begin{aligned} & Q_{\text{pre}}^{\text{WTG}}(k-1+\Delta t/t_{itv}) \\ & = \frac{1-e^{-\Delta t/T_s}}{1-e^{-t_{itv}/T_s}} Q_{\text{pre}}^{\text{WTG}}(k) + \frac{e^{-\Delta t/T_s} - e^{-t_{itv}/T_s}}{1-e^{-t_{itv}/T_s}} Q_{\text{pre}}^{\text{WTG}}(k-1) \end{aligned} \tag{9.14}$$

At last we come to output variables, despite of PCC voltage, voltages of WTGs and SVGs are also concerned for they are compactly correlated with system voltage security. With above, the predicted system voltage value can be derived with the network sensitivity matrix shown as $\frac{\partial V_{\text{PCC}}}{\partial P_{\text{WTG}}}$, $\frac{\partial V_{\text{PCC}}}{\partial Q_{\text{WTG}}}$, $\frac{\partial V_{\text{PCC}}}{\partial Q_{\text{SVG}}}$, $\frac{\partial V_{\text{WTG}}}{\partial P_{\text{WTG}}}$, $\frac{\partial V_{\text{WTG}}}{\partial Q_{\text{WTG}}}$, $\frac{\partial V_{\text{WTG}}}{\partial Q_{\text{SVG}}}$, $\frac{\partial V_{\text{SVG}}}{\partial P_{\text{WTG}}}$, $\frac{\partial V_{\text{SVG}}}{\partial Q_{\text{WTG}}}$ and $\frac{\partial V_{\text{SVG}}}{\partial Q_{\text{SVG}}}$. Although the values could be got by solving load flow equations, the sensitivity method's precision is acceptable and since it's linear the complexity of MPC optimization is largely reduced.

$$\begin{aligned} & V_{\text{PCC}}^{\text{pre}}(k) - V_{\text{PCC}}^{\text{real}}(0) \\ & = \frac{\partial V_{\text{PCC}}}{\partial P_{\text{WTG}}} \left[P_{\text{WTG}}^{\text{pre}}(k) - P_{\text{WTG}}^{\text{real}}(0)\right] \\ & \quad + \frac{\partial V_{\text{PCC}}}{\partial Q_{\text{WTG}}} \left[Q_{\text{WTG}}^{\text{pre}}(k) - Q_{\text{WTG}}^{\text{real}}(0)\right] \\ & \quad + \frac{\partial V_{\text{PCC}}}{\partial Q_{\text{SVG}}} \left[Q_{\text{SVG}}^{\text{pre}}(k) - Q_{SVG}^{\text{real}}(0)\right]. \end{aligned} \tag{9.15}$$

$$\begin{aligned} & V_{\text{WTG}}^{\text{pre}}(k) - V_{\text{WTG}}^{\text{real}}(0) \\ & = \frac{\partial V_{\text{WTG}}}{\partial P_{\text{WTG}}} \left[P_{\text{WTG}}^{\text{pre}}(k) - P_{\text{WTG}}^{\text{real}}(0)\right] \\ & \quad + \frac{\partial V_{\text{WTG}}}{\partial Q_{\text{WTG}}} \left[Q_{\text{WTG}}^{\text{pre}}(k) - Q_{\text{WTG}}^{\text{real}}(0)\right] \\ & \quad + \frac{\partial V_{\text{WTG}}}{\partial Q_{\text{SVG}}} \left[Q_{\text{SVG}}^{\text{pre}}(k) - Q_{SVG}^{\text{real}}(0)\right]. \end{aligned} \tag{9.16}$$

$$
\begin{aligned}
&V_{\mathrm{SVG}}^{\mathrm{pre}}(k) - V_{\mathrm{SVG}}^{\mathrm{real}}(0) \\
&\quad = \frac{\partial V_{\mathrm{SVG}}}{\partial P_{\mathrm{WTG}}}\left[P_{\mathrm{WTG}}^{\mathrm{pre}}(k) - P_{\mathrm{WTG}}^{\mathrm{real}}(0)\right] \\
&\qquad + \frac{\partial V_{\mathrm{SVG}}}{\partial Q_{\mathrm{WTG}}}\left[Q_{\mathrm{WTG}}^{\mathrm{pre}}(k) - Q_{\mathrm{WTG}}^{\mathrm{real}}(0)\right] \\
&\qquad + \frac{\partial V_{\mathrm{SVG}}}{\partial Q_{\mathrm{SVG}}}\left[Q_{\mathrm{SVG}}^{\mathrm{pre}}(k) - Q_{SVG}^{\mathrm{real}}(0)\right].
\end{aligned}
\tag{9.17}
$$

$V_{\mathrm{PCC}}^{\mathrm{real}}$, $V_{\mathrm{WTG}}^{\mathrm{real}}$, $V_{\mathrm{SVG}}^{\mathrm{real}}$, $Q_{\mathrm{WTG}}^{\mathrm{real}}$ and Q_{SVG}^{real} are real time measurements of voltage and reactive power. Actually, real-time measurements of voltage in prediction model work as feedback correction in MPC scheme.

In order to assure system security, voltages should be kept in acceptable bounds, which are express as $V_{\mathrm{PCC}}^{\min}$, $V_{\mathrm{PCC}}^{\max}$, $V_{\mathrm{WTG}}^{\min}$, $V_{\mathrm{WTG}}^{\max}$, $V_{\mathrm{SVG}}^{\min}$ and $V_{\mathrm{SVG}}^{\max}$ below:

$$
\begin{cases}
V_{\mathrm{PCC}}^{\min} \le V_{\mathrm{PCC}}^{\mathrm{pre}}(k) \le V_{\mathrm{PCC}}^{\max} \\
V_{\mathrm{WTG}}^{\min} \le V_{\mathrm{WTG}}^{\mathrm{pre}}(k) \le V_{\mathrm{WTG}}^{\max} \\
V_{\mathrm{SVG}}^{\min} \le V_{\mathrm{SVG}}^{\mathrm{pre}}(k) \le V_{\mathrm{SVG}}^{\max}
\end{cases}.
\tag{9.18}
$$

9.3 System Side Voltage Control

9.3.1 Synergic Preventive Voltage Control in CCVC

The major novel aspect of CCVC is the SCOPF-based preventive control method. As discussed in Sect. 9.1.2, cascading failures typically occur within 2–3 s. Once triggered, there is very little time to take effective control measures. It is therefore highly desirable to implement system-wide preventive control to ensure that the wind power pool area functions within normal operating conditions even when contingencies occur. In conventional system-wide voltage control strategies such as CSVC, only the current operating constraints are considered. The system may be held in a status that satisfies all the operating constraints, which we define it as the normal state. However, once a contingency happens, for instance, when a wind farm or a transmission line is out of operation because of fault, the reminder of the grid may not be able to still satisfy the operating constraints. We consider this not to be safe. In a cascading failure, prior to the fault, the wind power pool area was actually working in a normal but not safe state, which means that a localized fault may lead to voltage violations at other wind farms, and ultimately induce a cascading outage. Via preventive control, the system is optimized so that it should always operate in a normal and safe state, which not only satisfies all the current operating limitations but is also without violations under all the potential post-contingency scenarios. The SCOPF-based model can achieve system-wide preventive control, as described by

$$\begin{array}{lll} \min\limits_{u^0} & f(x^0, u^0) & (a) \\ s.t. & g_0(x^0, u^0) = 0 & (b) \\ & g_k(x^k, u^0) = 0 & (c) \\ & \underline{x} \le x^0 \le \bar{x} & (d) \\ & \underline{x}^k \le x^k \le \bar{x}^k & (e) \\ & \underline{u} \le u^0 \le \bar{u} & (f) \\ & k = 1, \ldots, N_C & \end{array} \tag{9.19}$$

where, x denotes the state variables (voltage magnitude V_i and voltage phase angle θ_i for node i), and u denotes the control variables. The superscript k refers to the value under the kth contingency and $k = 0$ refers to the pre-contingency base-case. N_C is the number of contingencies to be considered. Function $g(x, u) = 0$ is the load flow equation. The SCOPF model seeks an optimized control u^0 satisfying the constraints (9.19-f). u^0 will be performed on the base-case and maintained unchanged when the kth contingency occurs. So through constraints (9.19-b) and (9.19-c), we can calculate the states of the base-case x^0 and x^k under the kth post-contingency condition, which are required to satisfy the operation constraints (9.19-d) and (9.19-e). If both the base-case constraints ((9.19-b) and (9.19-d)) and the post-contingency constraints under all the pre-defined contingencies ((9.19-c) and (9.19-e)) are satisfied, then x^0 could be regarded as normal and safe. Within all the feasible solutions, u^0 with the minimum objective value is the final optimized solution.

Here u_0 includes $\{\boldsymbol{Q}_w^0, \boldsymbol{Q}_g^0, \boldsymbol{P}_w^0\}$, where $\boldsymbol{P}_w^0$ and $\boldsymbol{Q}_w^0$ are the active and reactive power outputs of the wind farms, respectively, and $\boldsymbol{Q}_g^0$ refers to the reactive power output of the conventional generators and compensators in the same area. The objective function is then

$$\min_{\{\boldsymbol{Q}_w^0, \boldsymbol{Q}_g^0, \boldsymbol{P}_w^0\}} \quad -w_1 f_1 + w_2 f_2 \tag{9.20}$$

where

$$\min_{\{\boldsymbol{Q}_w^0, \boldsymbol{Q}_g^0, \boldsymbol{P}_w^0\}} \quad -w_1 f_1 + w_2 f_2 \tag{9.21}$$

$$f_1 = \sum_{i=1}^{N_w} P_{w,i}^0 \tag{9.22}$$

$$f_2 = P_{Loss} = \sum_{(i,j) \in NL} (P_{ij} + P_{ji}) \tag{9.23}$$

There are two parts in the objective function: the first part with the higher-priority weighting w_1 is to maximize the MW output of all the N_w wind farms; the second part, with lower-priority weight w_2, represents transmission losses, where P_{ij} and P_{ij} denote the active power of the branch (i,j) on the start node i and the end node j, respectively. NL refers to the set of the branches.

In this paper, the contingencies to be considered are the N-1 of the wind farms. Suppose there are N_w wind farms in the area, the kth wind farm N-1 is described as:

$$\begin{cases} P^k_{w,i} = 0 & i = k \\ P^k_{w,i} = P^0_{w,i} & i \neq k \\ Q^k_{w,i} = 0 & i = k \\ Q^k_{w,i} = Q^0_{w,i} & i \neq k \end{cases} \quad i = 1, \ldots, N_w \tag{9.24}$$

For (9.19-d) and (9.19-e), we are most concerned with the voltage magnitude limitations, especially the wind farms' voltage that should be limited to avoid the WTGs being tripped off on base-case as well as the N-1 contingencies.

Note that the active power outputs of wind farms, $\boldsymbol{P}^0_w$, is also included in the optimization together with the reactive power outputs. In some cases, optimizing only $\boldsymbol{Q}^0_w$ and $\boldsymbol{Q}^0_g$ may not be sufficient to find a feasible solution to satisfy all of the pre- and post-contingency constraints. So $\boldsymbol{P}^0_w$ must be curtailed. For this reason, f_1 is modeled in the objective function to maximize the wind farms' active power outputs to accommodate as much wind power as possible, of course, in the premise of ensuring the safe operation both on pre- and post-contingency situations.

The SCOPF-based model is considerably more complicated than the conventional OPF, for the constraints of SCOPF are nearly $N_C + 1$ times than that of a traditional OPF. A Benders decomposition based method [15] is adopted to solve equations, where the $N_C + 1$ contingencies (including the pre-contingency base-case) are formed as sub-problems respectively. According to the optimized u^0, the reference value for the wind farms could be calculated and released to the WFVCs.

9.3.2 *System Side Voltage Security Region*

9.3.2.1 Definition of Two-Level Static Voltage Security Region

Nowadays, wind AVC systems are widely used in real wind farms which keeps the voltage magnitude of PCC bus of each wind farm within a range so as to mitigate the cascading tripping of centralized multiple wind farms as well as guarantee the secure operation condition of wind units.

In order to seek a feasible voltage range of PCC bus of each wind farm, the detailed static voltage security region of multiple wind farms is desired to set up for further automatic voltage control. However, in practical operation and control, the detailed knowledge of one wind farm, e.g., the topology of wind farm and reactive compensators, cannot be obtained by the other wind farms, so the two-level topology structure should be intensively modeled, respectively.

For each wind farm, the voltage security region is desired to guarantee the voltage magnitude of each wind unit within a certain range so as to protect wind units against serious over-voltage and low-voltage problems, which is called wind-farm-side voltage security region.

More important, in order to mitigate the cascading trips, the voltage security region of multiple wind farms should be considered on system wide to coordinate each wind farm's operation so as to ensure security both under normal operating conditions and N-1 contingencies. If an operating point is in the normal security region, but out of the N-1 region, this means that cascading is probably triggered by the first trip. Thus, even if the current operating status is normal, it is not secure enough. So we name it system-wide voltage security region.

Bearing in mind that voltage is an algebraic state variable and reactive power is a control variable, to guarantee that the voltage remains within the security region, the reactive power must be restricted. Therefore, the static voltage security region can be actually expressed as a set of constraints limiting the reactive power of each wind farm for system wide and each unit for wind farm wide to maintain the bus voltage in the secure range. As a result, the reactive power compensators in each wind farm must be studied in detail, based on the following considerations.

(i) The reactive power output of each wind unit normally ranges from −500 to +500 kVar for one 1.5-MW DFIG wind unit.
(ii) As for reactive power compensation equipment of each wind farm, they can be divided into two types: fast response compensators, such as SVC, STATCOM and slow response compensators, such as capacitor banks. In real wind farms of Northern China, there are few fast response compensators, so we only consider the capacitor banks.

However, the voltage security region should be reconsidered with the system's settings changed by the optimal reactive power flow from the tertiary voltage control (TVC), which aims to minimize the total network losses via regulating reactive power output of slow response compensators and the time frame is usually 5–15 min. Therefore, a quasi-steady-state (QSS) model is adopted to compute the voltage security region.

Different from thermal generators, the wind power is stochastic and intermittent, which leads the static voltage security region to be uncertain. Therefore, it is promising to take a robust static voltage security region into consideration.

9.3.2.2 Method for Robust Two-Level Static Voltage Security Region

For the QSS model, the sensitivity-based linearized model instead of primary non-linear power flow equations is employed for the proposed two-level static voltage security region. With respective to Jacobi matrix of power flow under normal condition, we obtain the following system of linear equations. From the system-wide viewpoint, only PCC bus of each wind farm needs to be considered as represented as (9.25). As for the wind farm side, both PCC buses and each wind unit bus should be taken into account by (9.26) and (9.27).

$$\Delta U_{PCC,w} = \sum_{i=1}^{N_w} \left(H_{w,i}\Delta P_i + S_{w,i}\Delta Q_i\right) \tag{9.25}$$

$$\Delta U_{PCC,w} = \sum_{i=1}^{N_u} \left(h_{w,i}\Delta p_i + s_{w,i}\Delta q_i\right) \tag{9.26}$$

$$\Delta u_{w,j} = \sum_{i=1}^{N_u} \left(h_{j,i}\Delta p_i + s_{j,i}\Delta q_i\right) \tag{9.27}$$

where ΔP, ΔQ are total active/reactive power derivation of wind farms; $\Delta \boldsymbol{p}$, $\Delta \boldsymbol{q}$ are active/reactive power derivation of wind units; $\Delta \boldsymbol{U}_{PCC}$ and $\Delta \boldsymbol{u}_w$ are the voltage derivation of PCC bus and wind unit, respectively.

(i) System-wide robust voltage security region

For system-wide voltage security region, we aim to find the reactive power range of each wind farm that can guarantee voltage security for all PCC buses. Under N-1 contingency, suppose wind farm w is tripped, the active and reactive power produced by wind units drops to zero, while the connected transmission lines and the slow switch-off of capacitance banks remain unchanged, such that

$$U_{PCC,i}^{\min} \le U_{PCC,i}^{0} + \sum_{w=1}^{N_w} S_{i,w}^{0}\left(Q_w - Q_w^{0}\right) + \sum_{w=1}^{N_w} H_{i,w}^{0}\Delta P_w \le U_{PCC,i}^{\max} \tag{9.28}$$

$$U_{PCC,i}^{\min} \le U_{PCC,i}^{s} + \sum_{w=1}^{N_w} S_{i,w}^{s}\left(Q_w - Q_w^{s}\right) + \sum_{w=1}^{N_w} H_{i,w}^{s}\Delta P_w \le U_{PCC,i}^{\max} \tag{9.29}$$

$$\forall \Delta P_w \in \left[\Delta P_w^{\min}, \Delta P_w^{\max}\right] \tag{9.30}$$

$$Q_w^{\min} \le Q_w \le Q_w^{\max} \tag{9.31}$$

However, the stochastic wind power inevitably makes the voltage security region uncertain, and it can be known from (9.28)–(9.31) that the system-wide voltage security region is an uncertain polyhedron where the voltage region in shadow area tells that the voltage may not be always secure due to the uncertain wind generation. In order to eliminate the impact of these uncertainties, the robust voltage security region should be considered by choosing the inner-most polyhedron.

Firstly, the uncertain polyhedron can be expressed as

$$[\Omega] = \left\{ \boldsymbol{Q}_w | \boldsymbol{U}_{PCC}^{\min} \le \boldsymbol{A}\boldsymbol{Q}_w + \boldsymbol{B}\Delta\boldsymbol{P}_w \le \boldsymbol{U}_{PCC}^{\max}, \forall \Delta\boldsymbol{P}_w \in \left[\Delta\boldsymbol{P}_w^{\min}, \Delta\boldsymbol{P}_w^{\max}\right] \right\} \tag{9.32}$$

where

$$\boldsymbol{B}\Delta\boldsymbol{P}_w = \left[\boldsymbol{B}^+ \Delta\boldsymbol{P}_w^{\min} + \boldsymbol{B}^- \Delta\boldsymbol{P}_w^{\max}, \boldsymbol{B}^- \Delta\boldsymbol{P}_w^{\min} + \boldsymbol{B}^+ \Delta\boldsymbol{P}_w^{\max}\right] \tag{9.33}$$

For convenience, let

$$\begin{aligned} [\vartheta^-, \vartheta^+] &= \boldsymbol{B}\Delta\boldsymbol{P}_w \\ \vartheta^- &= \boldsymbol{B}^+ \Delta\boldsymbol{P}_w^{\min} + \boldsymbol{B}^- \Delta\boldsymbol{P}_w^{\max} \\ \vartheta^+ &= \boldsymbol{B}^- \Delta\boldsymbol{P}_w^{\min} + \boldsymbol{B}^+ \Delta\boldsymbol{P}_w^{\max} \end{aligned} \tag{9.34}$$

Then, the system-wide robust voltage security region gives

$$\Omega = \left\{ \boldsymbol{Q}_w | \left(\boldsymbol{U}_{PCC}^{\min} - \vartheta^- \le \boldsymbol{A}\boldsymbol{Q}_w \le \boldsymbol{U}_{PCC}^{\max} - \vartheta^+ \right) \cap \boldsymbol{Q}_w^{\min} \le \boldsymbol{Q}_w \le \boldsymbol{Q}_w^{\max} \right\} \tag{9.35}$$

Note that the reactive power range of each wind farm is coupled by the above robust voltage security region Ω. However, it is usually difficult for one wind farm to achieve all the detailed knowledge of the other wind farms. Keep in mind that the elements in reactive power/voltage sensitive matrix H is always positive. Therefore, this robust voltage security region should be decoupled by finding an inner rectangle region within the polyhedron Ω, which yields (9.36). The $\left[\boldsymbol{Q}_w^-, \boldsymbol{Q}_w^+\right]$ is the decoupled reactive power rectangle range of each wind farm for robust voltage security region. According to the interval arithmetic, (9.36) can be further transformed into (9.37).

$$\overline{\Omega} = \left\{ \left[\boldsymbol{Q}_w^-, \boldsymbol{Q}_w^+\right] \middle| \left\{ \begin{array}{l} \boldsymbol{A}\boldsymbol{Q}_w \subseteq \left[\boldsymbol{U}_{PCC}^{\min} - \vartheta^-, \boldsymbol{U}_{PCC}^{\max} - \vartheta^+\right] \\ \left[\boldsymbol{Q}_w^-, \boldsymbol{Q}_w^+\right] \subseteq \left[\boldsymbol{Q}_w^{\min}, \boldsymbol{Q}_w^{\max}\right] \\ \forall \boldsymbol{Q}_w \in \left[\boldsymbol{Q}_w^-, \boldsymbol{Q}_w^+\right] \end{array} \right. \right\} \tag{9.36}$$

$$\overline{\Omega} = \left\{ \left[\boldsymbol{Q}_w^-, \boldsymbol{Q}_w^+\right] \middle| \left\{ \begin{array}{l} \boldsymbol{A}\boldsymbol{Q}_w^+ \le \boldsymbol{U}_{PCC}^{\max} - \vartheta^+ \\ \boldsymbol{A}\boldsymbol{Q}_w^- \ge \boldsymbol{U}_{PCC}^{\min} - \vartheta^- \\ \boldsymbol{Q}_w^{\min} \le \boldsymbol{Q}_w^- \le \boldsymbol{Q}_w^+ \le \boldsymbol{Q}_w^{\max} \end{array} \right. \right\} \tag{9.37}$$

Thus, the decoupled reactive power range can be obtained from the largest rectangle within the polyhedron Ω by

$$\max \quad \sum_{w=1}^{N_w} \frac{Q_w^+ - Q_w^-}{Q_w^{\max} - Q_w^{\min}} \tag{9.38}$$

$$\text{subject to:} \quad \boldsymbol{A}\boldsymbol{Q}_w^+ \le \boldsymbol{U}_{PCC}^{\max} - \vartheta^+ \tag{9.39}$$

$$AQ_w^- \geq U_{PCC}^{\min} - \vartheta^- \tag{9.40}$$

$$Q_w^{\min} \leq Q_w^- \leq Q_w^+ \leq Q_w^{\max} \tag{9.41}$$

(ii) Wind-farm-side robust voltage security region

For wind-farm-side voltage security region, the voltage magnitude of each wind unit in a wind farm should be strictly within secure range with the consideration of uncertainties. Thus, for wind farm w (w = 1, …, N_w), the voltage security region can be similarly expressed as

$$u_{w,i}^{\min} \leq u_{w,i} + \sum_{j=1}^{N_{w,u}} s_{w,i,j}^0 \left(q_{w,j} - q_{w,j}^0\right) + \sum_{j=1}^{N_{w,u}} h_{w,i,j}^0 \Delta p_{w,j} \leq u_{w,i}^{\max} \tag{9.42}$$

$$\forall \Delta p_{w,j} \in \left[\Delta p_{w,j}^{\min}, \Delta p_{w,j}^{\max}\right] \tag{9.43}$$

$$q_{w,j}^{\min} \leq q_{w,j} \leq q_{w,j}^{\max} \tag{9.44}$$

In the matrix form, we have

$$[\Theta_w] = \left\{ q_w | u^{\min} \leq Cq_w + D\Delta p_w \leq u^{\max}, \forall \Delta p_w \in \left[\Delta p_w^{\min}, \Delta p_w^{\max}\right] \right\} \tag{9.45}$$

where

$$D\Delta p_w = \left[D^+ \Delta p_w^{\min} + D^- \Delta p_w^{\max}, D^- \Delta p_w^{\min} + D^+ \Delta p_w^{\max}\right] \tag{9.46}$$

For convenience, let

$$\begin{aligned} \left[\varsigma_w^-, \varsigma_w^+\right] &= D\Delta p_w \\ \varsigma_w^- &= D^+ \Delta p_w^{\min} + D^- \Delta p_w^{\max} \\ \varsigma_w^+ &= D^- \Delta p_w^{\min} + D^+ \Delta p_w^{\max} \end{aligned} \tag{9.47}$$

Then, the wind-farm-side robust voltage security region gives

$$\Theta_w = \left\{ q_w | \left(u_w^{\min} - \varsigma_w^- \leq Cq_w \leq u_w^{\max} - \varsigma_w^+\right) \cap q_w^{\min} \leq q_w \leq q_w^{\max} \right\} \tag{9.48}$$

9.3.2.3 Hierarchically Alternate Coordination Method

Based on the proposed robust voltage security region, the original problem is actual an optimization problem that aims to find a voltage range of PCC bus of each wind farm, within which the two-level robust voltage security region {Ω, Θ1, Θ2, …, Θ*Nw*} is not empty.

In order to solve this problem, the two-level voltage security region should work together, it can be observed that the total active and reactive power limits as well as voltage each wind farm should be assigned by each wind farm, whereas the total reactive power of each wind farm should be restricted by the system-wide coordination of multiple wind farms. Especially, it should be pointed out that this two-level voltage security condition can be implemented by a hierarchically alternate coordination method between system and each wind farm which is described as follows.

At first, the wind-farm-side robust voltage security region is obtained by each wind farm with respect to its own independent information. Moreover, the total reactive power of each wind farm is restricted by the system-wide, i.e., $[Q^{min}, Q^{\max}]$ by (9.50). Therefore, total active power disturbance, total reactive power range and voltage magnitude region of PCC bus of wind farm w can be calculated by

(i) Total active power disturbance

$$\Delta P_w^{\min} = \min \sum_{j=1}^{N_{w,u}} \Delta p_{w,j}\, ,\, \Delta P_w^{\max} = \max \sum_{j=1}^{N_{w,u}} \Delta p_{w,j} \tag{9.49}$$

(ii) Total reactive power range

$$\begin{aligned} Q_w^{\max} &= \max_{\boldsymbol{q}_w} \sum_{j=1}^{N_u} \Delta q_{w,i} + Q_{C,w}^0 & Q_w^{\min} &= \min_{\boldsymbol{q}_w} \sum_{j=1}^{N_u} q_{w,i} + Q_{C,w}^0 \\ s.t. &\quad \boldsymbol{q}_w \in \Theta_w & s.t. &\quad \boldsymbol{q}_w \in \Theta_w \\ &\boldsymbol{Q}_w^- \le \sum_{j=1}^{N_u} q_{w,j} \le \boldsymbol{Q}_w^+ & &\boldsymbol{Q}_w^- \le \sum_{j=1}^{N_u} q_{w,j} \le \boldsymbol{Q}_w^+ \end{aligned} \tag{9.50}$$

(iii) Voltage magnitude region of PCC bus

$$\begin{aligned} U_{PCC,w}^{\max} &= \min_{\Delta \boldsymbol{p}} \max_{\boldsymbol{q}_w} \quad U_{PCC,w}^0 + \sum_{j=1}^{N_u} s_{w,j}\left(q_j - q_j^0\right) + \sum_{j=1}^{N_u} h_{w,j}\Delta p_j \\ s.t. &\quad \boldsymbol{q}_w \in \Theta_w, \quad \boldsymbol{Q}_w^- \le \sum_{j=1}^{N_u} q_{w,j} \le \boldsymbol{Q}_w^+, \forall \Delta \boldsymbol{p}_w \in \left[\Delta \boldsymbol{p}_w^{\min}, \Delta \boldsymbol{p}_w^{\max}\right] \end{aligned} \tag{9.51}$$

$$
\begin{aligned}
& U_{PCC,w}^{\min} = \max_{\Delta p} \min_{q_w} \quad U_{PCC,w}^{0} + \sum_{j=1}^{N_u} s_{w,j}\left(q_j - q_j^0\right) + \sum_{j=1}^{N_u} h_{w,j}\Delta p_j \\
& s.t. \quad \boldsymbol{q}_w \in \Theta_w, \quad \boldsymbol{Q}_w^- \le \sum_{j=1}^{N_u} q_{w,j} \le \boldsymbol{Q}_w^+, \forall \Delta \boldsymbol{p}_w \in \left[\Delta \boldsymbol{p}_w^{\min}, \Delta \boldsymbol{p}_w^{\max}\right]
\end{aligned} \tag{9.52}
$$

Note, (9.51) and (9.52) are actually a kind of "max-min" or "min-max" robust optimization models, which can be easily transformed into a tractable simple quadratic model by

$$
\begin{aligned}
& U_{PCC,w}^{\max} = \max_{q_w} \quad U_{PCC,w}^{0} + \sum_{j=1}^{N_u} s_{w,j}\left(q_j - q_j^0\right) + \sigma_w^- \\
& s.t. \quad \boldsymbol{q}_w \in \Theta_w, \quad Q_w^- \le \sum_{j=1}^{N_u} q_{w,j} \le Q_w^+, \forall \Delta \boldsymbol{p}_w \in \left[\Delta \boldsymbol{p}_w^{\min}, \Delta \boldsymbol{p}_w^{\max}\right]
\end{aligned} \tag{9.53}
$$

$$
\begin{aligned}
& U_{PCC,w}^{\min} = \min_{q_w} U_{PCC,w}^{0} + \sum_{j=1}^{N_u} s_{w,j}\left(q_j - q_j^0\right) + \sigma_w^+ \\
& s.t. \quad \boldsymbol{q}_w \in \Theta_w, \quad \boldsymbol{Q}_w^- \le \sum_{j=1}^{N_u} q_{w,j} \le \boldsymbol{Q}_w^+, \forall \Delta \boldsymbol{p}_w \in \left[\Delta \boldsymbol{p}_w^{\min}, \Delta \boldsymbol{p}_w^{\max}\right]
\end{aligned} \tag{9.54}
$$

where

$$
\begin{aligned}
\left[\sigma_w^-, \sigma_w^+\right] &= \sum_{j=1}^{N_u} h_{w,j}\Delta p_j \\
\sigma_w^- &= \sum_{j=1}^{N_u} \left(h_{w,j}^- \Delta p_j^{\max} + h_{w,j}^+ \Delta p_j^{\min}\right) \\
\sigma_w^+ &= \sum_{j=1}^{N_u} \left(h_{w,j}^- \Delta p_j^{\min} + h_{w,j}^+ \Delta p_j^{\max}\right)
\end{aligned} \tag{9.55}
$$

Furthermore, the system-wide robust voltage security region will be updated by (9.38)–(9.41) with respect to the new information of each wind farm from (9.49) to (9.52). Meanwhile, the updated system-wide robust voltage security region will generate a new total active power disturbance, total reactive power range and voltage magnitude region of PCC bus of each wind farm. The hierarchically alternate coordination of the two robust voltage security region leads to the feasible voltage range of PCC bus of each wind farm. Obviously, the voltage range of PCC bus is tightened during the alternate iteration process. In order to measure this deflation of i-th and $(i + 1)$-th iteration, it yields η for the convergence rule, such that the hierarchically alternate method is stop when η is smaller than a preset value.

Table 9.1 The flowchart of the proposed method

Flowchart of the proposed method	
Step 0	Given the preset convergence precision ε and system normal condition
Step 1	Calculate η by (9.56) and if η is smaller than ε, **Stop**; otherwise, go to **Step 2**
Step 2	Calculate the wind-farm-side robust voltage security region of each wind farm through (9.49)–(9.52), with the harvest of $\left[\boldsymbol{Q}_w^{\min}, \boldsymbol{Q}_w^{\max}\right]$, $\left[\Delta\boldsymbol{P}_w^{\min}, \Delta\boldsymbol{P}_w^{\max}\right]$ and $\left[\boldsymbol{U}_w^{\min}, \boldsymbol{U}_w^{\max}\right]$(i.e., {TR$_w$, TD$_w$, VM$_w$})
Step 3	Send the information of each wind farm to the master level system and calculate the system-wide robust voltage security region through (9.38)–(9.41), with the harvest of $\left[\boldsymbol{Q}_w^-, \boldsymbol{Q}_w^+\right]$(i.e., {CR$_w$})
Step 4	Send the information $\left[\boldsymbol{Q}_w^-, \boldsymbol{Q}_w^+\right]$ to the slave-level system of each wind farm, and go to **Step 1**

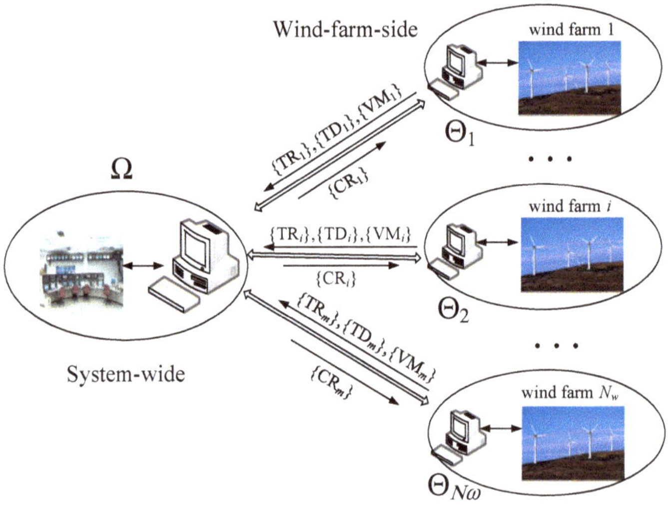

Fig. 9.7 Information exchange between system-wide workstation and wind-farm-side workstation

$$\eta = \max_{w} \left| \frac{\left(U_{PCC,w}^{\max,(k)} - U_{PCC,w}^{\min,(k)}\right) - \left(U_{PCC,w}^{\max,(k-1)} - U_{PCC,w}^{\min,(k-1)}\right)}{\left(U_{PCC,w}^{\max,(k)} - U_{PCC,w}^{\min,(k)}\right)} \right| \times 100\% \quad (9.56)$$

A procedure of the proposed hierarchically alternate coordination method for the two-level robust voltage security region is presented in Table 9.1.

To calculate this two-level robust voltage security region in centralized wind power integration, a master workstation is installed at system-wide control center and a slave workstation is installed at each wind farm, as illustrated in shown in Fig. 9.7. Each slave workstation uploads its own specific information to the master workstation, where the statuses of all wind farms are coordinated closely and compute the optimal strategy and send the information back to respective wind farms.

9.4 Field Application

The Wind-AVC system designed and developed by this paper was implemented in Zhangbei Wind Power Base of Northern China. Until July of 2013, 24 wind farms, including 1589 WTGs with the total generating capacity of 2379 MW, and 50 SVCs/SVGs with the total regulation capacity of +/−1000 Mvar, have been closed-loop controlled by the Wind-AVC system. Some results from real-life application are shown in the Section.

To evaluate the control performance of the close-loop control system, 2 days with similar operating conditions [11] and wind power generation, as shown in Fig. 9.8, were selected to be compared. Considering the intermittent characteristics, it is difficult to find 2 days with identical wind power output. As shown in Fig. 9.8, the two days' wind power were very similar before 15:00, whereas later, the day without control was with higher wind power output than the day with control, however, they were still with a similar trend.

Figure 9.9 shows the comparison of the high-side voltage of several typical wind farms in the area. It is clear that the voltage fluctuation was greatly reduced thanks to the Wind-AVC. The voltage profiles with control were much smoother, especially when the wind power rose rapidly shortly after noon, some very sharp voltage drops were observed on the day without Wind-AVC, while in contrast, the voltage could still be kept flat and smooth with Wind-AVC.

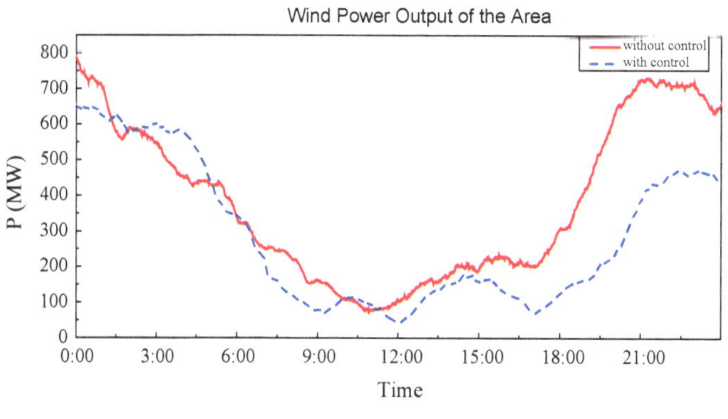

Fig. 9.8 Wind power output before and after the implementation of Wind-AVC. The *solid red curve* shows the active power on 13 Nov. 2011 without Wind-AVC. The *blue dashed curve* shows the active power on 18 Oct. 2012 following the implementation of Wind-AVC

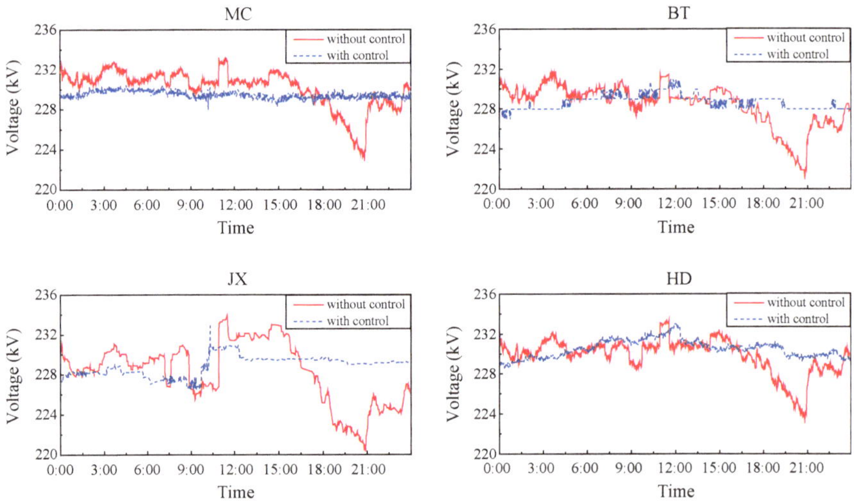

Fig. 9.9 Voltage curves comparison of typical wind farms with and without Wind-AVC

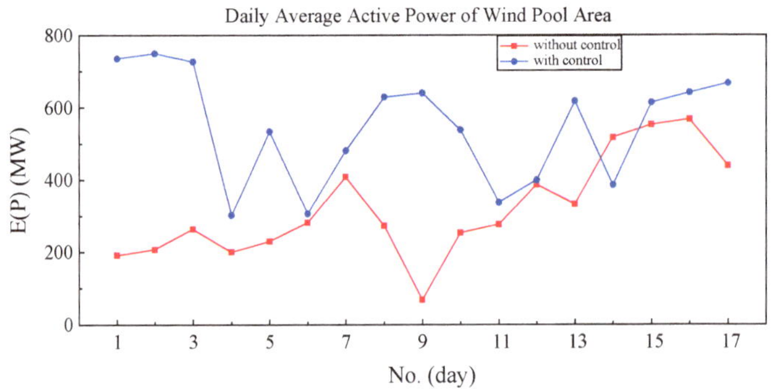

Fig. 9.10 The daily average active power of the wind power pool area over a 17-day period with and without Wind-AVC

Long-term comparison was also carried out, in which we selected two half-months. The first was from Aug 1st to 17th, 2012, when the wind area was during the low-wind season prior to the implementation of Wind-AVC, and the second was from Oct 15th to 31th, 2012, when the wind area was during the strong-wind season and the Wind-AVC was closed-loop controlled. The daily average wind power output comparison is shown as Fig. 9.10. Note that the wind power outputs of the days with Wind-AVC were obviously much higher than that of the days without control. This means that, it is likely that control was more challenging due to the greater wind power generated. However, as shown in Fig. 9.11 and Table 9.2, the voltage fluctuations had a smaller average standard deviation as well as lower peak-to-valley deviation with the closed-loop control of Wind-AVC.

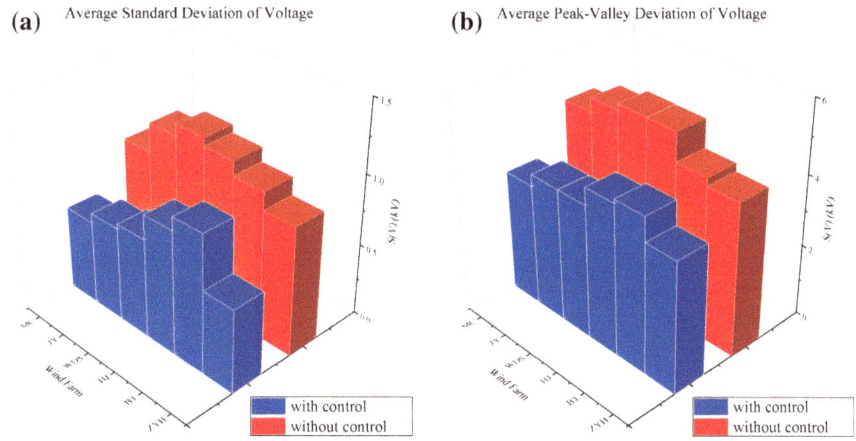

Fig. 9.11 Voltage fluctuation comparison of the wind pool area for long-term test with and without Wind-AVC

Table 9.2 Voltage fluctuation comparison of the wind farms for Long-term test with and without AVC

Wind farms	Average standard deviation			Average peak-valley deviation		
	Without control (kV)	With control (kV)	Reduction (%)	Without control (kV)	With control (kV)	Reduction (%)
ZYH	0.91	0.59	34.70	4.28	3.67	14.18
BT	1.03	0.97	5.63	4.50	4.35	3.17
HJ	1.13	0.87	22.47	5.28	4.24	19.60
WDS	1.18	0.70	41.13	5.29	3.75	29.03
JY	1.10	0.66	39.88	5.11	3.75	26.60
MC	0.90	0.60	33.59	4.70	3.42	27.26

Importantly, thanks to the autonomous-synergy voltage control scheme, the voltage distributions in the wind pool area are more reasonable and the cascading trip risk is also reduced. The Wind-AVC system is proved to be one of the important reasons to improve the operation security and reliability.

References

1. http://en.wikipedia.org/wiki/Wind_power_in_the_People's_Republic_of_China
2. M. Gang, J. Wang, G. Yan, Cascading trip-off of doubly-fed induction generators from grid at near full-load condition in a wind farm. Autom. Electr. Power Syst. **35**(22), 35–40 (2011)
3. D. Li, L. Jia, X. Xu, et al., Cause and countermeasure analysis on wind turbines' trip-off from grid. Autom. Electr. Power Syst. **35**(22), 41–44 (2011)

4. X. Ye, Z. Lu, Y. Qiao, et al., A primary analysis on mechanism of large scale cascading trip-off of wind turbine generators. Autom. Electr. Power Syst. **36**(8), 35–40 (2012)
5. J.P. Paul, J.Y. Leost, J.M. Tesseron, Survey of the secondary voltage control in France: present realization and investigations. IEEE Trans. Power Syst. **2**, 505–511 (1987)
6. M. Ilic-Spong, J. Christensen, K.L. Eichorn, Secondary voltage control using pilot point information. IEEE Trans. Power Syst. **3**(2), 660–668 (1988)
7. P. Lagonotte, J.C. Sabonnadiere, J.Y. Leost, J.P. Paul, Structural analysis of the electrical system: application to secondary voltage control in France. IEEE Trans. Power Syst. **4**(2), 479–486 (1989)
8. H. Lefebvre, D. Fragnier, J.Y. Boussion, P. Mallet, M. Bulot, Secondary coordinated voltage control system: feedback of EDF, in *Proceedings of the 2000 IEEE Power Engineering Society Summer Meeting*, pp. 290–295 (2000)
9. S. Corsi, M. Pozzi, C. Sabelli, A. Serrani, The coordinated automatic voltage control of the Italian transmission grid—Part I: Reasons of the choice and overview of the consolidated hierarchical system. IEEE Trans. Power Syst. **19**(4), 1723–1732 (2004)
10. S. Corsi, M. Pozzi, M. Sforna, G. Dell'Olio, The coordinated automatic voltage control of the Italian transmission grid—Part II: Control apparatuses and field performance of the consolidated hierarchical system. IEEE Trans. Power Syst. **19**(4), 1733–1741 (2004)
11. H. Sun, Q. Guo, B. Zhang, et al., An adaptive zone-division-based automatic voltage control system with applications in China. IEEE Trans. Power Syst. **28**(2), 1816–1828 (2013)
12. Q. Guo, H. Sun, M. Zhang, J. Tong, B. Zhang, B. Wang, Optimal voltage control of PJM smart transmission grid: study, implementation, and evaluation. IEEE Trans. Smart Grid **4**(3), 1665–1674 (2013)
13. F. Li, W. Qiao, H. Sun, H. Wan, J. Wang, Y. Xia, X. Zhao, P. Zhang, Smart transmission grid: vision and framework. IEEE Trans. Smart Grid **1**(2), 168–177 (2010)
14. K. Torchyan, M.S. Elmoursi, W. Xiao, Adaptive secondary voltage control for grid interface of large scale wind park, in *IEEE PowerTech Conferecne, Grenoble, France* (2013)
15. Y. Liu, Z. Chen, Voltage sensitivity based reactive power control on VSC-HVDC in a wind farm connected hybrid multi-infeed HVDC system, in *IEEE PowerTech Conferecne, Grenoble, France* (2013)
16. E. Heredia, D. Kosterev, M. Donnelly, Wind hub reactive resource coordination and voltage control study by sequence power flow, in *IEEE Power and Energy Society General Meeting, Vancouver, Canada* (2013)
17. T. Zheng, S. Jiao, K. Ding, L. Lin, A coordinated voltage control strategy of wind farms based on sensitivity method, in *IEEE PowerTech Conferecne, Grenoble, France* (2013)
18. B. Zhang, H. Zhang, H. Sun, et al., Interaction and coordination between multiple control centers: development and practice, in *CIGRE 2006, SC C2 System Control and Operation, SC C2-302*
19. H. Sun, B. Zhang, W. Wu, Q. Guo, Family of energy management system for smart grid, in *IEEE PES Innovative Smart Grid Technologies Europe, Berlin, Germany, 13–18 Oct 2012* (2012)
20. Q. Guo, H. Sun, Y. Liu, et al., Distributed automatic voltage control framework for large-scale wind integration in China, in *IEEE Power and Energy Society General Meeting, San Diego, USA* (2012)
21. L. Yuan, J.D. McCalley, Decomposed SCOPF for improving efficiency. IEEE Trans. Power Syst. **24**(1), 494–495 (2009)
22. N. Xiang, S. Wang, E. Yu, An application of estimation-identification approach of multiple bad data in power system state estimation. IEEE Trans. PAS **PAS-103**(2) (1984)
23. B.M. Zhang, K.L. Lo, A recursive measurement error estimation identification method for bad data analysis in power system state estimation. IEEE Trans. Power Syst. **PWRS-6**(1), 191–197 (1991)
24. G.P. Azevedo, C.S. Souza, B. Feijó, Enhancing the human-computer interface of power system applications. IEEE Trans. Power Syst. **11**(2), 646–653 (1996)

25. T.J. Overbye, J.D. Weber, Visualization of power system data, in *Proceedings of the 33rd Hawaii International Conference on System Sciences* (2000), pp. 1228–1234
26. R. Klump, W. Wu, G. Dooley, Displaying aggregate data, interrelated quantities, and data trends in electric power systems, in *Proceedings of the 36th Hawaii International Conference on System Sciences* (2003)
27. C. Jia, S. Hongbin, T. Lei, et al., Three-dimensional visualization technique in power system control centers and its real-time applications. Autom. Electr. Power Syst. **32**(6), 20–24 (2008) (in Chinese)

Chapter 10
Risk Averse Security Constrained Stochastic Congestion Management

Abbas Rabiee, Alireza Soroudi and Andrew Keane

Nomenclature

For quick reference, the main notations used throughout the chapter are stated in this section. For quick reference, the main notations used throughout the chapter are stated in this section.

Sets:

NB	Set of system buses.
NB_G	Set of system buses with generation units.
NB_{CM}	Set of load buses participating in PSCCM.
NG	Set of generators.
NG_{CM}	Set of generators participating in PSCCM.
NB_E	Set of load buses participating in emergency control.
NG_E	Set of generators participating in emergency control.
NG_b	Set of generation units connected to bus b.
NS	Set of post-contingency scenarios.
NL	Set of branches (transmission lines and transformers).

A. Rabiee
Faculty of Engineering, Department of Electrical Engineering,
University of Zanjan, Zanjan, Iran
e-mail: rabiee@znu.ac.ir

A. Soroudi (✉) · A. Keane
School of Electrical, Electronic and Communications Engineering,
University College Dublin, Dublin, Ireland
e-mail: alireza.soroudi@ucd.ie

A. Keane
e-mail: andrew.keane@ucd.ie

P. Du et al. (eds.), *Integration of Large-Scale Renewable Energy into Bulk Power Systems*, Power Electronics and Power Systems,
DOI 10.1007/978-3-319-55581-2_10

Indices:

i	Index for generation units running from 1 to *NG*.
s	Index for scenarios running from 1 to *NS*.
b/j	Index for bus (node) number running from 1 to *NB*.
ℓ	Index for transmission lines running from 1 to *NL*.

Variables:

$\Delta P_{G_i}^{up/dn,bc}$	Active power increment/decrement in generator i due to CM at the bc.
$\Delta P_{G_i,s}^{up/dn}$	Corrective active power increment/decrement in generator i at scenario s.
$\Delta \widehat{P}_{G_i,s}^{up/dn}$	Emergency active power increment/decrement in generator i at scenario s.
$(P/Q)_{L_b}^{bc}$	Active/reactive load in bus b at the bc.
$(P/Q)_{L_b,s}$	Active/reactive load in bus b at scenario s.
$\Delta(P/Q)_{L_b}^{bc}$	Active/reactive voluntarily load reduction in bus b due to CM at the bc.
$\Delta(P/Q)_{L_b,s}^{LC}$	Active/reactive involuntarily load reduction in bus b due to CM at scenario s.
$\Delta(\widehat{P}/\widehat{Q})_{L_b,s}^{LC}$	Active/reactive involuntarily load reduction in bus b as an emergency control at scenario s.
$S_{\ell}^{bc}(V,\theta)$	Apparent power flowing through ℓ^{th} branch at bc.
$S_{\ell,s}(V,\theta)$	Apparent power flowing through ℓ^{th} branch at scenario s.
$(P/Q)_{G_i}^{bc}$	Modified active power generation by generator i at bc.
$P_{G_i,s}$	Modified active power generation by generator i at scenario s.
V_b^{bc}/θ_b^{bc}	Voltage magnitude/angle of bus b at the bc.
$V_{b,s}/\theta_{b,s}$	Voltage magnitude/angle of bus b at scenario s.
$\widehat{V}_{b,s}/\widehat{\theta}_{b,s}$	Voltage magnitude/angle of bus b at scenario s, when corrective controls fail.

Parameters:

$\mu_{G_i}^{up/dn,bc}$	Cost of active power generation increase/decrease (re-dispatch) at the bc (\$/MW).
$\mu_{G_i,s}^{up/dn}$	Cost of corrective active power generation increase/decrease (re-dispatch) at scenario s (\$/MW).
$\widehat{\mu}_{G_i,s}^{up/dn}$	Cost of emergency active power generation increase/decrease (re-dispatch) at scenario s (\$/MW).
$\mu_{L_b}^{bc}/\mu_{L_b,s}^{P}$	Cost of voluntarily load reduction in bus b at bc/scenario s (\$/MW).
$\mu_{L_b,s}^{LC}$	Cost of involuntarily load reduction in bus b at scenario s (\$/MW).
$\widehat{\mu}_{L_b,s}^{LC}$	Cost of involuntarily load reduction in bus b at scenario s as an emergency control (\$/MW).

$(P/Q)_{L_b}^{fr}$	Forecasted values of active/reactive load demand.
$P_{G_i}^{sch}$	Initial (before PSCCM) schedule of active power generation by unit i.
$\Delta P_{G_i}^{up/dn,max}$	Maximum active power increment/decrement in generator i for CM purposes.
$\Delta P_{G_i,s}^{up/dn,max}$	Maximum active power increment/decrement in generator i at scenario s.
$\Delta \widehat{P}_{G_i,s}^{up/dn,max}$	Maximum active power increment/decrement in generator i as an emergency control at scenario s.
$\Delta(P/Q)_{L_b}^{max,bc}$	Maximum limit on the active/reactive voluntarily load reduction in bus b due to CM at the bc.
$\Delta(P/Q)_{L_b,s}^{max}$	Maximum limit on the active/reactive voluntarily load reduction in bus b due to CM at scenario s.
$\Delta(P/Q)_{L_b,s}^{LC,max}$	Maximum limit on the active/reactive involuntarily load reduction in bus b due to CM at scenario s.
$\Delta(\widehat{P}/\widehat{Q})_{L_b,s}^{LC,max}$	Maximum limit on the active/reactive involuntarily load reduction in bus b as an emergency control at scenario s.
$V_b^{max/min}$	Minimum/maximum voltage magnitude at bus b.
S_ℓ^{max}	Maximum power carrying capacity of branch ℓ.
$(P/Q)_{G_i}^{\min/\max}$	Minimum/maximum active/reactive power output of generator i.
$Y_{bj}^{bc}/\phi_{bj}^{bc}$	Magnitude/angle of bj^{th} element of network admittance matrix at bc.
$Y_{bj,s}/\phi_{bj,s}$	Magnitude/angle of bj^{th} element of network admittance matrix at scenario s.
π_s/π_{bc}	Probability at scenario s/base-case (bc) scenario.
π_c	Probability of post-contingency corrective control successfully implemented.
RU_{G_i}/RD_{G_i}	Ram-up/dn rate of generation unit i.
$\tau_{bc}/\tau_{CM}/\tau_E$	The lead-time for execution of PSCCM in bc/scenario s/ emergency state s.
$I_{G_i,s}$	On/off status of generation unit i, at scenario s.

Risk associated variables and parameters:

η	Auxiliary variable to define CVaR.
ζ	Conditional value at risk (CVaR).
$E(.)$	Expected value operator.
β	Weighting factor to indicate the risk neutral ($\beta = 0$) to risk averse ($\beta = 1$) strategies.

10.1 Introduction

10.1.1 Motivation and Problem Description

In a deregulated electricity market, transmission lines are working closer to their limits for getting a more economic operation point. The amount of the electricity that can be transmitted along a transmission line is constrained by thermal limit, voltage limit and stability limits. Different participants of the energy market are responsible for their own profits and trying to maximize their profits employing the bidding strategies. In a day-ahead pool based electricity market, the amount of the accepted offered power and requested demands are determined based on the received bids in a day before of actual energy transactions. Additionally, the actual wind power generation might be different with predicted schedule. In other words, the wind power generation is subject to uncertainty. This structure frequently results in a congestion problem [1, 2]. In a congestion management framework, the system operator determines the required changes in the cleared market outputs that relieves the system secure operation with minimum cost.

10.1.2 Literature Review

Three forms of the congestion management approaches have been proposed in literature [3]. In the first approach, centralized optimization along with optimal power flow (OPF) programs are used for relieving the congestion. Price signals derived from anticipated market resolution are used to deter congestion by constraining scheduled generator output before real time operation in the second approach. In the third approach, the effect of bilateral contracts between producer and consumer on the congestion of transmission lines are checked and allowed or disallowed based on their effects on congestion. Methods based on the OPF are the most significant techniques for congestion management in a power system considering transmission and operational constraints [1, 4, 5]. In [4], a multi-objective framework for minimizing the cost of congestion management and line overloading index has been proposed and solved using multi-objective particle swarm optimization (MOPSO) method. Congestion management considering voltage stability limit is proposed in [1] using AC OPF. The congestion management schemes used in England and Wales, Norway, Sweden, PJM, and California systems are analyzed and DC power flow based unified framework has been proposed in [5]. A review of different congestion management approaches can be found in [3]. In [6], a combination of demand response (DR) and flexible alternating current transmission system (FACTS) devices, i.e., TCSC and SVC, are used for reliving congestion of the transmission lines. A zonal/cluster-based congestion management approach has been proposed in [7], where the generators with strongest and nonuniform distribution of sensitivity indexes in the most sensitive zones are selected for rescheduling. A new approach for short term forecasting congestion and other DC-OPF variables has been presented in [8]. The impact of load variation, the size of a power supplier and random failures on market power have been investigated in [9]. A market-based transmission planning is proposed in [10] which considers some power system uncertainties and selects the final plan after risk assessment of all solutions. The investment cost, operation cost, congestion cost, load curtailment cost, and cost caused by system unreliability are all considered. In [11], a Monte Carlo method was proposed to enable the TSO to evaluate, for each possible choice of the total transfer capacity (TTC) limit among areas, the maximum probability of congestion in a market framework, thus selecting the limit corresponding to the acceptable risk level. The network security is probabilistically treated in [12] for determining the optimal network reinforcement. The relevant contingencies are selected and optimized using the available preventive and corrective control actions (in form of special protection schemes (SPS) and demand response). Reference [1] proposes a congestion management method which is based on online conditions obtained from power flow equations. It not only guarantees a stable operating point but also provides a minimum required distance to voltage collapse conditions. The Reactive power generation of wind turbines and demand response (DR) flexibility are used in [2] to mitigate the distribution network congestion. Information Gap Decision Theory [13] is used to address uncertainty of wind generation generation. However the network contingencies are ignored.

10.1.3 Contribution

In this chapter, a stochastic security constrained congestion management (PSCCM) framework is proposed. The proposed approach modifies the base-case (normal) operation point of the system taking into account the effects of probable severe contingencies as well as wind power generation uncertainties. The proposed PSCCM is a two stage stochastic programming problem, in which at its first stage modifications such as generation re-dispatch, load reduction and voltage set points re-adjustments are applied on the base-case operation point, which is also named as preventive controls. In its second stage, the proper control actions which are necessary for restoration of the system at the post-contingency condition, are determined. These control actions are also called corrective controls. It determines the optimal preventive/corrective control actions while minimizing the corresponding costs. In order to avoid expensive controls especially in some post-contingency scenarios, a well established risk index named as conditional value at risk (CVaR) is also included in the PSCCM model, which facilitates compromise decisions for system operator This risk measure is chosen because it is convex and draws a balance between the cost objective function and the diversification of risk across the preventive and corrective actions.

10.1.4 Chapter Organization

The remainder of this chapter is organized as follows: Sect. 10.2 deals with uncertainty and risk modeling. Section 10.3 presents detailed description of the problem and the formulation of the proposed PSCCM approach. Simulation results and extensive discussions are presented in Sect. 10.4 and finally, Sect. 10.5 summarizes the findings of this work.

10.2 Uncertainty and Risk Modeling

The decision making process in engineering problem is defined as finding the optimal values of a set of decision variables in order to minimize (maximize) an objective function. It is evident that the optimal values of these variables depend on the input data as well as the problem's constraints. It is not always possible for decision maker to have the precise values of these parameters in decision making process. In other words, some input parameters are subject to uncertainty. Based on the uncertainty nature, there are several methods to deal with the uncertain parameters such as Information gap decision theory [14, 15], fuzzy logic [16], robust optimization [17] and probabilistic techniques [18]. In this chapter, a probabilistic technique is used to handle the uncertainty of contingencies. One of the well known methods is scenario based modeling. In this method, the uncertain parameter is described using a probability density function (PDF). Then the continuous PDF is divided into predefined scenarios (Z_s) with specific probabilities (π_s). Without loss of generality, a minimization problem is discussed and explained. Suppose an objective function $f(X, Z_s)$ is to be minimized. The decision variable are represented by X and the uncertain parameters are described using Z_s.

$$\min_{X} E(X) \tag{10.1a}$$

$$E(X) = \sum_{s \in \Omega_s} \pi_s f(X, Z_s) \tag{10.1b}$$

$$\mathbf{H}(X, Z_s) \leq 0, \quad \forall s \in \Omega_s \tag{10.1c}$$

$$\mathbf{G}(X, Z_s) = 0, \quad \forall s \in \Omega_s \tag{10.1d}$$

where Ω_s is the set of all possible realization of uncertain parameters Z_s. $\mathbf{H}, \mathbf{G}$ are the inequality and equality constraints. The decision making procedure described in (10.1), has two important shortcomings as follows:

1. The decision variable vector (X) is independent of scenarios so it should be conservatively chosen to cope with every possible scenario in Ω_s.
2. It only minimizes the expected value of the objective function f. Several distribution of f can produce the same expected value. This might increase the risk of decision maker.

In order to add the flexibility of decision making process (and overcome the first shortcoming), two stage stochastic decision making is used here. The second shortcoming can be removed using risk measures. There are several risk measures proposed to be used in scenario based decision making under uncertainty like variance, shortfall probability, expected shortfall, value at risk and conditional value at risk (CVaR). In this chapter, CVaR (ζ) is used as the risk measure to be optimized along with expected value of cost because of its significant advantages over other risk measures, numerical efficiency and stability of calculations [19]. The procedure for minimization is as follows: In this approach, the decision variables are divided into two categories. The first variable set (Ψ), represents the "**Here and now**" decisions. These variables are determined before knowing the exact value of uncertain parameters. The second variable set (ς_s), represents the "**Wait and see**" decisions. These variables are determined after the realization of uncertain parameters. So $X \in \{\Psi, \varsigma_s\}$.

$$\min_{X,\eta} OF = (1-\beta) \times E(X) + \beta \times \zeta \tag{10.2}$$

$$\zeta = \eta + \frac{1}{1-\epsilon} \sum_{s \in \Omega_s} \pi_s \max(f(X, Z_s) - \eta, 0) \tag{10.3}$$

where ζ is computed as the expected cost in the $(1-\epsilon) \times 100\%$ worst scenarios (or CVaR). η is the value at risk. The considered uncertainties in this work are contingencies which endanger the safe operation of power system. The failure of lines and generating units are taken into account.

10.3 The Proposed Framework for PSCCM

Practical power systems are subjected to occurrence of various contingencies, such as transmission lines, generators and transformers outage. It is obvious that the occurrence of severe contingencies, strongly affects the stability and (in more general sense) the security of system. Hence, it is necessary to identify such critical contingencies, and modify the operation point of the system in base-case and post-contingency states against these vulnerable incidents. The control actions performed prior to happening of contingencies, i.e. in base-case state, are called preventive controls, and those initiated after occurrence of any contingency is named as corrective controls [20]. It is evident that, corrective control actions are much more expensive than preventive controls, due to the fact that the former initiate at the emergency state, and consequently they are essential for preserving the stability of system and maintaining the balanced energy flow in the network.

10.3.1 Basic Description of the Proposed Framework

If the system operator (SO) is able to modify the base-case operating point of the system properly, it is possible to reduce the costs of employing corrective controls, and hence the system will be operated both economically and more safely. Thus, the SO should detect critical contingencies, following the clearance of market and determination of the tentative operation point of the system in the base-case. Then, by manipulating this operation point subject to the identified critical contingencies, a modified operation point is obtained, which is secure against most of these contingencies and the corresponding corrective controls for any of these contingencies are obtained along with the preventive controls for the base-case condition. Thus the proposed framework for this aim is as follows.

- **Step 0** (Initialization): Market settlement and determination of the initial operation point (with known system topology, active power generation, voltage set points, load levels, etc.
- **Step 1** (Contingency screening): Perform contingency analysis on the initial operation point obtained in **Step 0** and determine the list of severe (critical) contingencies
- **Step 2** (PSCCM): Run the proposed PSCCM model (which is described in the following section) and determine the optimal values for both preventive/corrective control actions, along with the modified base-case operation point.
- **Step 3** (re-checking the obtained modified operation point): run **Step 1** again for the modified base-case operation point and check that if any new critical contingencies created or not. If yes, add these new contingencies to the previous list and update the list of severe contingencies. Then, go to **Step 2**, and modify the base-case operation point subject to the updated contingency list by determination the proper preventive/corrective control actions. Otherwise, terminate and declare the required modifications, to the participants in this program (such as GenCos, retailers, customers, etc.).

10.3.2 Formulation of PSCCM

The objective function of the congestion management (to be minimized) is defined as the expected total cost (ETC) of incurred up/down power adjustments as follows:

$$\min_{\bar{U}} ETC = \pi_{bc} \times TC_{bc} + \sum_{\forall s \neq bc} \pi_s \times TC_{C,s} \tag{10.4}$$

where:

$$TC_{bc} = \sum_{i \in NG_{CM}} \mu_{G_i}^{up/dn,bc} \Delta P_{G_i}^{up/dn,bc} + \sum_{b \in NB_{CM}} \mu_{L_b}^{bc} \Delta P_{L_b}^{bc} \tag{10.5}$$

$$TC_{C,s} = \left\{ \begin{array}{c} \sum_{i \in NG_{CM}} \mu_{G_i,s}^{up/dn} \Delta P_{G_i,s}^{up/dn} + \\ \sum_{b \in NB_{CM}} \left\{ \mu_{L_b,s}^{P} \Delta P_{L_b,s} + \mu_{L_b,s}^{LC} \Delta P_{L_b,s}^{LC} \right\} \end{array} \right\} \tag{10.6}$$

Equations (10.5) and (10.6) are base-case and post-contingency total costs, respectively. The first term in both (10.5) and (10.6) is the total payment for up/down adjustments of power generation and the second term is the expected payment for reduction of demands. Objective function (10.4) should be minimized subject to the following constraints:

(a) Active/reactive power flow constraints at base-case (bc) ($\forall b \in NB$):

$$\sum_{i=1}^{NG_b} P_{G_i}^{bc} - P_{L_b}^{bc} = V_b^{bc} \sum_{j=1}^{N_B} V_j^{bc} Y_{bj}^{bc} \cos(\theta_b^{bc} - \theta_j^{bc} - \phi_{bj}^{bc}) \tag{10.7}$$

$$\sum_{i=1}^{NG_b} Q_{G_i}^{bc} - Q_{L_b}^{bc} = V_b^{bc} \sum_{j=1}^{N_B} V_j^{bc} Y_{bj}^{bc} \sin(\theta_b^{bc} - \theta_j^{bc} - \phi_{bj}^{bc}) \tag{10.8}$$

(b) Active/reactive power generation/demand, voltage and line flow limits at bc:

$$P_{G_i}^{bc} = P_{G_i}^{sch} + \Delta P_{G_i}^{up,bc} - \Delta P_{G_i}^{dn,bc} \qquad \forall i \in NG_{CM} \tag{10.9}$$

$$P_{L_b}^{bc} = P_{L_b}^{fr} - \Delta P_{L_b}^{bc} \qquad \forall b \in NB \tag{10.10}$$

$$Q_{L_b}^{bc} = Q_{L_b}^{fr} - \Delta Q_{L_b}^{bc} \qquad \forall b \in NB \tag{10.11}$$

$$P_{G_i}^{min} \le P_{G_i}^{bc} \le P_{G_i}^{max} \qquad \forall i \in NG \tag{10.12}$$

$$Q_{G_i}^{min} \le Q_{G_i}^{bc} \le Q_{G_i}^{max} \qquad \forall i \in NG \tag{10.13}$$

Also, $\forall i \in NG_{CM}$:

$$0 \le \Delta P_{G_i}^{up} \le \Delta P_{G_i}^{up,max} \tag{10.14}$$

$$0 \le \Delta P_{G_i}^{dn} \le \Delta P_{G_i}^{dn,max} \tag{10.15}$$

$$\Delta P_{G_i}^{up,max} = min(P_{G_i}^{max} - P_{G_i}^{sch}, RU_{G_i} \times \tau_{bc}) \tag{10.16}$$

$$\Delta P_{G_i}^{dn,max} = min(P_{G_i}^{sch} - P_{G_i}^{min}, RD_{G_i} \times \tau_{bc}) \tag{10.17}$$

And,

$$0 \le \Delta P_{L_b}^{bc} \le \Delta P_{L_b}^{max,bc} \qquad \forall b \in NB_{CM} \tag{10.18}$$

$$0 \le \Delta Q_{L_b}^{bc} \le \Delta Q_{L_b}^{max,bc} \qquad \forall b \in NB_{CM} \tag{10.19}$$

$$V_b^{min} \le V_b^{bc} \le V_b^{max} \qquad \forall b \in NB \tag{10.20}$$

$$\left| S_\ell^{bc}(V, \theta) \right| \le S_\ell^{max} \qquad \forall \ell \in NL \tag{10.21}$$

(c) Active/reactive power flow constraints at post-contingency scenarios ($\forall b \in NB$):

$$\sum_{i=1}^{NG_b}(P_{G_i,s}I_{G_i,s}) - P_{L_b,s} = \tag{10.22}$$

$$V_{b,s}\sum_{j=1}^{N_B}V_{j,s}Y_{bj,s}\cos(\theta_{b,s}-\theta_{j,s}-\phi_{bj,s})$$

$$\sum_{i=1}^{NG_b}(Q_{G_i,s}I_{G_i,s}) - Q_{L_b,s} = \tag{10.23}$$

$$V_{b,s}\sum_{j=1}^{NB}V_{j,s}Y_{bj,s}\sin(\theta_{b,s}-\theta_{j,s}-\phi_{bj,s})$$

(d) Active/reactive power generation/demand, voltage and line flow limits at post-contingency scenarios:

$$P_{G_i,s} = (P_{G_i}^{bc} + \Delta P_{G_i,s}^{up} - \Delta P_{G_i,s}^{dn}) \times I_{G_i,s} \qquad \forall i \in NG_{CM} \tag{10.24}$$

$$P_{L_b,s} = P_{L_b}^{bc} - \Delta P_{L_b,s} - \Delta P_{L_b,s}^{LC} \qquad \forall b \in NB \tag{10.25}$$

$$Q_{L_b,s} = Q_{L_b}^{bc} - \Delta Q_{L_b,s} - \Delta Q_{L_b,s}^{LC} \qquad \forall b \in NB \tag{10.26}$$

$$(P_{G_i}^{min}I_{G_i,s}) \leq P_{G_i,s} \leq (P_{G_i}^{max}I_{G_i,s}) \qquad \forall i \in NG \tag{10.27}$$

$$(Q_{G_i}^{min}I_{G_i,s}) \leq Q_{G_i,s} \leq (Q_{G_i}^{max}I_{G_i,s}) \qquad \forall i \in NG \tag{10.28}$$

and, $\forall i \in NG_{CM}$:

$$0 \leq \Delta P_{G_i,s}^{up} \leq (\Delta P_{G_i,s}^{up,max} I_{G_i,s}) \tag{10.29}$$

$$0 \leq \Delta P_{G_i,s}^{dn} \leq (\Delta P_{G_i,s}^{dn,max} I_{G_i,s}) \tag{10.30}$$

$$\Delta P_{G_i,s}^{up,max} = I_{G_i,s} \times min(P_{G_i}^{max} - P_{G_i}^{bc}, RU_{G_i} \times \tau_{CM}) \tag{10.31}$$

$$\Delta P_{G_i,s}^{dn,max} = I_{G_i,s} \times min(P_{G_i}^{bc} - P_{G_i}^{min}, RD_{G_i} \times \tau_{CM}) \tag{10.32}$$

Also,

$$0 \leq \Delta P_{L_b,s} \leq \Delta P_{L_b,s}^{max} \qquad \forall b \in NB_{CM} \tag{10.33}$$

$$0 \leq \Delta Q_{L_b,s} \leq \Delta Q_{L_b,s}^{max} \qquad \forall b \in NB_{CM} \tag{10.34}$$

$$0 \leq \Delta P_{L_b,s}^{LC} \leq \Delta P_{L_b,s}^{LC,max} \qquad \forall b \in NB \tag{10.35}$$

$$0 \leq \Delta Q_{L_b,s}^{LC} \leq \Delta Q_{L_b,s}^{LC,max} \qquad \forall b \in NB \tag{10.36}$$

$$V_b^{min} \leq V_{b,s} \leq V_b^{max} \qquad \forall b \in NB \tag{10.37}$$

$$\left|S_{\ell,s}(V,\theta)\right| \leq S_\ell^{max} \qquad \forall \ell \in NL \tag{10.38}$$

It is worth noting that, the involuntarily load reduction is undesirable for customers and hence in the proposed PSCCM, it is admissible only in the emergency post-contingency states not in base-case operation. The objective function of PSCCM is modified as follows, considering the effect risk in the through CVaR index.

$$\min_{X,\eta} OF = (1-\beta)\times ETC + \beta \times \zeta \tag{10.39}$$

$$\zeta = \eta + \frac{1}{1-\epsilon}\sum_{\hat{s}\in\Omega_{\hat{s}}} \pi_{\hat{s}} \max(TC_{\hat{s}} - \eta, 0) \tag{10.40}$$

where $\Omega_{\hat{s}} = \{NS\} \cup \{bc\}$, is the overall set of all scenarios (including the base-case and post-contingency scenarios). The above PSSCM model is solved to determine the following control variables: $\overline{U} = U_{bc} \cup \sum_{s} U_s$, where: $U_{bc} = \left\{\Delta P_{G_i}^{up/dn,bc}, \Delta(P/Q)_{L_b}^{bc}, V_b^{bc}(\forall b \in NB_G)\right\}$, and $U_s = \left\{\begin{matrix}\Delta(P/Q)_{G_i,s}^{up}, V_{b,s}(\forall b \in NB_G), \\ Delta(P/Q)_{L_b,s}, \Delta(P/Q)_{L_b,s}^{LC}\end{matrix}\right\}$. U_{bc} is also referred as first stage (or here and now) decision variable and U_s is named as second stage or wait and see decision variable in the stochastic programming.

10.3.3 Corrective Controls Failure

Based on the approach proposed in [21], we considered the case where the corrective control actions fail to dependably respond following occurance of severe contingencies. In this regard, two cost terms are defined for post-contingency actions, namely the cost of the successful corrective controls (which is given by (10.6)), and another term which is the cost of emergency remedial actions when the scheduled corrective control actions fail to respond. It is assumed that the corrective controls are successful with the probability of π_c, and hence the probability of corrective actions failure is $1 - \pi_c$. Thus the cost of base case, successful corrective actions (in scenario s) are calculated from (10.5) and (10.6), whereas the cost of remedial emergency controls, $TC_{E,s}$, is calculated as follows:

$$TC_{E,s} = \left\{\sum_{i\in NG_E} \hat{\mu}_{G_i,s}^{up/dn} \Delta\hat{P}_{G_i,s}^{up/dn} + \sum_{b\in NB_E} \hat{\mu}_{L_b,s}^{LC} \Delta\hat{P}_{L_b,s}^{LC}\right\} \tag{10.41}$$

It is worth noting that, in the case of corrective control failure, emergency controls are activated to restore a feasible post-contingency operation point. In this chapter, it is assumed that the load curtailment (as the last but as a quick and effective) remedial action is activated in the emergency state (i.e. when the corrective controls fail). Besides, some fast response generation units (such as gas or hydro turbo-generators) are responsible for emergency load balancing service in the network. Hence, the cost of emergency controls defined by (10.42), consists of two terms, for load curtailment and load emergency balancing service by a specified set of generators (NG_E). Therefore, by considering the probability of corrective control failure, the ETC is now calculated as follows.

$$\min_{\bar{U}} ETC = \pi_{bc} \times TC_{bc} + \sum_{\forall s \neq bc} \pi_s \times \left(\pi_c TC_{C,s} + (1-\pi_c) TC_{E,s}\right)$$

In this case, the OF (10.39) should be minimized subject to the base case and two sets of post-contingency constrains. One set of post contingency constraints are those related to the successful corrective controls, and the other set describes the behavior of the system when the corrective controls fail and hence the emergency controls activated. Thus, (10.39) should be minimized subject to (10.7)–(10.39), and the following constraints of post-contingency emergency states.

(a) Active/reactive power flow constraints at post-contingency scenarios, when the corrective controls fail to respond ($\forall b \in NB$):

$$\sum_{i=1}^{NG_b} (\widehat{P}_{G_i,s} I_{G_i,s}) - \widehat{P}_{L_b,s} = V_{b,s} \sum_{j=1}^{N_B} \widehat{V}_{j,s} Y_{bj,s} \cos(\widehat{\theta}_{b,s} - \widehat{\theta}_{j,s} - \phi_{bj,s}) \tag{10.42}$$

$$\sum_{i=1}^{NG_b} (\widehat{Q}_{G_i,s} I_{G_i,s}) - \widehat{Q}_{L_b,s} = \widehat{V}_{b,s} \sum_{j=1}^{NB} \widehat{V}_{j,s} Y_{bj,s} \sin(\widehat{\theta}_{b,s} - \widehat{\theta}_{j,s} - \phi_{bj,s}) \tag{10.43}$$

(b) Active/reactive power generation/demand, voltage and line flow limits at post-contingency scenarios, when the corrective controls fail to respond:

$$\widehat{P}_{L_b,s} = P^{bc}_{L_b} - \Delta\widehat{P}^{LC}_{L_b,s} \quad \forall b \in NB \tag{10.44}$$

$$\widehat{Q}_{L_b,s} = Q^{bc}_{L_b} - \Delta\widehat{Q}^{LC}_{L_b,s} \quad \forall b \in NB \tag{10.45}$$

Also, $\forall i \in NG_E$

$$\widehat{P}_{G_i,s} = (P^{bc}_{G_i} + \Delta\widehat{P}^{up}_{G_i,s} - \Delta\widehat{P}^{dn}_{G_i,s}) \times I_{G_i,s} \tag{10.46}$$

$$0 \le \Delta\widehat{P}^{up}_{G_i,s} \le (\Delta\widehat{P}^{up,max}_{G_i,s} I_{G_i,s}) \tag{10.47}$$

$$0 \le \Delta\widehat{P}^{dn}_{G_i,s} \le (\Delta\widehat{P}^{dn,max}_{G_i,s} I_{G_i,s}) \tag{10.48}$$

$$\Delta\widehat{P}^{up,max}_{G_i,s} = I_{G_i,s} \times min(P^{max}_{G_i} - P^{bc}_{G_i}, RU_{G_i} \times \tau_E) \tag{10.49}$$

$$\Delta\widehat{P}^{dn,max}_{G_i,s} = I_{G_i,s} \times min(P^{bc}_{G_i} - P^{min}_{G_i}, RD_{G_i} \times \tau_E) \tag{10.50}$$

And,

$$0 \le \Delta\widehat{P}^{LC}_{L_b,s} \le \Delta\widehat{P}^{LC,max}_{L_b,s} \quad \forall b \in NB_E \tag{10.51}$$

$$0 \le \Delta\widehat{Q}^{LC}_{L_b,s} \le \Delta\widehat{Q}^{LC,max}_{L_b,s} \quad \forall b \in NB_E \tag{10.52}$$

$$V_b^{min} \leq \widehat{V}_{b,s} \leq V_b^{max} \qquad \forall b \in NB \tag{10.53}$$

$$\left|\widehat{S}_{\ell,s}(\widehat{V},\widehat{\theta})\right| \leq S_\ell^{max} \qquad \forall \ell \in NL \tag{10.54}$$

10.3.4 Contingency Selection Criterion

By investigating the system condition following the occurrence of any contingency (operational constraints non-satisfaction), it is possible to identify critical outages. Hence, in this chapter the critical contingencies are selected based on the following procedure: run post-contingency AC power flow, subject to the outage of generation units or transmission lines (single outage), considering the reactive power limits of all generation units. The following situations may happen:

- If the power flow is diverged for a specified contingency, the corresponding outage is identified as a critical contingency.
- If the post-contingency load flow converged, but some of the important operational limits, such as line flow or voltage limits are violated, the contingency is also a critical one.
- If the post contingency state is normal (i.e. load flow converged and no operational limit is violated), the corresponding contingency is invulnerable.

By executing the above procedure, the most severe contingencies which need to preventive/corrective control actions, are identified and the list of contingencies which should be considered in the PSCCM model, is constructed.

10.3.5 Wind Power Generation Uncertainty Modeling

In this section the impact of volatile wind power generation is also considered in the proposed PSCCM. The variation of wind power generation is an uncertain parameter which can be modeled probabilistically using historical data records of wind speed [16, 18]. In this work, variation of wind speed, v, is modeled using Weibull probability density function (PDF).

$$PDF(v) = (\frac{k}{\lambda})\,(\frac{\lambda}{v})^{k-1} exp[-(\frac{v}{\lambda})^k] \tag{10.55}$$

The generated power of a wind turbine in terms of wind speed is approximated as follows [16]:

$$P_b^w(v) = \begin{cases} 0 & \text{if } v \leq v_{in}^c \text{ or } v \geq v_{out}^c \\ \frac{v - v_{in}^c}{v_{rated}^c - v_{in}^c} P_{b,r}^w & \text{if } v_{in}^c \leq v \leq v_{rated} \\ P_{b,r}^w & \text{else} \end{cases} \tag{10.56}$$

where v_{in}^c, v_{rated}^c, and v_{out}^c are the cut-in, rated and cut-off speed of wind turbine, respectively. $P_{b,r}^w$ denotes rated power of the wind turbine installed at bus b. More accurate relations could also be used instead of the linear $P - V$ relation for the interval $v_{in}^c \leq v \leq v_{rated}$. Using the technique described in [16], the PDF of wind speed is divided into several intervals, and the probability of falling into each interval is calculated. Each interval is given a mean value which is further used. It is assumed that the outage of elements and wind power generation scenarios are independent so the scenarios are mixed to attain the whole set of scenarios as follows.

$$\pi_s = \pi_w \times \pi_l \tag{10.57}$$

where π_w and π_l are the probabilities of the w-th wind and the l-th outage scenarios, respectively. The total number of scenarios, i.e., NS, will be $l_n \times w_n$, where w_n, l_n are the number of wind and load states.

10.4 Simulation Results

10.4.1 Studied System

The proposed PSCCM model is examined on the IEEE RTS 24-bus system. This system consists of 33 generators, and 38 transmission lines. The data of this system including the data of loads, generating units and transmission lines are given in [22]. The reason why this test system is adopted for numerical examination of the proposed approach is that standard reliability data such as the failure rate and forced outage rate (FOR) of equipments are available for this test system, which could be attained from [23]. The following modifications are made on the original data given in [22].

1. The load level used in this chapter is 10% higher than the original value for this system, given in [22].
2. Maximum limit of active power generation capacity is also assumed to be 10% higher than the original data [22].
3. The lower limit of voltage of bus B_6 is assumed to be 0.85 pu in scenario s_3.

The proposed PSCCM model is implemented in GAMS [24] environment, and solved by CONOPT solver running on an Intel®Xeon™CPU E5-1620 3.6 GHz PC with 8 GB RAM. The cost of active power generation re-dispatch, voluntarily and involuntarily load reductions, for base case and post-contingency states are also given in [25]. The initial operation point (prior to any modification by the proposed PSCCM approach) in terms of active/reactive power generation, consumption and bus voltage magnitudes, is given in [25]. Due to the fact that almost all load points consume active and reactive power simultaneously, curtailing any load will result in reduction of both active and reactive powers together. In this chapter, it is assumed that load reduction in the base-case and each post-contingency probable scenario, is based on its initial power factor. Also, it is assumed that 10% of the initial load of each bus participates in voluntarily load reduction both in base-case and post-contingency scenarios. Hence, the remaining load in each bus, may be subjected to involuntarily load curtailment. Besides, it is assumed that τ_{bc} and τ_{CM} are 5 and 15 min, respectively. ϵ value in (10.40) is assumed to be 5%. For the initial point given in [25], by applying all single outages (i.e. the outage of 38 transmission lines and transformers, and 33 generation units) the post-contingency load flow is performed. Then, for each outage, the maximum value of the ratio of apparent power flowing from any branch to its corresponding nominal value is obtained (as a simple and exact index which directly reflects occurrence of congestion), which is named as contingency index. If this index is greater than 1 for a specified outage, it means this outage leads to congestion (i.e. at least one of the available branches encounters with congestion, subject to the considered contingency) and hence, it is vulnerable contingency, otherwise it is safe (or invulnerable) incident. Figure 10.1 shows the value of this index for the outage of different branches and Fig. 10.2 depicts it for the outage of generation units. It is evidently observed from these figures that for the initial operating point given in [25], the outage of following elements are critical: Transmission lines L_5, L_7, L_{10}, L_{27} and generators G_7 and G_8. It is observed that, the outage of line L_{10} is the most severe single outage event, which also leads to divergence of load flow equations in the post-contingency operation state. Also, generators G_7 and G_8 are similar with the same ratings and forced outage rate (FOR) and located in a common bus, hence their effects are aggregated and one scenario considered for them in the proposed PSCCM. Thus, there are totally 5 post-contingency critical scenarios which should be regarded in the proposed PSCCM. Table 10.1 gives the probabilities associated with these scenarios. The probability of failure i is calculated assuming that component i fails and the rest of the components continue the safe operation.

The proposed PSCCM approach is solved for different risk levels (i.e. from $\beta = 0$ at the risk neutral strategy (RNS) to $\beta = 1$ at the risk averse strategy (RAS)).

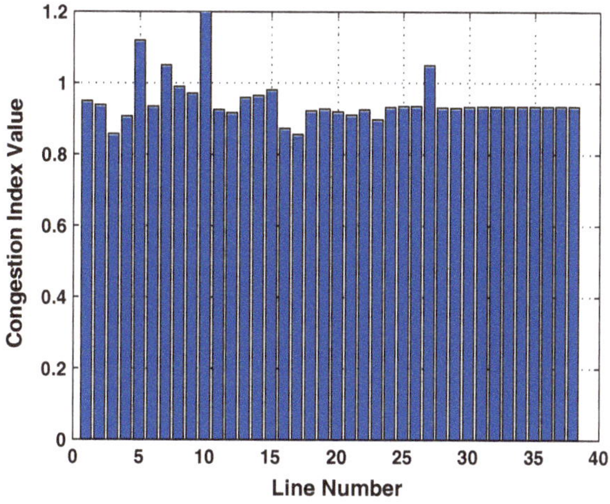

Fig. 10.1 The value of congestion index for the outage of different transmission lines (before applying PSCCM)

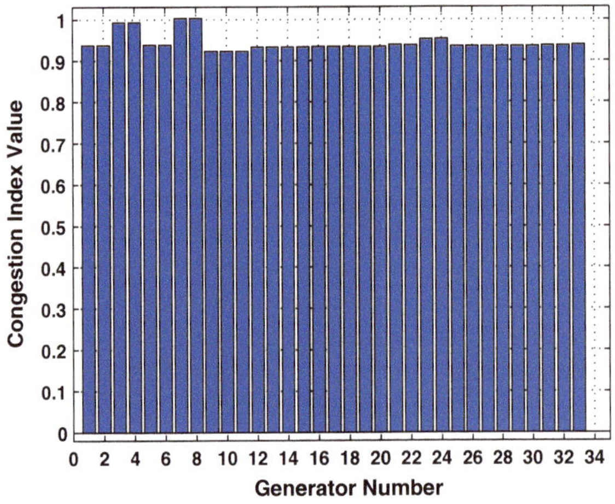

Fig. 10.2 The value of congestion index for the outage of different generation units (before applying PSCCM)

Table 10.1 Probability of base-case and post-contingency scenarios

Scenario	Contingency	Probability
Base-case (bc)	Safe contingencies	0.771233
	No contingency	0.218925
Critical contingencies	s_1 (L_5)	0.000524
	s_2 (L_7)	0.001676
	s_3 (L_{10})	0.001261
	s_4 (L_{27})	0.000493
	s_5 (G_7)	0.039835

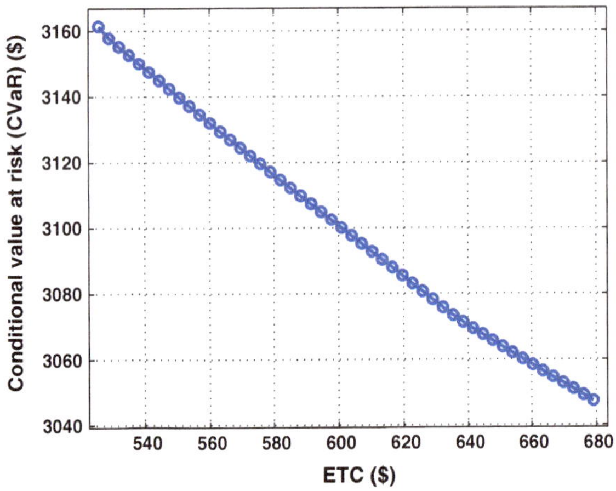

Fig. 10.3 The variation of CVaR versus ETC (corrective controls failure not considered)

Table 10.2 Optimal solutions and their attributes

Solution	β ($)	ETC ($)	ζ ($)	μ_{ETC}	μ_ζ	$min(\mu_{ETC}, \mu_\zeta)$	Solution	β ($)	ETC ($)	ζ ($)	μ_{ETC}	μ_ζ	$min(\mu_{ETC}, \mu_\zeta)$
sol_1	0	525.855	3161.688	1	0	0	sol_{26}	0.51	604.09	3097.833	0.49	0.559	0.49
sol_2	0.02	528.984	3158.078	0.98	0.032	0.032	sol_{27}	0.531	607.22	3095.404	0.469	0.581	0.469
sol_3	0.041	532.114	3155.478	0.959	0.054	0.054	sol_{28}	0.551	610.349	3092.974	0.449	0.602	0.449
sol_4	0.061	535.243	3152.878	0.939	0.077	0.077	sol_{29}	0.571	613.479	3090.545	0.428	0.623	0.428
sol_5	0.082	538.372	3150.28	0.918	0.1	0.1	sol_{30}	0.592	616.608	3088.117	0.408	0.645	0.408
sol_6	0.102	541.502	3147.681	0.898	0.123	0.123	sol_{31}	0.612	619.737	3085.689	0.387	0.666	0.387
sol_7	0.122	544.631	3145.083	0.878	0.145	0.145	sol_{32}	0.633	622.867	3083.261	0.367	0.687	0.367
sol_8	0.143	547.761	3142.486	0.857	0.168	0.168	sol_{33}	0.653	625.996	3080.834	0.347	0.708	0.347
sol_9	0.163	550.89	3139.89	0.837	0.191	0.191	sol_{34}	0.674	629.126	3078.407	0.326	0.73	0.326
sol_{10}	0.184	554.02	3137.294	0.816	0.214	0.214	sol_{35}	0.694	632.255	3075.981	0.306	0.751	0.306
sol_{11}	0.204	557.149	3134.698	0.796	0.236	0.236	sol_{36}	0.714	635.385	3073.555	0.285	0.772	0.285
sol_{12}	0.225	560.278	3132.103	0.775	0.259	0.259	sol_{37}	0.735	638.514	3071.526	0.265	0.79	0.265
sol_{13}	0.245	563.408	3129.509	0.755	0.282	0.282	sol_{38}	0.755	641.643	3069.644	0.245	0.806	0.245
sol_{14}	0.265	566.537	3127.023	0.735	0.304	0.304	sol_{39}	0.776	644.773	3067.797	0.224	0.823	0.224
sol_{15}	0.286	569.667	3124.588	0.714	0.325	0.325	sol_{40}	0.796	647.902	3065.951	0.204	0.839	0.204
sol_{16}	0.306	572.796	3122.154	0.694	0.346	0.346	sol_{41}	0.816	651.032	3064.105	0.183	0.855	0.183
sol_{17}	0.327	575.925	3119.72	0.673	0.368	0.368	sol_{42}	0.837	654.161	3062.26	0.163	0.871	0.163
sol_{18}	0.347	579.055	3117.286	0.653	0.389	0.389	sol_{43}	0.857	657.291	3060.415	0.142	0.887	0.142
sol_{19}	0.367	582.184	3114.853	0.632	0.41	0.41	sol_{44}	0.878	660.42	3058.57	0.122	0.903	0.122
sol_{20}	0.388	585.314	3112.421	0.612	0.432	0.432	sol_{45}	0.898	663.549	3056.725	0.102	0.92	0.102
sol_{21}	0.408	588.443	3109.988	0.592	0.453	0.453	sol_{46}	0.918	666.679	3054.88	0.081	0.936	0.081
sol_{22}	0.429	591.573	3107.557	0.571	0.474	0.474	sol_{47}	0.939	669.808	3053.036	0.061	0.952	0.061
sol_{23}	0.449	594.702	3105.125	0.551	0.496	0.496	sol_{48}	0.959	672.938	3051.192	0.04	0.968	0.04
sol_{24}	0.469	597.831	3102.694	0.53	0.517	**0.517**	sol_{49}	0.98	676.067	3049.349	0.02	0.984	0.02
sol_{25}	0.49	600.961	3100.263	0.51	0.538	0.51	sol_{50}	1	679.13	3047.544	0	1	0

Figure 10.3 presents the values of ETC and CVaR for all risk levels (i.e. from $\beta = 0$ to $\beta = 1$) in the above interval. Table 10.2 summarizes the detailed results corresponding to RNS and RAS.

10.4.2 *PSCCM Implementation Without Considering Wind Power Generation*

In this section, it is assumed that the network is operated without any wind farms and hence the uncertainty of wind power generation is neglected. The optimal values of control variables for both RNS and RAS are given in this section.

A. Risk neutral strategy (RNS)

In this case, by setting $\beta = 0$, CVaR is neglected, and hence the focus is on the minimization of expected total cost (i.e. ETC). The ETC, is obtained \$525.855 in this case. The expected cost of base-case preventive controls (i.e. $\pi_{bc} TC_{bc}$) is \$383.365, whereas the total expected cost of post-contingency modifications (corrective actions), is \$142.490. Also, CVaR is \$3161.688 in this case. Active power generation schedules for generation units and the re-dispatches carried out on the initial generation schedules in base-case (bc) are depicted in Fig. 10.4. Besides, the optimal re-dispatches in the post-contingency scenarios are given in Fig. 10.5. These modifications are made on the base-case optimal schedules given in Fig. 10.4. It is observed from Fig. 10.5 that only in scenarios $s_2 - s_4$ active power re-dispatch is scheduled, which are the most severe post-contingency scenarios. The optimal voltage magnitudes of generator buses in base case and post-contingency scenarios are given in Table 10.3.

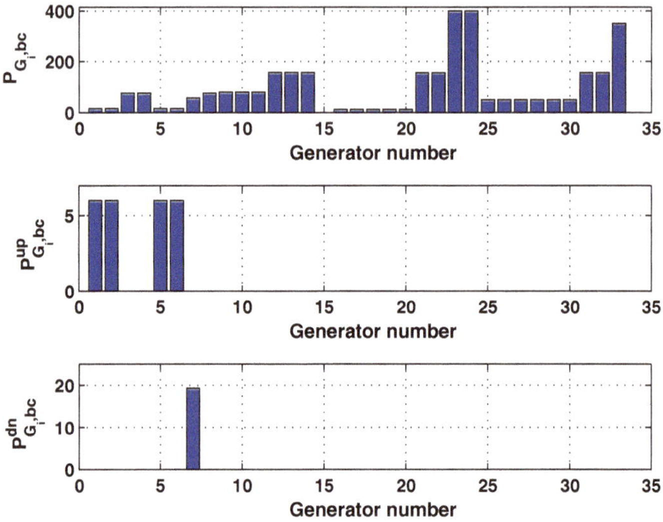

Fig. 10.4 Active power generation schedules and re-dispatches (MW), in bc for RNS

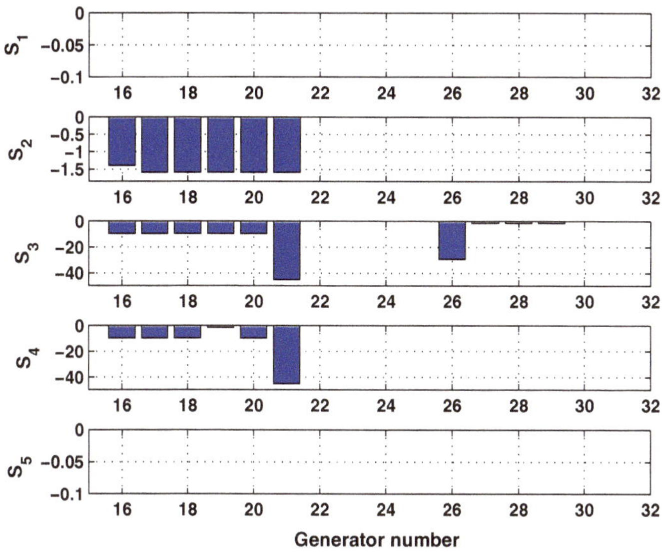

Fig. 10.5 Active power generation re-dispatches (MW), in post-contingency scenarios for RNS

Table 10.3 Optimal voltages of generator buses at base-case and post-contingency scenarios in RNS (in pu)

Node	bc	s_1	s_2	s_3	s_4	s_5
B_1	1.050	1.027	1.050	1.050	1.050	1.044
B_2	1.050	1.025	1.050	1.046	1.050	1.043
B_7	0.983	1.025	1.050	1.050	1.050	1.050
B_{13}	0.958	0.970	1.050	1.050	1.050	1.050
B_{14}	1.007	1.010	1.050	1.050	1.050	1.050
B_{15}	0.990	0.991	1.037	1.040	1.020	1.039
B_{16}	0.994	0.997	1.035	1.044	1.024	1.043
B_{18}	0.991	1.010	1.034	1.050	1.014	1.050
B_{21}	1.002	1.025	1.034	1.050	1.014	1.050
B_{22}	1.050	1.050	1.025	1.041	1.005	1.050
B_{23}	1.001	1.000	1.050	1.050	1.050	1.050

In this case, no load reduction is needed in base-case (i.e. $\Delta P^{bc}_{L_b} = \Delta Q^{bc}_{L_b} = 0, \forall b$), and the amount of voluntarily and involuntarily load reductions in all post-contingency scenarios are described in Table 10.4. It is observed from this table that, $\Delta P^{LC}_{L_b,s}$ is only scheduled in bus B_3 at scenario s_4 (i.e. subject to the outage of line L_{27}), and in the bus B_6 at scenario s_3 (outage of line L_{10}), which is vital for prevention of post contingency congestion in the transmission system, as well as feasible post-contingency operation of the system. The expected values of voluntarily and

Table 10.4 Voluntarily and involuntarily load reductions at different scenarios in RNS, (MW)

Node	Scenario	$\Delta P_{L_b,s}$	$\Delta P^{LC}_{L_b,s}$
B_3	s_2	19.800	0.000
B_3	s_4	19.800	39.698
B_4	s_4	8.140	0.000
B_4	s_5	2.768	0.000
B_6	s_3	14.960	99.956
B_7	s_5	13.750	0.000
B_9	s_2	0.843	0.000
B_9	s_4	19.250	0.000
B_9	s_5	19.250	0.000
B_{20}	s_5	14.080	0.000

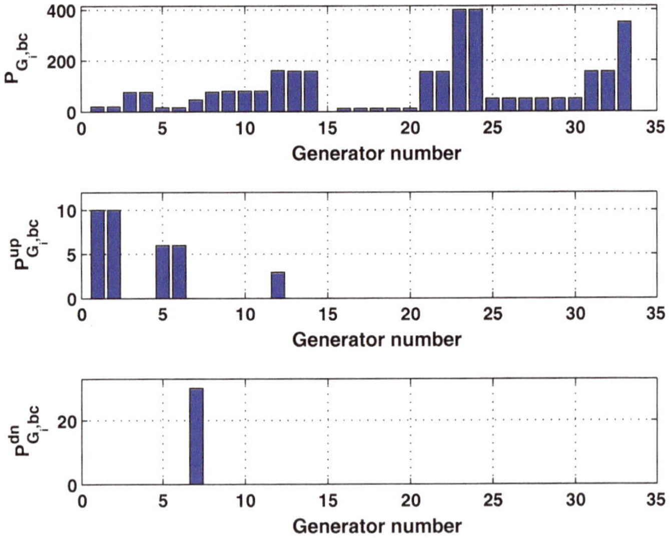

Fig. 10.6 Active power generation schedules and re-dispatches (MW), in bc for RAS

involuntarily active power reduction are 0.463 MW and 0.033 MW respectively, in this case.

B. Risk averse strategy (RAS)

In this case, CVaR is fully considered and the aim of PSCCM is to minimize the value of CVaR for $\beta = 1$. The value of ETC is obtained \$679.130, in which 80.852% of this cost (i.e. \$549.089) belongs to base-case preventive control actions and the remaining 19.148% (i.e. \$130.041) is the expected cost of corrective controls in post-contingency scenarios. The value of CVaR is obtained \$3047.544 in this case.

The optimal schedule of active power generation, along with the corresponding re-dispatches with respect to the initial point, are depicted in Fig. 10.6 for this case.

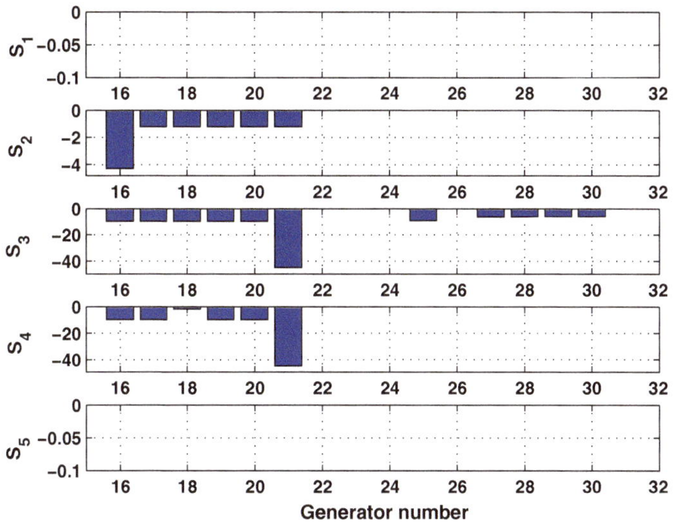

Fig. 10.7 Active power generation re-dispatches (MW), in post-contingency scenarios for RAS

Table 10.5 Optimal voltages of generator buses at base-case and post-contingency scenarios in RAS (in pu)

Node	bc	s_1	s_2	s_3	s_4	s_5
B_1	1.050	1.019	1.050	1.050	1.050	1.044
B_2	1.050	1.017	1.050	1.046	1.050	1.043
B_7	1.050	0.989	1.050	1.050	1.050	1.050
B_{13}	0.956	0.979	1.050	1.050	1.050	1.050
B_{14}	1.007	1.015	1.050	1.050	1.050	1.050
B_{15}	0.988	0.991	1.037	1.040	1.020	1.039
B_{16}	0.991	1.002	1.035	1.044	1.024	1.043
B_{18}	0.991	0.988	1.034	1.050	1.014	1.050
B_{21}	1.002	0.994	1.034	1.050	1.014	1.050
B_{22}	1.050	1.039	1.025	1.041	1.005	1.050
B_{23}	0.989	1.050	1.050	1.050	1.050	1.050

Also, in this case the optimal values of necessary corrective re-dispatches in post-contingency scenarios are given in Fig. 10.7, which shows necessary re-dispatches in scenarios $s_2 - s_4$. The optimal voltage magnitudes of generator buses in the base-case and post-contingency scenarios are given in Table 10.5. Besides, the schedules of both voluntarily and involuntarily load curtailments in different post-contingency scenarios are given in Table 10.6. In this case, the expected values of voluntarily and involuntarily active power reduction are 0.366 MW and 0.033 MW respectively.

Table 10.6 Voluntarily and involuntarily load reductions at different scenarios in RAS, (MW)

Node	Scenario	$\Delta P_{L_b,s}$	$\Delta P_{L_b,s}^{LC}$
B_3	s_2	19.800	0.000
B_3	s_4	19.800	39.827
B_4	s_4	8.140	0.000
B_6	s_3	14.960	100.176
B_7	s_5	13.750	0.000
B_9	s_2	1.653	0.000
B_9	s_4	19.250	0.000
B_9	s_5	11.169	0.000
B_{20}	s_5	14.080	0.000

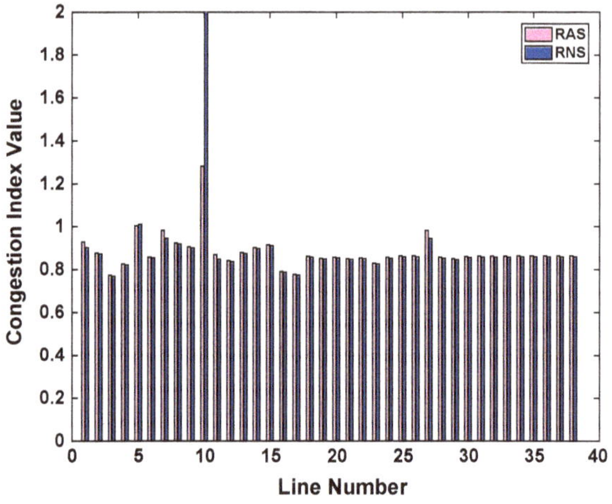

Fig. 10.8 Congestion index value for the outage of different transmission lines (after applying PSCCM)

C. Discussion on the obtained results

As it is aforementioned, the aim of PSCCM is to modify the base-case operation point of the system in a way that if any probable severe contingency happens, the system in the post contingency state also remains in stable region, and the operational constraints, such as line flow limits and voltage limits are met. As it is observed from the above numerical results in RNS and RAS, this goal is achieved by means of base-case preventive control actions and post-contingency corrective controls, which are specified for any probable post-contingency scenario.

For the optimal control actions obtained by PSCCM in the base-case state, again a contingency analysis should be performed through the procedure described in Sect. 10.3.4, in order to ensure that no new contingency is created following the modifications performed on the base-case operation point. Figures 10.8 and 10.9 show the contingency analysis results for the modified base-case in both RNS and RAS. In Fig. 10.8 the value of congestion index versus the single outage (i.e. outage of one

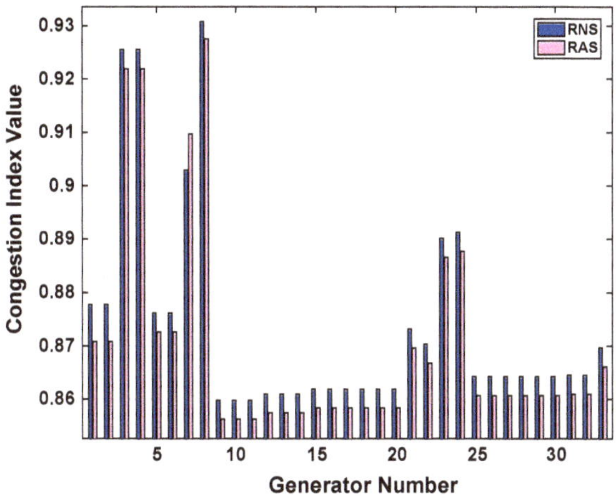

Fig. 10.9 Congestion index value for the outage of different generation units (after applying PSCCM)

Table 10.7 Total costs with and without PSCCM ($)

Scenario	Without PSCCM	With PSCCM	
		RNS	RAS
bc	0.000	387.175	554.547
s_1	405.621	0.000	0.000
s_2	4910.402	3931.629	4090.598
s_3	309552.743	260501.640	261072.895
s_4	98940.136	87358.167	87616.934
s_5	9504.279	6257.781	4836.091
Expected preventive costs	0.000	383.365	549.089
Expected corrective costs	187.391	142.489	130.041
ETC	187.391	525.855	679.130

element) of all 38 branches is shown and compared for both RNS and RAS. Also, Fig. 10.9 gives its values for the single outage of all 33 generation units in RNS and RAS. It is evidently observed from Fig. 10.8 that for the modified base-case no new contingency is created, and only the outage of lines L_5 (in RNS) and L_{10} (in both RNS and RAS) are also remain in the list of vulnerable contingencies, which could be managed accordingly by the corresponding optimal post-contingency corrective actions. Besides, it is observed from Fig. 10.8 that the RAS leads to a feasible post-contingency state for the outage of line L_{10}, which was the most severe contingency in the initial condition, and at the corresponding post-contingency operation point only one of the transmission lines experiences 28% over load (but the load flow is feasible). Also, it is observed from Fig. 10.9 that for the modified base-case operation point, the system becomes secure against the single outage of all generation units. This reflects the fact that the PSCCM improves security of the system, since the modified base-case is secured against more contingencies.

In order to investigate the economical benefits of the proposed PSCCM method, the initial (i.e. unmodified base-case) operation point of the system is subjected to the aforementioned 5 contingencies and at the post-contingency state, the cost of corrective controls are calculated. The obtained results are given in Table 10.7 which are compared with those achieved by PSCCM (in RNS and RAS). It is observed from this table that although the cost of preventive controls in the base-case is zero (i.e. no base-case modification performed), but the cost of post-contingency corrective control actions for each scenario is much higher than the corresponding cost obtained by the proposed PSCCM.

In this chapter, we obtained the Pareto optimal set for numerous values of beta. But, two critical solutions, i.e. the RNS (for $\beta = 0$) and RAS (for $\beta = 1$) are discussed and compared in the above. The main motivation for this, is to illustrate the effect of considering risk (through CVaR index) in the PSCCM. However, if one intended to select a compromise solution form the obtained Pareto optimal set, it depends to its priorities, i.e. whether the cost is more important to him/her or the risk. Also, in order to choose the best solution among the obtained Pareto optimal set, fuzzy satisfying approach could be utilized, where a membership function is defined for each Pareto optimal solution and the best compromise solution is obtained by comparison of the membership function of different solutions. This approach is described in [26]. In Table 10.2, the values of μ_{ETC} and μ_{ζ} are calculated as follows [26]:

$$\mu_{ETC} = \frac{ETC_{max} - ETC}{ETC_{max} - ETC_{min}} \tag{10.58}$$

$$\mu_{\zeta} = \frac{\zeta_{max} - \zeta}{\zeta_{max} - \zeta_{min}} \tag{10.59}$$

In order to find the best compromise solution which considers both risk and cost issues at the same time, the minimum value of μ_{ETC} and μ_{ζ} (i.e. $min(\mu_{ETC}, \mu_{\zeta})$) are calculated for each solution. It is worth to mention that for each solution, μ_{ETC} and μ_{ζ} are membership functions of ETC and CVaR, respectively. Now, between all solutions, the solution with maximum value of $min(\mu_{ETC}, \mu_{\zeta})$ is the best compromise solution. It could be observed from Table 10.2 that the best compromise solution is sol_{24}. Similar to sol_1 (i.e. the RNS) and sol_{50} (i.e. the RAS), the optimal values of preventive and corrective controls is obtained by for this solution running the PSCCM. The detailed results for this compromise solution are not presented in the chapter for the sake of brevity.

D. Considering the corrective controls failure

As it is explained in Sect. 10.3.3, post-contingency corrective controls may fail to respond. In this section, it is assumed that the corrective controls are successfully implemented with the probability of 80%, and hence their failure probabilty is 20% (i.e. $\pi_c = 0.80$). It is assumed that the generation units $G_{20} - G_{25}$ participate in the emergency load following service, when the corrective controls fail to respond. Variation of CVaR versus the ETC in this case for different values of β is depicted in Fig. 10.10.

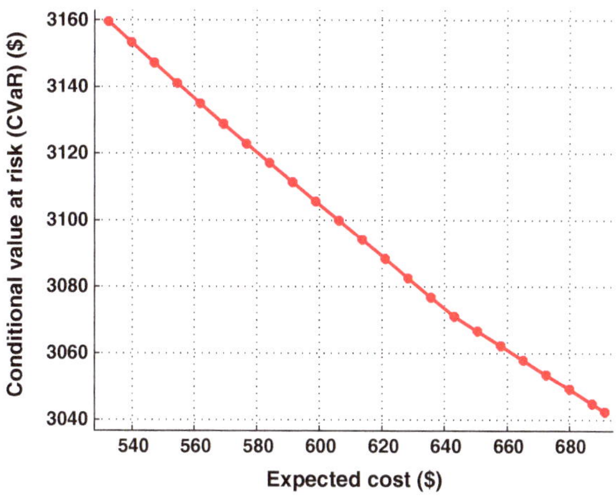

Fig. 10.10 The variation of CVaR versus ETC (corrective controls failure is considered)

In the RNS, the ETC is obtained \$532.493 and the CVaR is \$3159.793. In this case, the expected cost of base-case preventive controls (i.e. TC_{bc}) is \$381.318, whereas the expected cost of corrective controls (i.e. $TC_{C,s}$) is \$142.478 and the cost of remedial actions in the case of corrective controls failure (i.e. $TC_{E,s}$) is \$185.965. The ETC and CVaR in the RAS case are \$691.557 and \$3042.531, respectively. The values of TC_{bc}, $TC_{C,s}$ and $TC_{E,s}$ are \$554.185, \$129.583 and \$168.527, respectively.

Also, Figs. 10.11 and 10.12 show the active power re-dispatches in the post-contingency corrective controls and emergency state (where the corrective controls fail), in RNS and RAS, respectively. It is observed from Fig. 10.11 that in the RNS case the emergency load following service is activated by generation units G_{21} and G_{24} in scenario s_5, to compensate the outage of generator G_7, and in scenarios $s_2 - s_4$ these generation units are also participate to alleviate the effect of transmission lines outage.

Also, the scheduled voluntarily and involuntarily load reductions in post contingency states are given in Table 10.8. This table gives the optimal involuntary load reductions in different scenarios in the case of corrective controls failure are given for both RNS and RAS.

Finally, Table 10.9 summarizes the cost associated with the base-case preventive controls, post-contingency corrective controls, and post-contingency emergency controls, in different scenarios. By comparing the results given in Tables 10.7 and 10.9, one can see the effect of considering the failure of post-contingency corrective controls. It is observed from these tables that the costs (i.e. TC_{bc} and $TC_{C,s}$) does not change significantly when the effect of corrective controls failure is considered in the proposed PSCCM model.

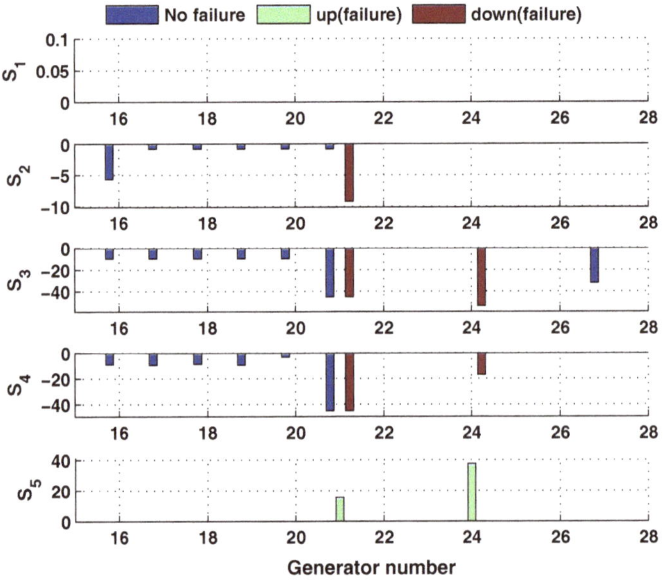

Fig. 10.11 Active power generation re-dispatches (MW), in post-contingency scenarios for RNS (corrective controls failure is considered)

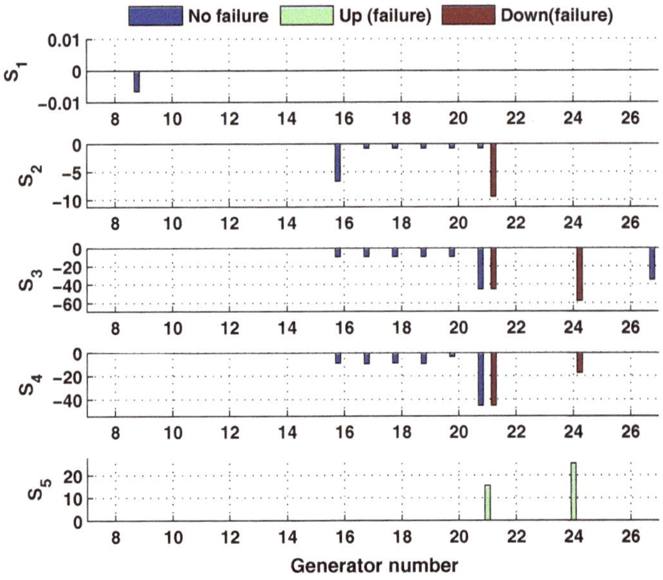

Fig. 10.12 Active power generation re-dispatches (MW), in post-contingency scenarios for RAS (corrective controls failure is considered)

Table 10.8 Voluntarily and involuntarily load reduction at different scenarios in RNS and RNS, (MW) (corrective controls failure is considered)

		RNS			RAS		
Node	Scenario	$\Delta P_{L_b,s}$	$\Delta P^{LC}_{L_b,s}$	$\Delta \hat{P}^{LC}_{L_b,s}$	$\Delta P_{L_b,s}$	$\Delta P^{LC}_{L_b,s}$	$\Delta \hat{P}^{LC}_{L_b,s}$
B_3	s_2	19.800	0.000	19.925	19.800	0.000	20.115
B_3	s_4	19.800	39.667	64.589	19.800	39.827	64.759
B_4	s_4	8.140	0.000	0.000	8.140	0.000	0.000
B_4	s_5	2.768	0.000	0.000	0.000	0.000	0.000
B_6	s_3	14.960	99.947	115.180	14.960	100.176	115.381
B_7	s_5	13.750	0.000	0.000	13.750	0.000	0.000
B_9	s_2	0.652	0.000	0.000	1.648	0.000	0.000
B_9	s_4	19.250	0.000	0.000	19.250	0.000	0.000
B_9	s_5	19.250	0.000	0.000	10.762	0.000	0.000
B_{20}	s_5	14.080	0.000	0.000	14.080	0.000	0.000

Table 10.9 Expected costs ($) in RNS and RAS, (considering the corrective controls failure)

	RNS		RAS	
Base-case scenario (bc)	385.1078799		559.6928562	
Expected preventive costs	381.3180032		554.1848752	
Post-contingency scenario	$TC_{C,s}$	$TC_{E,s}$	$TC_{C,s}$	$TC_{E,s}$
s_1	0.000	0.000	0.453	0.000
s_2	3912.804	35450.596	4101.336	35792.844
s_3	260502.421	294068.068	261093.180	294800.733
s_4	87318.512	125752.292	87629.259	126084.506
s_5	6257.781	8200.252	4783.571	6206.219
Expected corrective costs	142.478	185.965	129.583	168.528
ETC	532.493		691.556	

10.4.3 PSCCM Implementation Considering the Wind Power Generation Uncertainty

In this section, it is assumed that a wind farm (WF) is installed at bus *B*4 which generates intermittent power. This bus is selected arbitrary. In order to model the uncertainty of power generated by this WF, Weibull PDF is utilized here. The parameters of this PDF are as follows [27]: $k = 2.5034$ and $\lambda = 10.0434$. Using these parameters, five distinct scenarios are generated [27], which are summarized in Table 10.10. Similar to the former case, the results obtained for both RNS and RAS. The voltage magnitudes in generator nodes in different scenarios of wind power generation are depicted in Fig. 10.13 for the RNS. Also, active power generations by thermal units in RNS case are given in Fig. 10.14.

Table 10.10 Wind power generation scenarios and the associated probabilities

Scenarios	S_1	S_2	S_3	S_4	S_5
Wind generation (pu)	0	0.129	0.494	0.869	1.000
Probability (%)	6.89	20.44	40.48	19.92	12.27

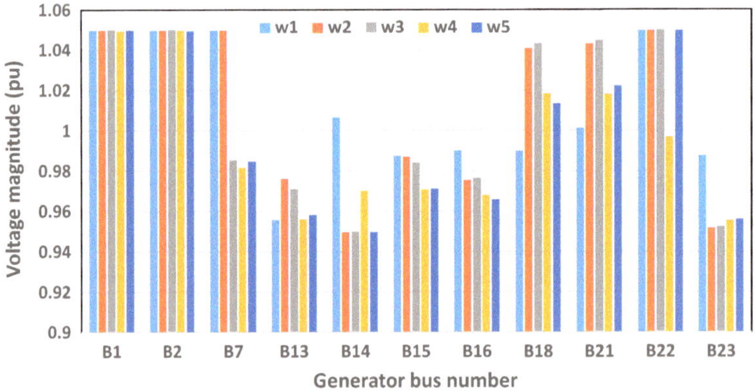

Fig. 10.13 Voltage magnitudes in generator nodes in different wind power generation scenarios (RNS)

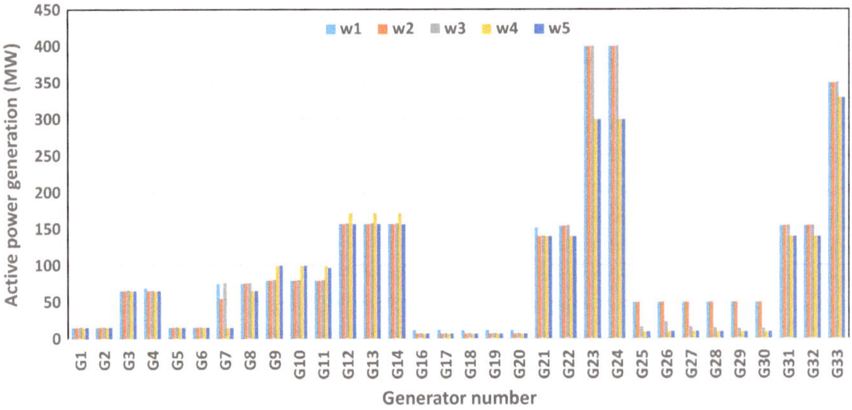

Fig. 10.14 Active power in generator nodes in different wind power generation scenarios (RNS)

In the case of RAS, for the considered wind power generation scenarios, the voltage magnitudes of generator nodes and the active power generation outputs of thermal units are given in Figs. 10.15 and 10.16, respectively.

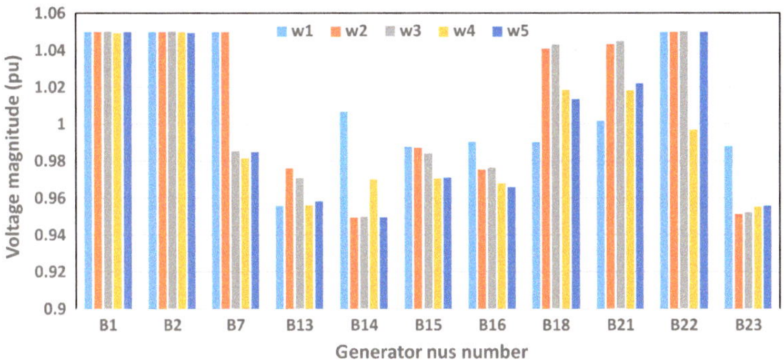

Fig. 10.15 Voltage magnitudes in generator nodes in different wind power generation scenarios (RAS)

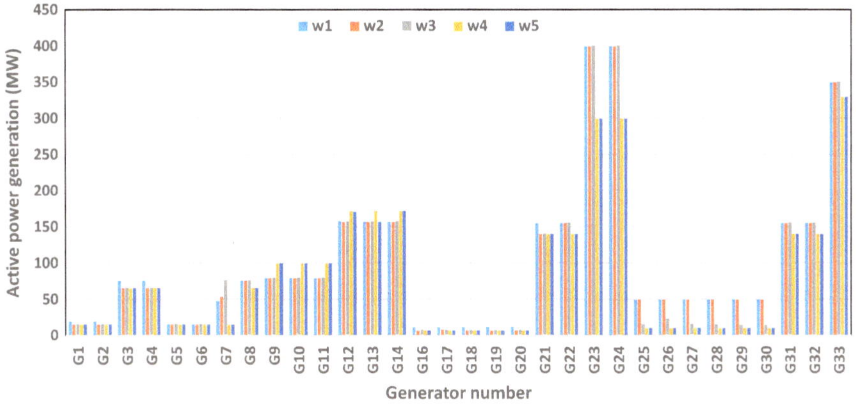

Fig. 10.16 Active power in generator nodes in different wind power generation scenarios (RAS)

Finally, Figs. 10.17, 10.18, 10.19 and 10.20 give the contingency analysis results for the modified base-case in both RNS and RAS at the presence of uncertain wind power generation. It is worth to note that in these figures the expected value of congestion index is calculated using the probability of each wind power generation scenario. In Fig. 10.17 the expected value of congestion index versus the single outage (i.e. outage of one element) of all 38 branches is presented and compared for RNS case. Besides, Fig. 10.18 gives the congestion index for the outage of generators in RNS. Also, Fig. 10.19 shows the expected value of congestion index for the single outage of all 38 branches in RAS. Finally, Fig. 10.20 depicts the expected value of

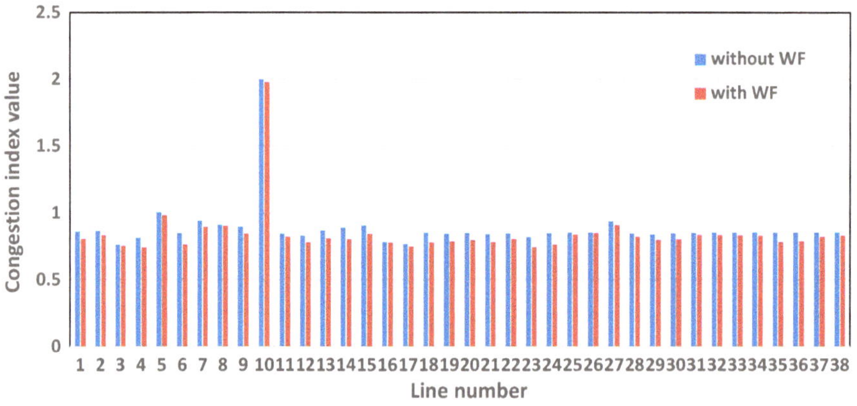

Fig. 10.17 Congestion index versus line outage (RNS)

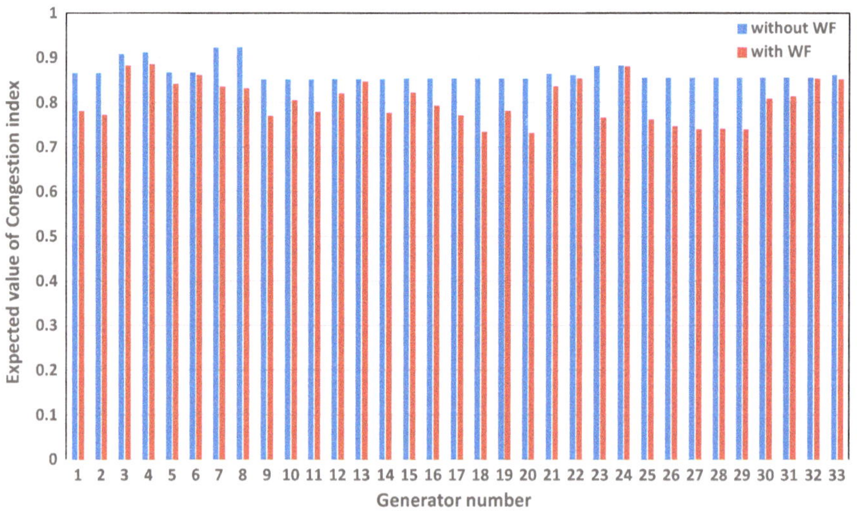

Fig. 10.18 Congestion index versus generator outage (RNS)

congestion index for the outage of all 33 generation units in RAS. In Figs. 10.17, 10.18, 10.19 and 10.20 the value of congestion index, without considering the wind power generation, is also shown in order to investigate the impact of wind power generations on congestion relief.

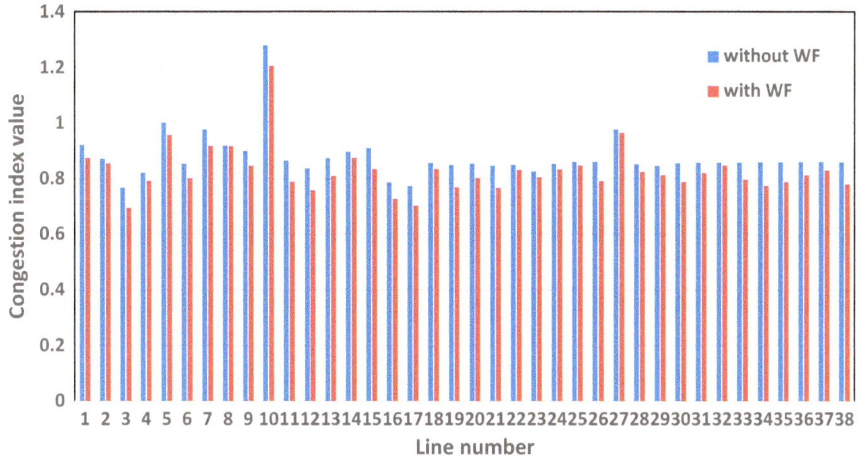

Fig. 10.19 Congestion index versus line outage (RAS)

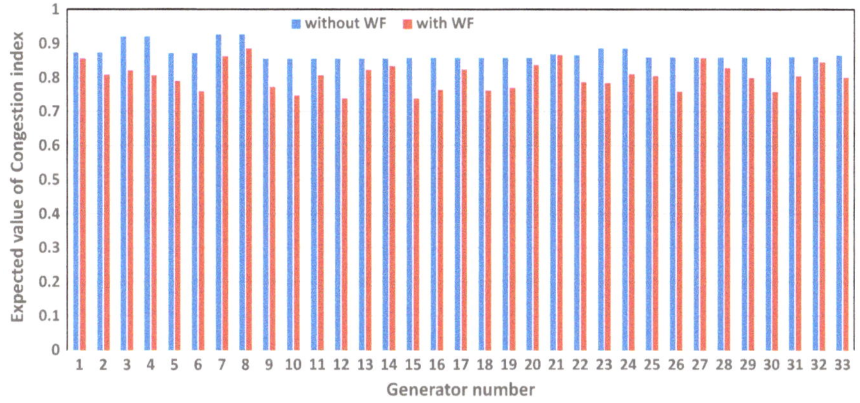

Fig. 10.20 Congestion index versus generator outage (RAS)

10.5 Conclusion

In this chapter, a probabilistic risk based methodology is proposed for a system operator in which a set of preventive and corrective actions are employed to avoid network congestion. It considers the uncertainties associated with various contingencies including transmission lines and generating units. The formulated problem was implemented on the IEEE RTS 24-bus system to demonstrate its effectiveness and suitability. Although a small-scale transmission network is used here, the results shows that it can also be applied to larger systems. Considering the proposed congestion management carried out, the general conclusions below are in order:

1. The value of ETC in RAS is higher than the corresponding ETC in RNS. Also, the expected cost of base-case preventive controls in RAS is higher than the corresponding cost in the RNS, while contrarily the expected cost of corrective controls in RAS is less than that that in RNS. This means that employing the CVaR as a risk measure, reduces the post-contingency corrective costs, but with the expense of increasing the base-case preventive costs.
2. Considering the possibility of corrective controls failure, makes the proposed PSCCM approach more generic. It is observed that including the cost of emergency remedial actions in the case of corrective controls failure, increases the associated expected costs of preventive and corrective controls, slightly. The previous conclusion is valid, i.e. including the CVaR decreases the corrective and emergency control costs in RAS with respect to the RNS.
3. In one hand the number of vulnerable contingencies decreased by modifications performed by PSCCM on the base-case operation point, on the other hand, these modifications lead to lower corrective control cost at the post-contingency scenarios.

Also, in order to extend this work, the following remarks are convenient:

1. Other effective control actions like start-up of fast response generators can further reduce the total corrective control costs.
2. Other uncertainty methodologies may be used to enhance the computational aspects of the proposed congestion management model like Information Gap Decision Theory (IGDT). Specially when no precise information is available about the PDF of uncertain parameters.
3. The current framework is based on day ahead scheduling. The more information is in hand, the more robust decisions can be made. This would require real time data regarding the network topology, demand and generation values. The smart grid technology [28, 29] can provide valuable real-time data regarding the network condition to better handle the uncertainties existing in the problem.
4. The current framework only considers the outage of lines and generating units as the uncertain parameters. In order to make the proposed approach closer to reality, considering other uncertainty resources (e.g. renewable resources) can be useful.

Acknowledgements This work was conducted in the Electricity Research Centre, University College Dublin, Ireland, which is supported by the Electricity Research Centres Industry Affiliates Programme (http://erc.ucd.ie/industry/). This material is based upon works supported by the Science Foundation Ireland, by funding Alireza Soroudi, under Grant No. SFI/09/SRC/E1780. The opinions, findings and conclusions or recommendations expressed in this material are those of the author(s) and do not necessarily reflect the views of the Science Foundation Ireland.

References

1. A. Conejo, F. Milano, R. Garcia-Bertrand, Congestion management ensuring voltage stability. IEEE Trans. Power Syst. **21**(1), 357–364 (2006)
2. C. Murphy, A. Soroudi, A. Keane, Information gap decision theory-based congestion and voltage management in the presence of uncertain wind power. IEEE Trans. Sustain. Energ. **7**(2), 841–849 (2016)
3. A. Kumar, S. Srivastava, S. Singh, Congestion management in competitive power market: a bibliographical survey. Electr. Power Syst. Res. **76**(13), 153–164 (2005)
4. J. Hazra, A. Sinha, Congestion management using multiobjective particle swarm optimization. IEEE Trans Power Syst. **22**(4), 1726 –1734 (2007)
5. E. Bompard, P. Correia, G. Gross, M. Amelin, Congestion-management schemes: a comparative analysis under a unified framework. IEEE Trans. Power Syst. **18**(1), 346–352 (2003)
6. A. Yousefi, T. Nguyen, H. Zareipour, O. Malik, Congestion management using demand response and facts devices. Int. J. Electr. Power Energ. Syst. **37**(1), 78–85 (2012)
7. A. Kumar, S. Srivastava, S. Singh, A zonal congestion management approach using real and reactive power rescheduling. IEEE Trans. Power Syst. **19**(1), 554–562 (2004)
8. Q. Zhou, L. Tesfatsion, C.-C. Liu, Short-term congestion forecasting in wholesale power markets. IEEE Trans. Power Syst. **26**(4), 2185–2196 (2011)
9. P. Wang, Y. Xiao, Y. Ding, Nodal market power assessment in electricity markets. IEEE Trans. Power Syst. **19**(3), 1373–1379 (2004)
10. M. Buygi, G. Balzer, H. Shanechi, M. Shahidehpour, Market-based transmission expansion planning. IEEE Trans. Power Syst. **19**(4), 2060–2067 (2004)
11. A. Berizzi, C. Bovo, M. Delfanti, M. Merlo, M. Pasquadibisceglie, A monte carlo approach for ttc evaluation. IEEE Trans. Power Syst. **22**(2), 735–743 (2007)
12. R. Moreno, D. Pudjianto, G. Strbac, Transmission network investment with probabilistic security and corrective control. IEEE Trans. Power Syst. **28**(4), 3935–3944 (2013)
13. P. Maghouli, A. Soroudi, A. Keane, Robust computational framework for mid-term techno-economical assessment of energy storage. IET Gener. Transm. Distrib. **10**(3), 822–831 (2016)
14. Y. Ben-Haim, *Info-Gap Decision Theory: Decisions Under Severe Uncertainty*. Academic Press (2006)
15. A. Rabiee, A. Soroudi, A. Keane, Information gap decision theory based opf with hvdc connected wind farms. IEEE Trans. Power Syst. **30**(6), 3396–3406 (2015)
16. A. Soroudi, Possibilistic-scenario model for dg impact assessment on distribution networks in an uncertain environment. IEEE Trans. Power Syst. **PP**(99), 1 (2012)
17. A. Soroudi, T. Amraee, Decision making under uncertainty in energy systems: state of the art. Renew. Sustain. Energ. Rev. **28**, 376–384 (2013)
18. A. Soroudi, Taxonomy of uncertainty modeling techniques in renewable energy system studies, in *Large Scale Renewable Power Generation* (Springer Singapore, 2014), pp. 1–17
19. R. Rockafellar, S. Uryasev, Conditional value-at-risk for general loss distributions. J. Bank. Finance **26**(7), 1443–1471 (2002)
20. Z. Feng, V. Ajjarapu, D. Maratukulam, A comprehensive approach for preventive and corrective control to mitigate voltage collapse. IEEE Trans. Power Syst. **15**(2), 791–797 (2000)
21. E. Karangelos, P. Panciatici, L. Wehenkel, Whither probabilistic security management for real-time operation of power systems? in *Bulk Power System Dynamics and Control—IX Optimization, Security and Control of the Emerging Power Grid (IREP), 2013 IREP Symposium*, Aug 2013, pp. 1–17
22. R. Zimmerman, C. Murillo-Sanchez, R. Thomas, Matpower: steady-state operations, planning, and analysis tools for power systems research and education. IEEE Trans. Power Syst. **26**(1), 12–19 (2011)
23. R. Billinton, S. Jonnavithula, A test system for teaching overall power system reliability assessment. IEEE Trans. Power Syst. **11**(4), 1670–1676 (1996)
24. A. Brooke, D. Kendrick, A. Meeraus, R. Raman, *GAMS: A User's Guide*. GAMS Development Corporation (1998)

25. RTSdata, https://sites.google.com/site/alirezasoroudi/datalf.xlsx, Technical report, Sept 2014
26. A. Rabiee, A. Soroudi, B. Mohammadi-Ivatloo, M. Parniani, Corrective voltage control scheme considering demand response and stochastic wind power. IEEE Trans. Power Syst. **29**(6), 2965–2973 (2014)
27. S. Wen, H. Lan, Q. Fu, D.C. Yu, L. Zhang, Economic allocation for energy storage system considering wind power distribution. IEEE Trans. Power Syst. **30**(2), 644–652 (2015)
28. P. Vrba, V. Marik, P. Siano, P. Leitao, G. Zhabelova, V. Vyatkin, T. Strasser, A review of agent and service-oriented concepts applied to intelligent energy systems. IEEE Trans. Ind. Inform. **10**(3), 1890–1903 (2014)
29. P. Siano, C. Cecati, H. Yu, J. Kolbusz, Real time operation of smart grids via fcn networks and optimal power flow. IEEE Trans. Ind. Inform. **8**(4), 944–952 (2012)

Index

P. Du et al. (eds.), *Integration of Large-Scale Renewable Energy into Bulk Power Systems*, Power Electronics and Power Systems,
DOI 10.1007/978-3-319-55581-2

MIX
Papier aus verantwortungsvollen Quellen
Paper from responsible sources
FSC® C105338

If you have any concerns about our products,
you can contact us on
ProductSafety@springernature.com

In case Publisher is established outside the EU,
the EU authorized representative is:
Springer Nature Customer Service Center GmbH
Europaplatz 3, 69115 Heidelberg, Germany

Printed by Libri Plureos GmbH
in Hamburg, Germany